CISM COURSES AND LECTURES

The series presents lecture notes, monographs, edited works and proceedings in the field of Mechanics, Engineering, Computer Science and Applied Mathematics.
Purpose of the series is to make known in the international scientific and technical community results obtained in some of the activities organized by CISM, the International Centre for Mechanical Sciences.

INTERNATIONAL CENTRE FOR MECHANICAL SCIENCES

COURSES AND LECTURES - No. 452

MECHANICS AND THERMOMECHANICS OF RUBBERLIKE SOLIDS

EDITED BY

GIUSEPPE SACCOMANDI
UNIVERSITY OF LECCE

RAYMOND W. OGDEN
UNIVERSITY OF GLASGOW

Springer-Verlag Wien GmbH

This volume contains 108 illustrations

Originally published by Springer-Verlag Wien New York in 2004
SPIN 10990797

In order to make this volume available as economically and as rapidly as possible the authors' typescripts have been reproduced in their original forms. This method unfortunately has its typographical limitations but it is hoped that they in no way distract the reader.

ISBN 978-3-211-21251-6 ISBN 978-3-7091-2540-3 (eBook)
DOI 10.1007/978-3-7091-2540-3

PREFACE

Natural rubber is obtained from the milky secretion (latex) of various plants, but today the only commercial sources of natural rubber are the tree Hevea brasiliensis and the shrub known as guayule (Parthenium argentatum). Hevea brasiliensis is a tropical tree native to the Amazon Basin in Brazil and adjoining countries, whereas guayule is native to the arid regions of Mexico and the south west of the United States.

When the rubber tree is tapped we obtain latex. It has a milky colour and if it is allowed to flow onto a hand, for instance, it will quickly coagulate and form a ribbon or strand. These strands can be rubbed together to form a ball that is capable of bouncing. Unless the latex is preserved it will rapidly degrade and normally ammonia is used as a preservative. In this raw state latex is perfectly safe and can be used as an adhesive. Hevea latex is a polydisperse system in which rubber particles constitute 30–45% of its volume.

Native Americans used rubber for the production of sports equipment, as incense and lip balm, and also for the production of religious figurines. Still today Amazon Indians use it to smear over their feet to wear a kind of boot.

Of all the wonderful tales brought back by Christopher Columbus in 1496 after his second voyage to the New World, none was stranger than the tale of a ball used by the people of Haiti that bounced. Although they did not realize it, Columbus and his crew were the first Europeans to see rubber. The Aztecs and other Central American nations used rubber balls during a game named Chaah or Tlachtlic. The game was a mixture of soccer, volleyball and basketball in which two teams of six players attempted to hit the ball through one of the stone rings attached to a wall.

Rubber did not really attract the interest of Europeans until reports about it were made (1736-51) to the French Academy of Sciences by Charles de la Condamine and François Fresneau and also after this date, initially, the new material was merely a scientific curiosity. The British scientist Joseph Priestley remarked on its usefulness for rubbing pencil marks from paper in 1770. Other applications gradually developed, notably for waterproofing shoes and clothing, and the first rubber factory in the world was established near Paris in 1803, the first in England by Thomas Hancock in 1820, and in the 1830s John Haskins and Edward Chaffee established the first rubber-goods factory in the United States.

At the time rubber had many weaknesses. It softened with heat and hardened with cold; it was tacky, odorous, and perishable. In 1834 the German chemist Friedrich Ludersdorf and the American chemist Nathaniel Hayward discovered that the addition of sulphur to gum rubber lessened or eliminated the stickiness of finished rubber goods. In 1839 the American inventor Charles Goodyear, using the findings of the two chemists, discovered that cooking rubber with sulphur, in a process called vulcanisation, removed the gum's unfavourable properties. Vulcanized rubber has increased strength and elasticity and greater resistance to changes in temperature than unvulcanized rubber; it is impermeable to gases, and resistant to abrasion, chemical action, heat, and electricity. Vulcanized rubber also exhibits high frictional resistance on dry surfaces and low frictional resistance on water-wet surfaces. The vulcanization process remains fundamentally the same as it was in 1839. With the invention of the bicycle, the motorcar and the pneumatic rubber tire, demand for rubber grew rapidly. By 1900 more than 40,000 tons were used each year. Today the world's natural rubber use is over 4,500,000 tons per annum.

There are also artificially produced substances that resemble natural rubber in essential chemical and physical properties. Such substances, usually called elastomers, are produced by chemical reactions, known as condensation or polymerization, of certain unsaturated hydrocarbons. The basic materials of synthetic rubber are monomers: compounds of relatively low molecular weight that form the building units of huge molecules called polymers. After fabrication elastomers are also cured by vulcanization. The origin of synthetic-rubber technology can be traced back to 1860, when the British chemist Charles Hanson Greville Williams (1829–1910) determined that natural rubber was a polymer of the monomer isoprene. Much effort was directed during the next 70 years to synthesizing rubber in the laboratory by using isoprene as the monomer. Other monomers were also investigated, and during World War I German chemists polymerized dimethylbutadiene to produce an elastomer called methyl rubber, which was of limited use. A breakthrough in synthetic-rubber research did not occur, however, until about 1930, when the American chemist Wallace Hume Carothers (1896–1937) and the German scientist Hermann Staudinger (1881–1965) placed polymer chemistry on a theoretically sound scientific basis. The work of these scientists contributed greatly to the present-day knowledge that polymers are huge molecules with chain-like structures, arising from chemical reactions of large numbers of monomers, and that elastomers can be prepared from monomers other than isoprene. More than a dozen major classes of synthetic rubber are made from raw material derived from petroleum, coal, oil, natural gas, and acetylene. Many of them are copolymers, i.e. polymers consisting of more than one monomer. By changing the composition it is possible to achieve particular properties desired for specific applications. Today 70% of all rubber used is synthetic, but despite this fact natural rubber continues to be in high demand due to its better elasticity, resilience and heat transfer properties.

The first experiments on the elasticity of natural rubber were carried out by John Gough in 1802 and the results were published in the Memoirs of the Literary and Philosophical Society of Manchester, Second Series (Volume I, 1805, 288–295). The title of this paper was: 'A description of a property of caoutchouc or Indian rubber with some reflections on the cause of elasticity of this substance'. Two experiments were conducted: the first concerned the evolution of heat on stretching rubber; the second the counterintuitive effect of contraction of stretched rubber upon heating. Other experiments had been done previously to determine the chemical composition of rubber. In 1930 the polymer physicists W. Kuhn, E. Guth and H. Mark developed the first mathematical model to explain the elasticity of rubber. Progress in the quantitative modelling of polymeric materials was made mainly in the period from the 1930s to the 1960s by polymer chemists such as P.J. Flory and L.R.G. Treloar, but also by applied mathematicians, notably R.S. Rivlin. The approach used by polymer chemists and physicists is usually based at the molecular level, whereas applied mathematicians adopt a phenomenological approach based on the theories of continuum mechanics and nonlinear elasticity.

Today interest in rubber-like materials it is still very much alive. Rubber is used mainly to produce tires (68%), but also to manufacture other goods and products; for example, footwear (5%), engineering devices (7.8%), latex products (8%), adhesives (3.2%) and pharmaceutical articles, floor covering, sports goods and toys. The development of special 'smart' rubber-like materials such as electro-active or magneto-sensitive rubbers has introduced the possibility of new and exciting applications; for example, active suspensions for vehicles and artificial muscles. Moreover, the mechanics of rubber-like materials have some similarities to those of soft tissue (the materials of arteries, ligaments, tendons, etc.). The biomechanics of soft tissues is a multidisciplinary field of research, but it requires a good background of knowledge from our experience of analysing the mechanics of rubber-like materials.

Within this framework it is very important to have a complete and rational approach to the mathematical modelling of the mechanics of rubber-like materials. Models for rubber-like materials may be based on the molecular or the phenomenological approach. The aim of CISM advance course on the Mechanics and Thermomechanics of Rubberlike Solids, held in Udine during the week September 30 to October 4, 2002, was to provide a unique state-of-the art multidisciplinary overview of rubber mechanics through the interactions of topics ranging from the mathematical modelling of material response (elastic and inelastic) at both the molecular and phenomenological levels, analysis of the governing differential equations and boundary-value problems, computational methods for the solution of these problems and experimental techniques for the determination of the mechanical properties of rubber, with particular reference to industrial applications and the needs of industry for new developments in material modelling and analysis.

The course included seven lectures by Burak Erman (Sabanci University, Turkey) on the classical molecular theories of rubber elasticity and bioelastomers, and six lectures by Paul Buckley (University of Oxford, UK) on experimental techniques. Giuseppe Saccomandi (University of Lecce, Italy) devoted five lectures to the phenomenological theory of nonlinear elasticity, and Ray Ogden (University of Glasgow, UK) delivered five lectures on inelastic theories, including pseudoelasticity and viscoelasticity. Numerical methods were discussed in seven lectures by Sanjay Govindjee (University of California at Berkeley, USA), while some industrial applications were highlighted in three lectures by Franco Bettinali (ENEL s.p.a., Italy). The course was attended by 35 students with different background (engineering science, material science, mathematics and physics), from all over the world, both from universities and from companies in the rubber industry. The students contributed to a very interesting discussion afternoon with several short talks on topics ranging from fatigue of rubber to the mechanics of elastic structures.

As always the CISM staff and the Rector M.G. Velarde have been very helpful and professional in coping with all the technical needs of the course. We think that the lecture notes provide, not only a bird's eye overview of this interdisciplinary subject but also show that the subject remains very active and offers many interesting challenges from the perspectives of both pure and applied research.

Giuseppe Saccomandi
Ray Ogden

CONTENTS

Experimental Methods for Rubberlike Solids

Paul Buckley

Department of Engineering Science, University of Oxford, Oxford, UK

Abstract. Rubberlike solids are typified by a highly elastic response to stress, combined with a low Young's modulus typically of order 1 MPa. They are amorphous polymers, crosslinked (chemically or physically) and at temperatures above their glass transition. Filler particles are often present. In experimental studies, length scales and time scales are usually chosen to ensure the measured response is that of a homogeneously deforming continuum, and protocols recognize the possibility of stress-induced structural changes. To characterise the response in three dimensions, experiments are conducted with various strain-states. These can be achieved with a variety of experimental configurations, including purpose-built biaxial testing machines. Each experiment explores a path in principal stretch space. Since rubberlike solids show some deviation from ideal elasticity, it is also necessary to characterise the dependence of their response on the time-sequence of strain or stress. For small strains the response is then linear viscoelastic, giving creep, stress relaxation and hysteresis effects, related through the linear theory. At larger strains, and especially in the presence of reinforcing fillers, characteristic deviations from linearity are observed – the well-known Payne and Mullins effects, for which stress-induced structural breakdown is at least partially responsible. Experiments are also used to validate numerical models for prediction of non-uniform deformations. It is advantageous to employ simple geometries, preferably where there is a known analytical solution to provide a benchmark.

1 Introduction to Rubberlike Solids

1.1 Natural Rubber

Rubber is a remarkable material. It occurs naturally and serves a huge range of purposes in the modern world. So much so, the last century saw the production of many synthetic substitutes with similar characteristics, to meet specific objectives. We employ the term *rubberlike* here to indicate all these materials. To learn just what this means, let us consider what is special about rubber.

Since the term *rubber* is often applied generically to refer to all rubberlike materials, we shall refer here to the original, naturally occurring, material of this type as *natural rubber.* This occurs in the sap of the tree Hevea Braziliensis, from which it is usually harvested. Chemically, it is a form of the polymer 1,4-polyisoprene which may be represented by its monomer unit

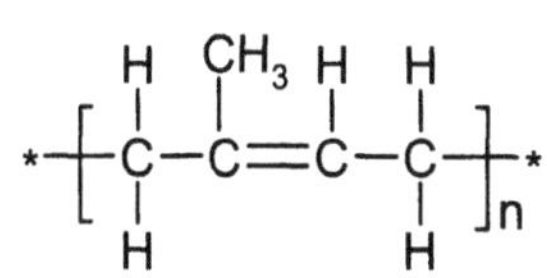

The double bond prevents rotation of the right-hand half of the monomer relative the left-hand half, and in natural rubber the three central bonds are arranged in this fashion

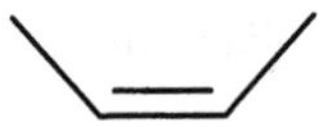

and the polymer is known as *cis* 1,4-polyisoprene.

The special physical properties of natural rubber stem from the high degree of flexibility of this molecule, resulting from the ease of rotation about the three single C-C bonds per monomer. Thus, at room temperature the molecules are highly mobile, and the polymer as created is in the liquid state. It is amorphous and above its glass transition temperature (which is -70°C). However, entanglement of the very long molecules (typically containing about 5000 isoprene monomer units) makes it highly viscous. For practical applications the liquid is turned into a solid, a process known as "curing", usually by reacting it with around one percent of sulphur. The double bonds are susceptible to attack, and the sulphur reacts with some of these on each molecule, to form sulphur bridges or "crosslinks" between them. The result is known as "vulcanised rubber". Provided there is sufficient crosslinking - usually there are at least 20 crosslinks per molecule – the viscosity of the liquid becomes effectively infinite and the resulting material, although soft, is a solid. *Figure 1* shows shear modulus G versus temperature T for natural rubber crosslinked with different percentages of sulphur.

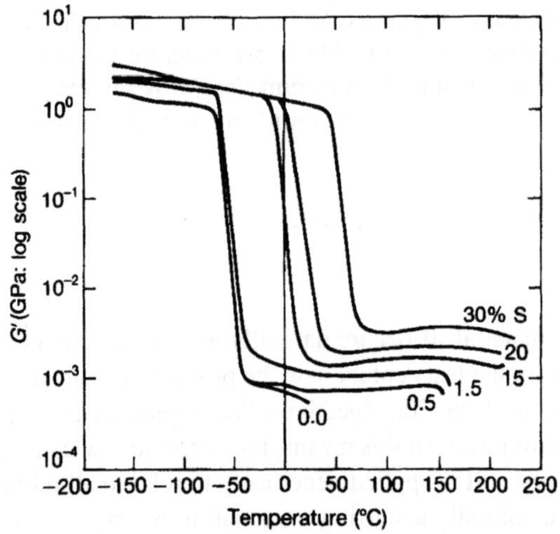

Figure 1. Shear modulus of natural rubber crosslinked with various percentages of sulphur. Characteristic "rubberlike" behaviour at room temperature is associated with the lightest degrees of crosslinking (0, 0.5 and 1.5%), giving a shear modulus of order 1 MPa. From McCrum, Buckley and Bucknall (1997).

It can be seen that for 0, 0.5 and 1.5% sulphur the shear modulus is of order 1 MPa at room temperature. This is the form in which natural rubber is normally used, and the low shear modulus is therefore one of the distinctive features of a "rubberlike" material. At the higher percentages of sulphur shown, crosslinking is so severe, and molecular mobility so reduced, that the glass transition temperature T_g is raised to room temperature or above, and we have a much harder, glassy, material known as ebonite.

Further distinctive features of the mechanical response of lightly crosslinked natural rubber are revealed in *Figure 2*, showing nominal[1] stress σ_n versus nominal[1] strain ε_n for tensile strain cycles up to 200% extension. Note that, when extensions are as large as this, we shall see later it is helpful to express deformation in terms of the stretch

$$\lambda = \frac{l}{L}, \tag{1}$$

where l and L are current and original lengths respectively. It is simply related to the usual definition of nominal strain:

$$\varepsilon_n = \lambda - 1 \ . \tag{2}$$

Figure 2 shows natural rubber to be highly resilient: on unloading, the strain returns to zero on removal of stress. Furthermore, it shows very little hysteresis: the energy dissipated per unit volume in one load-unload cycle (enclosed area of the stress-strain curve) is a small fraction of the energy expended in reaching the maximum load (area under the stress-strain curve).

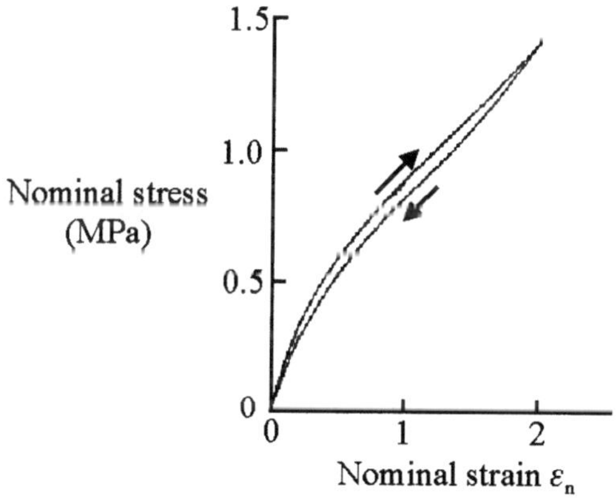

Figure 2. Load/unload cycle to 200% nominal strain in a sample of crosslinked natural rubber. Data from Lindley (1964).

The remaining distinguishing feature of natural rubber is the achievement of extremely large elongations-to-break. This can be seen in *Figure 3*, showing stress-strain loops (nominal stress versus nominal strain again) for load/unload cycles to various maximum strains. Each loop refers to a new specimen. Note the extreme extensibility of the material: in the case shown, even stretching by 600% is not sufficient to cause failure. There are two other interesting features of these curves. Note that the hysteresis does not remain low – the loops become fatter as the maximum strain increases – and note the characteristic increase in gradient at large strains – the phenomenon of strain-stiffening.

[1] By "nominal" we mean: calculated on the original specimen dimensions at the start of the test.

Thus we see the key characteristics of crosslinked natural rubber are:

- low shear stiffness (shear modulus typically 0.1-10 MPa)
- high resilience
- high extensibility.

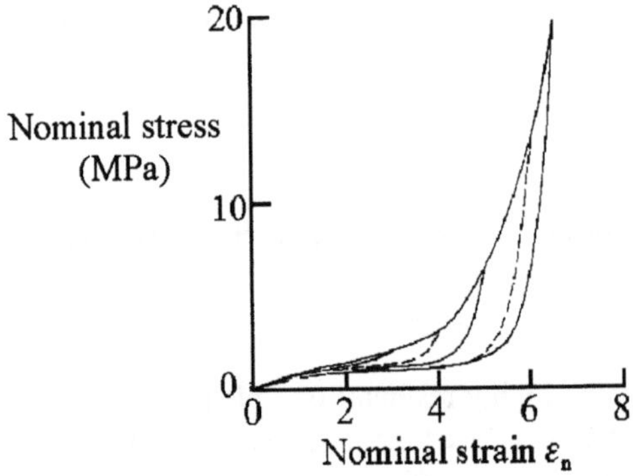

Figure 3. First load/unload cycles to various maximum strains in specimens of crosslinked natural rubber. Data from Lindley (1964).

1.2 Other rubberlike polymers

There is a huge range of products where the properties of natural rubber are highly desirable: e.g. car tyres, engine mountings, inflatable boats, to name a few. It is fortunate, therefore, that such properties are not confined to just cis-1,4 polyisoprene, but are also shared by several other polymers at ambient temperature. As for natural rubber, these polymers are amorphous and above their glass transitions at room temperature, and are able to be crosslinked. It is the whole group of polymers exhibiting such behaviour that we call the *rubberlike solids*. They are also known as *elastomers*. In addition to polyisoprene, some of the other homopolymers frequently encountered in the rubberlike (or elastomeric) state are

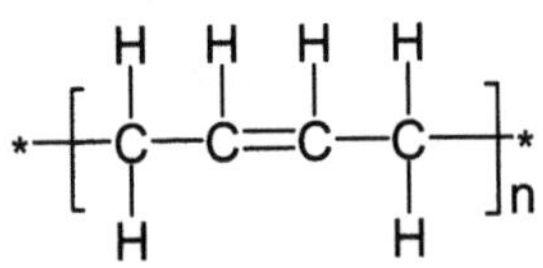
1,4-polybutadiene

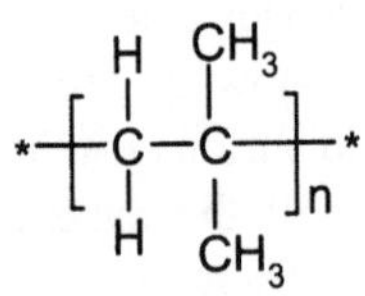
polyisobutylene
(butyl rubber)

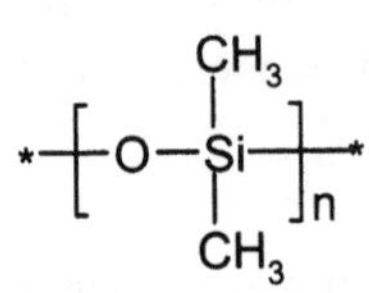
poly(dimethyl siloxane)
(silicone rubber)

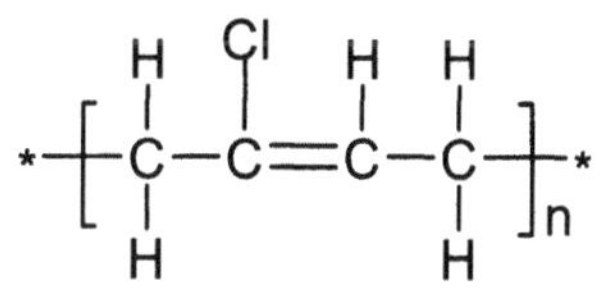

1,4-polychloroprene

In addition, common rubberlike polymers include random copolymers of polybutadiene with

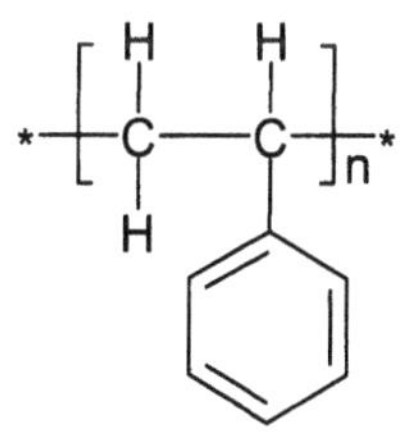

polystyrene

to produce styrene-butadiene rubber (SBR), and of

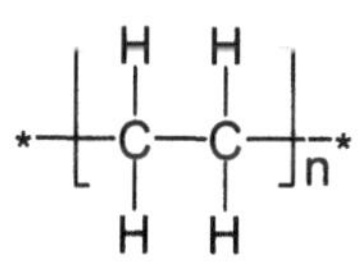

polyethylene with

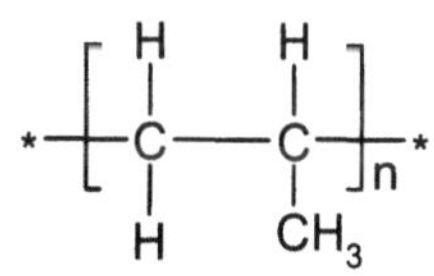

polypropylene

to produce ethylene-propylene rubber (EPR).

In all such cases, solid-like behaviour results from chemical crosslinking of the molecules – either by sulphur or other crosslinking chemistry (e.g. peroxides). This does introduce some practical engineering problems. Any product to be formed from a crosslinked polymer must be shaped, for example in a mould, before the crosslinking reaction occurs. The manufacturing operation therefore involves reaction chemistry, with all the attendant problems of potential toxicity and safe storage of chemicals etc. Also, once formed, the product cannot be recycled without first destroying the molecular network that was created[2]. These problems are overcome in the following class of materials.

1.3 Thermoplastic Rubberlike Solids

In these materials the conditions for rubberlike behaviour are met, but connectivity between the molecules is achieved by physical means, i.e. secondary bonding such as van der Waals bonds or hydrogen bonds. In such cases, the mobile molecular chains essential for rubbery behaviour are combined in block copolymers with less mobile molecular segments that cohere to form relatively rigid domains, thereby acting as physical crosslinks. Thus each molecule contains "soft" (rubberlike) segments connected to "hard" segments. The two types of segment are chosen to be immiscible, leading to a two-phase microstructure. Provided the ratio of soft segments to hard

[2] The accumulation of used rubber car tyres, that cannot be easily recycled, is currently a vexing environmental problem.

segments is high enough, the structure consists of small hard regions (glassy or crystalline) where the hard segments have agglomerated, in a rubbery matrix formed by the soft segments – see the sketch in *Figure 4*. Such materials are known as "thermoplastic elastomers". They offer significant manufacturing advantages over the chemically crosslinked elastomers, since they are easily formed into shape by the following simple sequence.

1. Heat to above the softening temperature of the hard phase (glass transition or melting temperature).
2. Form into shape by moulding or extrusion.
3. Cool to solidify the hard phase.

They are also easily recycled, by re-softening, shaping and cooling, in the same fashion as other thermoplastic polymers. The drawback of these materials, however, is their inferior resilience as compared to, for example, natural rubber. For any particular product requiring a rubberlike material, the manufacturer must weigh the relative importance of all these factors, including cost, in selecting the optimum material.

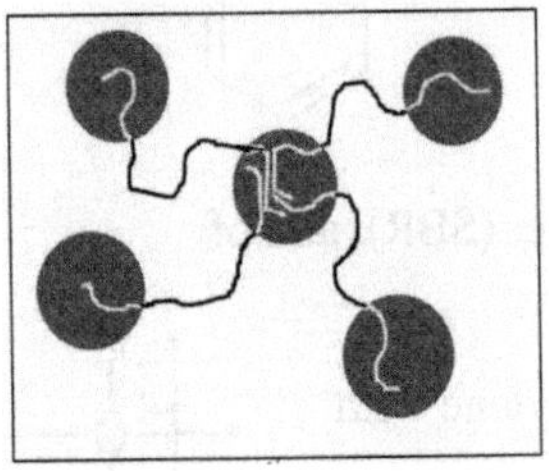

Figure 4. Schematic sketch of the two-phase structure of a thermoplastic elastomer. Four typical molecules are shown. Each has its ends embedded in hard domains, providing "physical crosslinks".

A good example of a thermoplastic elastomer is the styrene-butadiene-styrene class of block copolymers, with structure as given below

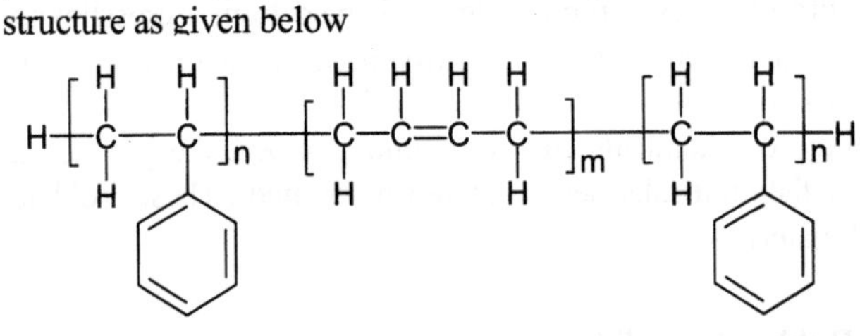

where typical values are $n \approx 100$ and $m \approx 800$: thus the molar ratio soft:hard is 4:1, ensuring that the rubbery polybutadiene forms the continuous matrix and polystyrene the discrete domains.

An important sub-set of the thermoplastic elastomers is the family of *polyurethane* elastomers, formed by reacting di-isocyanate molecules, of general structure $O{=}C{=}N\text{-}R_1\text{-}N{=}C{=}O$, with a macrodiol of general structure $HO\text{-}R_2\text{-}OH$, where R_2 is a flexible polymer chain of molecular weight typically 2000, and a chain extender such as ethylene glycol $HO\text{-}(CH_2)_2\text{-}OH$. This produces copolymers containing the urethane linkage after which these elastomers are named

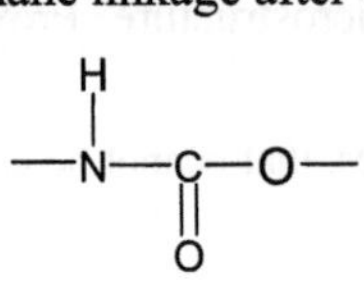

For example, a typical polyurethane elastomer consists of molecules with a repeat unit of the form:

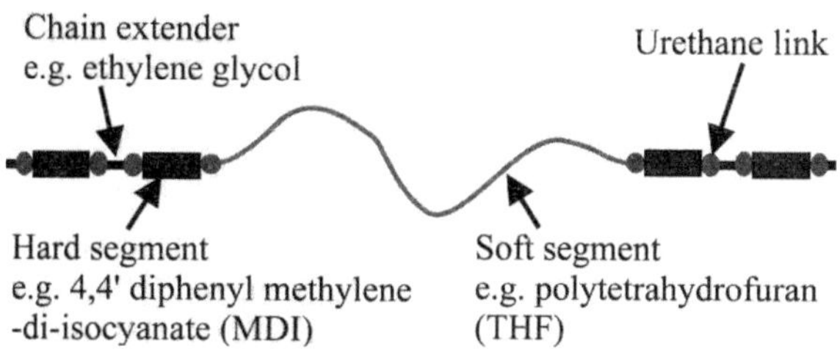

Phase segregation produces a microstructure similar to that shown in *Figure 4*. In the example above, assemblies of the hard MDI segments, hydrogen-bonded to their neighbours via the urethane links, form discrete domains, in a continuous phase of rubbery THF. The polyurethane elastomers form a particularly versatile class of rubber, since the properties can be tailored by varying the chemistry of hard and soft segments. The Handbook by Oertel (1994) provides a useful and authoritative source of information.

It is clear that the thermoplastic elastomers are rubbery-matrix composite materials, with two very desirable features for a composite: a fine dispersion of the reinforcing particles, and strong covalent bonding between the particles and matrix. In the case of the SBS copolymers, a typical hard particle diameter is 20 nm, while for the polyurethane it is even smaller, ca 5 nm. The effect on the mechanical response can be seen in data from the author's laboratory in *Figure 5*. This shows load/unload cycles for a polyurethane elastomer. Compare with *Figure 3* and note that the THF matrix at room temperature is in the rubbery state, with properties close to natural rubber. Clearly the presence of the hard segments has two significant effects. Firstly, the initial stiffness has risen: now we have Young's modulus $E \simeq 70$ MPa. Secondly there is greater hysteresis, presumably arising from some plastic flow within the hard domains or at their boundaries with the soft matrix. However, these materials are relatively soft and resilient, and hence they are useful in many applications where rubberlike properties are required. Note that a large proportion of the strain is recovered on unloading.

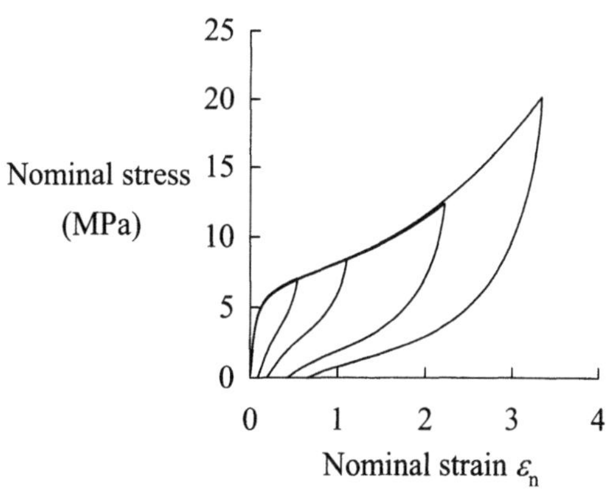

Figure 5. Load/ unload cycles for a polyurethane elastomer based on MDI hard segments and THF soft segments. Each cycle refers to a new specimen.

1.4 Filled Rubber

In practical applications, it is common for rubbery polymers to be reinforced by fine filler particles. The purpose is to improve particular properties such as abrasion resistance. For the polymers with a -C-C- backbone, the most widely used filler is carbon black. This form of powdered carbon is obtainable in a range of particle sizes: the most reinforcing action being achieved with the smallest particles, for a given proportion of filler. *Figure 6* shows how the particles of carbon black tend to cohere into aggregates. Typical particles may be as small as 20 nm wide, but the aggregated clumps of particles are significantly larger.

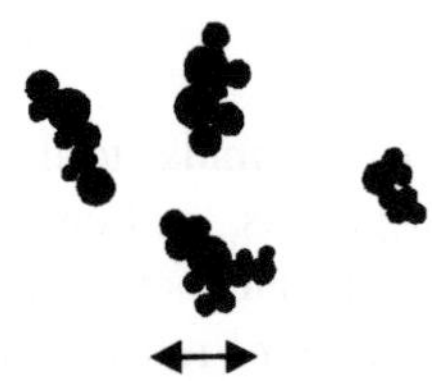

Figure 6. A sketch of the aggregated particles of carbon as dispersed in a sample of rubber. The scale bar indicates 100 nm.

The proportion of filler in a rubberlike solid is usually expressed in parts (usually by mass) expressed per hundred of rubber (phr). When carbon black is dispersed, for example, in natural rubber, and the rubber is then cured (i.e. crosslinked) we again end up with a form of composite material: relatively rigid reinforcing particles in a matrix of rubbery polymer. Although the particle-matrix bonding is not by covalent bonding, as in the case of the thermoplastic elastomers, there is significant physico-chemical interaction between the carbon and the rubber. The details remain a matter of lively debate. Certainly, the strength of the bond between particles and matrix is sensitive to the physical roughness and chemical state of the carbon surfaces. Medalia and Krauss (1994) provide a useful review of the issues.

Figure 7 shows load/unload cycles for a sample of carbon black filled natural rubber.

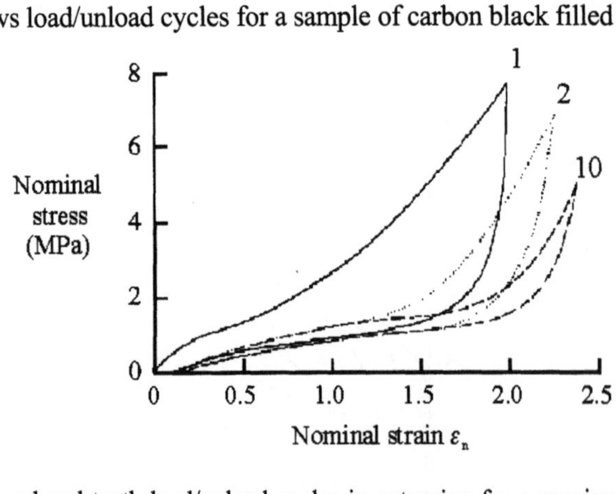

Figure 7. First, second and tenth load/unload cycles in extension for a specimen of natural rubber filled with 50 phr carbon black. Data from Lindley (1964).

Comparing the curves with those in *Figures 2* and *3*, we see that the carbon particles increase hysteresis. However, the reinforced rubber retains the low Young's modulus (typically 1-10 MPa), high recoverability of strain, and high extensibility associated with rubberlike solids.
Figure 7 also reveals a new feature, that seems to be common to rubberlike polymers, and is particularly pronounced in the presence of particulate reinforcements. The stress-strain response is sensitive to the prior straining sequence. Curves obtained on reloading show lower stiffness, lower hysteresis and higher recoverability of strain if the material has previously been deformed to a higher strain. This effect poses something of a dilemma for the experimenter (see next section); but it must certainly be recognised in the design of any programme of mechanical testing of rubberlike solids.

2 Mechanical and Thermomechanical Testing of Rubberlike Solids

2.1 Motivation and aims of experimental studies

Mechanical and thermomechanical testing of rubberlike solids impose particular requirements on the experimenter. Firstly, it is apparent from the previous section that rubberlike solids can withstand unusually large deformations without failure; for example a typical lightly crosslinked natural rubber can reach an extension in excess of 600%. Therefore experimental methods intending to explore the mechanical response fully must be able to accommodate such deformations. Thus the test machine must be able to accommodate correspondingly large displacements. Similarly, methods employed to determine the stress response must retain their validity across such a wide range of deformations.

Secondly, as we saw especially for the two-phase rubberlike materials, response depends on the prior stress/strain history. A complete investigation therefore requires the employment of varying histories, and in particular the use of non-monotonic loading.

Thirdly, another feature of the response, not yet mentioned, is that in common with other polymers, the observed hysteresis is associated with time/frequency-dependence in their mechanical response. As it happens, in unreinforcéd rubbery polymers the phenomenon is less marked than in many other polymers. Nevertheless, some applications hinge critically on material performance across the time/frequency domain, and test methods must be able to encompass this.

Fourthly, for engineering relevance, there is a need to recognise that performance in a product usually involves mechanical response in three dimensions. In the case of rubbery polymers undergoing large deformations, the correct three-dimensional analytical description of their finite strain response remains a matter of dispute. As yet, there is no widely accepted method for predicting the 3D response on the basis of response in a uniaxial test. Therefore test methods are required that will allow measurement of the response in two and three dimensions.

Taking all these considerations together, the challenge for the experimenter is considerable. Moreover, experiments are expensive. Costs include the following.

1. Amortization of the cost of the testing machine over a defined period.
2. Purchase of materials to be tested.
3. Cost of technician's time in preparation of test specimens (moulding of material samples to be tested, and cutting of test pieces).
4. Cost of machine operator's time in setting up and carrying out experiments.
5. Cost of analyst's time in processing output data from experiments.

6. Costs of reporting the results.
7. Overhead costs of running the laboratory.

A conservative estimate might be £500 for one rather straightforward experiment taking two hours on a standard test machine. So why do experiments?

It is impossible to escape the need to measure mechanical and thermomechanical response of materials such as elastomers. First and foremost there is the engineering need. There are many products employing these materials where mechanical or coupled thermomechanical behaviour is a critical determinant of performance. Manufacturers need the information for optimisation of materials and products. Secondly there is the scientific need. Although rubberlike solids are of such practical importance, the physical origins of many aspects of their behaviour remain poorly understood. Further progress will be possible only on the basis of more comprehensive information concerning their mechanical and thermomechanical response.

Here we shall consider experiments required to meet one or more of the following general objectives.

1. To generate information on the behaviour of the material under conditions of interest.
2. To compare the behaviour of the material with the predictions of physical or phenomenological theories, to test their validity.
3. To obtain the parameters in physical or phenomenological theories of the material behaviour.
4. To validate continuum models that claim to predict inhomogeneous fields of deformation, stress and temperature.

There are also other important reasons for testing rubberlike solids, such as the following.

5. Quality control in manufacturing.
6. Whole-product testing to predict service performance.

Procedures for 5 and 6, however, tend to be dominated by industry-specific requirements, and lie beyond the scope of the present book.

In design and execution of experimental programmes, the experimenter aims for the optimum balance of precision, accuracy and economy, appropriate to the application.

2.2 Principles of mechanical and thermomechanical testing

A systematic approach to achieving these goals begins with definitions of important terms.

The first step is to define the intended scope of the testing. Here we shall employ a continuum representation of rubberlike solids. But in reality these solids consist of large numbers of discrete molecules and possibly larger-scale filler particles. So we must specify a minimum length scale Λ for which we claim validity of the continuum approximation. In view of the typical scale of filler particles (see Section 1), an appropriate order of magnitude might be $\Lambda = 10\ \mu\text{m}$. In the continuum description employed here and in other chapters of this book, a material is assumed to be continuous on length scales greater than Λ. The aim of experimental methods is then to characterise the continuum properties of a homogeneous region of material larger than Λ. Clearly, these properties result from an averaging over physical processes occurring at length scales *smaller* than Λ. We assume therefore that, with respect to its influence on the continuum response, the instantaneous state of the microstructure is adequately represented by an ordered set S of internal tensor variables of various rank. As in other polymers, there is evidence for stress-induced structural damage in rubberlike solids: indeed this is one explanation for the different responses on successive load cycles (see *Figure 7*). For this reason it will be convenient to label $\dot{\mathcal{S}}$ the ordered set of time-derivatives of the elements of $\mathcal{S}$.

The second step is to define a reference, stress-free, configuration of the solid material, as an equilibrium state where it would remain unchanged indefinitely in the absence of stress or temperature change. At the origin of time $t = 0$, the material is assumed to exist in this reference configuration with $\mathcal{S} = \mathcal{S}_0$, say, and the elements of $\dot{\mathcal{S}}$ vanish.

The third step is to define quantities with which to describe the thermomechanical response of the material. This issue is non-trivial, given the need to accommodate the geometry of large deformations that are potentially achievable. Other chapters of this book discuss the matter in further detail. For the present, we shall be content with four quantities.

1. Temperature is expressed in terms of the *absolute temperature* T on the Kelvin scale.
2. Stress is expressed in terms of the *Cauchy stress* $\boldsymbol{\sigma}$, defined in terms of the traction vector $\boldsymbol{t}$ and the force $\Delta\boldsymbol{P}$ acting on an element of area Δa with unit normal $\boldsymbol{n}$ (in the *current*, rotated and strained configuration):

$$\boldsymbol{t} = \operatorname*{Lim}_{\Delta a \to 0} \frac{\Delta \boldsymbol{P}}{\Delta a} = \boldsymbol{\sigma n}. \tag{3}$$

3. Deformation is expressed generally in terms of the *deformation gradient* $\boldsymbol{F}$, defined in terms of the relation between the new position of a material point $\boldsymbol{x} = x_i\mathbf{e}_i$ and its original position $\boldsymbol{X} = X_i\mathbf{e}_i$ in the reference configuration - where $\mathbf{e}_i$ (i=1..3) form an orthonormal basis. A typical component of $\boldsymbol{F}$ is:

$$F_{ij} = \frac{\partial x_i}{\partial X_j}. \tag{4}$$

$\boldsymbol{F}$ does not represent the deformation objectively, as it includes a contribution from rigid body rotation. From its polar decomposition, however, $\boldsymbol{F}$ may always be expressed $\boldsymbol{F} = \boldsymbol{VR}$ where $\boldsymbol{R}$ is a rotation and $\boldsymbol{V}$ a subsequent rotation-free deformation. Objectivity requires that we express material response in terms of $\boldsymbol{V}$ but not $\boldsymbol{R}$. However it is more convenient to use instead a related quantity - the left Cauchy-Green deformation tensor $\boldsymbol{B}$:

$$\boldsymbol{B} = \boldsymbol{F}\boldsymbol{F}^{\mathrm{T}} = \boldsymbol{V}^2. \tag{5}$$

This is an appropriate descriptor when expressing experimental results in terms of the Cauchy stress, since it quantifies deformation in the *current* (as-rotated) configuration. Clearly, the undeformed configuration corresponds to $\boldsymbol{B} = \boldsymbol{F} = \mathbf{I}$, while a pure three-dimensional extension parallel to the reference axes $x_1 = \lambda_1 X_1$, $x_2 = \lambda_2 X_2$, $x_3 = \lambda_3 X_3$ corresponds to

$$\boldsymbol{F} = \begin{pmatrix} \lambda_1 & 0 & 0 \\ 0 & \lambda_2 & 0 \\ 0 & 0 & \lambda_3 \end{pmatrix}, \quad \boldsymbol{B} = \begin{pmatrix} \lambda_1^2 & 0 & 0 \\ 0 & \lambda_2^2 & 0 \\ 0 & 0 & \lambda_3^2 \end{pmatrix}, \tag{6}$$

where $\lambda_1, \lambda_2, \lambda_3$ are known as the *principal stretches*. It is clear that $\lambda_1^2, \lambda_2^2, \lambda_3^2$ are the eigen values of $\boldsymbol{B}$.

4. Also of interest in the thermomechanical response is that, as a consequence of deformation, the material in general acts as a net source or sink of heat. We define the specific heat source s as the rate of heat generation per unit mass.

To study the behaviour of the material (Objectives 1-4 above), a test sample with dimensions sufficiently exceeding Λ is exposed uniformly to the required independent variables, and the required dependent variables measured. The sample must be sufficiently large to enable these measurements to be made with the precision required. The process may be expressed schematically as in *Figure 8*. The material sample may be considered in engineer's parlance as a *system*, which is exposed to inputs – here shown as temperature and deformation – and producing outputs – here shown as heat and stress.

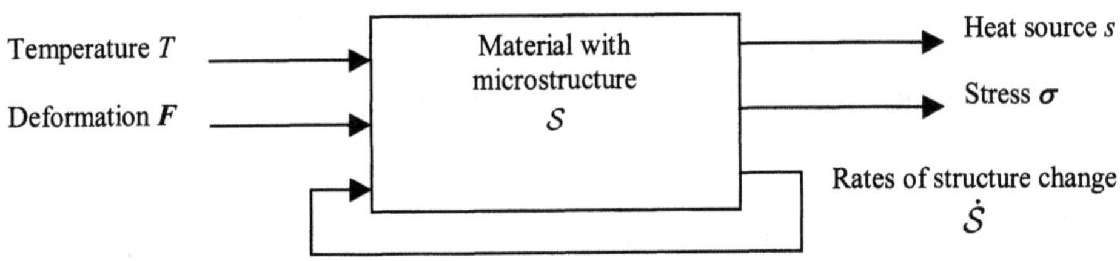

Figure 8. A test sample of material may be considered as a *system*, subjected to *inputs* temperature and deformation, and producing *outputs* heat and stress. The temperature and deformation may also cause evolution of the material microstructure.

The reference configuration corresponds to:

$$t = 0, \quad T = T_0, \quad \boldsymbol{B} = \mathbf{I}, \quad \boldsymbol{\sigma} = \mathbf{0}, \quad s = 0, \quad \mathcal{S} = \mathcal{S}_0. \tag{7}$$

Now consider a later time t. The sample has been subjected uniformly to a history of temperature $T(u)$ and deformation gradient $\boldsymbol{F}(u)$, where $0 \le u \le t$. It responds at time t with a stress $\boldsymbol{\sigma}(t)$, heat output rate $s(t)$ and, in general, with finite rates of evolution of the internal variables $\dot{\mathcal{S}}$. The latter possibility introduces feedback as shown in *Figure 8*. We shall assume here that the *only* agents able to produce stress, heat or change of internal variables are the deformation and temperature. Moreover, objectivity of the material description requires that it is only the non-rotational part of the deformation that intervenes in this respect. Hence the relevant measure of deformation here is $\boldsymbol{B}$ rather than $\boldsymbol{F}$. It follows that the most general form of the stress response may be written:

$$\boldsymbol{\sigma}(t) = \mathop{\mathcal{F}}_{u=0}^{t}\left\{T(u), \boldsymbol{B}(u); \mathcal{S}_0\right\} , \tag{8}$$

where $\mathcal{F}$ is a tensor functional of T and $\boldsymbol{B}$ over the time interval $0 \le u \le t$ that also depends on the (constant) elements of the set $\mathcal{S}_0$. Our decision to choose strain as an independent variable and stress as dependent variable is, in principle, arbitrary. But it has the advantage of being the form in which most constitutive models are expressed in this book and elsewhere. In practical experiments it is usually more convenient to impose most of the components of stress and to measure resulting strain than vice versa, suggesting an alternative form, equivalent to equation (8) thus:

$$B(t) = \underset{u=0}{\overset{t}{\mathcal{F}'}}\{T(u), \sigma(u); S_0\} \ , \tag{9}$$

where $\mathcal{F}'$ is a new tensor functional of T and σ depending also on S_0.

Recall that there are 6 independent components of strain and 6 of stress. Of these 12, we may arbitrarily impose only 6, since equations (8) and/or (9) provide 6 equations linking them. In fact, experiments are frequently carried out with mixed boundary conditions, where one or two of the components of strain are imposed, while the remaining five or four components of stress are imposed.

The practical implementation of this approach to testing, and the attendant difficulties, are illustrated by careful consideration of the widely-used tensile test, at constant rate of grip displacement. A similar analysis may be applied to other test methods. One half of a typical test specimen is sketched in *Figure 9*. The intention is to deform it isothermally at a constant rate of uniaxial extension: $\lambda_1(u) = \dot{\lambda}u$ where $\dot{\lambda}$ is a constant. The specimen has initial dimensions: $L \times B \times D$. These change to $l \times b \times d$ on deformation. We assume the specimen has uniform temperature and is in mechanical equilibrium.

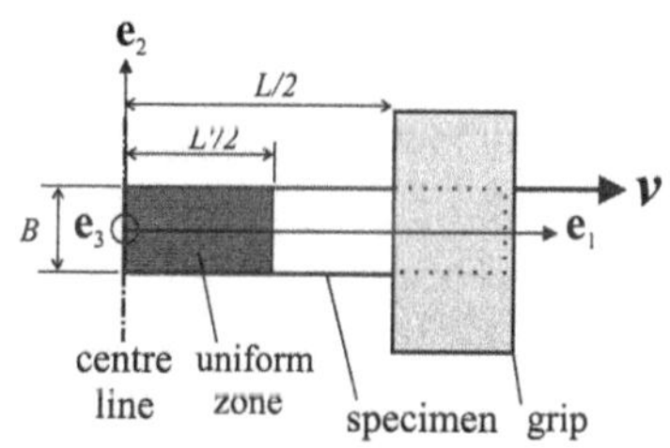

Figure 9. Schematic diagram of (one half of) a rectangular test specimen (length L between grips, width B, thickness D) at the start of a conventional tensile test. Velocity $\boldsymbol{v} = \dot{\lambda}(L/2)\mathbf{e}_1$ where $\dot{\lambda}$ is constant.

For simplicity, the specimen is shown without the usual waisted shape specified in many standard test protocols to minimise the probability of premature failure near the grip. Consider the boundary conditions and their implications. With the usual form of rigid grip, and invoking mirror symmetry about the plane $X_1 = 0$, such a test imposes the following mixed boundary conditions:

$$x_1 = 0, \ \boldsymbol{t} = -|\boldsymbol{t}|\mathbf{e}_1, \ \text{on face } X_1 = 0 \ , \tag{10}$$

$$x_1(u) = X_1\left(1 + \dot{\lambda}u\right), x_2 = X_2, \ \text{on face } X_1 = L/2 \ , \tag{11}$$

$$\boldsymbol{t} = \mathbf{0} \ \text{ on faces } X_2 = \pm B/2, \ X_3 = \pm D/2 \,. \tag{12}$$

Far from the grip ($X_1 \ll L/2$), these conditions ensure:

$$\sigma_{12} = \sigma_{23} = \sigma_{31} = \sigma_{22} = \sigma_{33} = 0 \,.$$

Hence it follows (from equilibrium) that σ_{11} is uniform. Within this *uniform zone* it follows therefore from equation (9) that the strain $\boldsymbol{B}$ is uniform, and if the material is isotropic it follows from the diagonality of stress $\boldsymbol{\sigma}$ that $\boldsymbol{B}$ (and hence $\boldsymbol{V}$) is also diagonal and there is no rotation.

Hence we deduce $\boldsymbol{F}$ is uniform and diagonal. Thus an experiment of this type imposes, within the uniform zone, five components of the stress as above, and one component of $\boldsymbol{F}$:

$$F_{11} = \lambda_1 = 1 + f(u) \tag{13}$$

and hence of $\boldsymbol{B}$. The function $1+f(u)$ will be approximately linear as intended, but the closeness of the approximation will depend on the response of the non-uniform region near the grip. As it is imperfectly defined by the experiment, it is necessary to measure it. The testing machine usually also measures the applied axial force P, and hence the nominal stress $\sigma_n = P/BD$. From the axial symmetry of the experiment in the uniform zone, it follows that $F_{22} = \lambda_2 = \lambda_3 = 1 - g(u)$. Measurement of the function $g(u)$ would complete characterisation of the response, since the three off-diagonal terms in $\boldsymbol{F}$ vanish. It is clear that the deformation measurements would need to be made within the uniform zone $-L'/2 < X_1 < L'/2$. The well-known St Venant Principle (for example, see Timoshenko and Goodier (1951)) gives an estimate of the size of the nonuniform zone to be avoided: $L/2 - L'/2 \approx B$ in the linear elastic, isotropic case. Conversion of the measured nominal stress into true (Cauchy) stress σ requires knowledge of λ_2, since it depends upon the new lateral dimensions b,d of the specimen:

$$\sigma = \frac{P}{bd} = \frac{\sigma_n}{\lambda_2^2}. \tag{14}$$

But fortunately rubberlike solids deform in tension almost at constant volume, and for many purposes therefore an adequate approximation for the volume ratio (new volume : old volume) is $J \equiv \det \boldsymbol{F} = \lambda_1\lambda_2\lambda_3 \simeq 1$. Then it is unnecessary to measure λ_2 directly, and instead the true stress may be found simply from $\sigma = \lambda_1 \sigma_n$.

2.3 Limiting Time-scales

It is important to recognise that there is not complete freedom in experiment design, if the assumptions of the systematic approach outlined above are to be justified. Specifically, there are physically-imposed constraints on the time-scales of experiments.

First consider the achievement of a zone of uniform stress within the sample. This is possible only if inertial forces vanish: i.e. if the material deforms in a state of mechanical equilibrium. The velocity of longitudinal stress waves in a solid is given in terms of the Young's modulus E and density ρ of the solid:

$$v = \sqrt{E/\rho}. \tag{15}$$

For a rubberlike solid this has an unusually low value of order 30 m/s, arising from the unusually low modulus. Thus the transit time for a longitudal stress wave to pass along the length of the specimen in *Figure 9* is $L/v \sim 1$ ms, for a typical specimen gauge length of 30 mm. Given that multiple (at least 10, say) wave reflections are required for equilibrium to be established, and assuming that we require to resolve nominal strains as small as 5×10^{-2} (i.e. of order 1% of a maximum nominal strain of 5), it follows that we are constrained to testing speeds $\dot{\lambda} < 5$ s^{-1}.

The intention for most experiments is that they be carried out under isothermal conditions. Thus the experimenter wishes to achieve a constant temperature T prior to the test, although this value may differ from the initial temperature T_0 in the reference configuration. If we neglect radiative heat transfer, the two modes of heat transfer available are conduction and convection.

They each have their own characteristic time-constants for completion. Again considering the rectangular specimen in *Figure 9*, these time-constants are respectively

$$\tau_{\text{cond}} = \frac{D^2}{4\alpha}, \quad \tau_{\text{conv}} = \frac{D\rho c}{2h}, \tag{16}$$

where α is the thermal diffusivity of the rubber, ρ is the density, and c the specific heat, while h is the heat transfer coefficient for convection on the faces $X_3 = \pm D/2$ of the specimen. With approximate typical values $D = 2$ mm, $\alpha = 10^{-7}$ m^{2}s^{-1}, $\rho = 1000$ kg m^{-3}, $c = 1000$ J kg^{-1} K^{-1}, $h = 10$ W m^{-2} K^{-1}, we obtain $\tau_{\text{cond}} \sim 10$ s, $\tau_{\text{conv}} \sim 100$ s. Thus in this case, to be sure of adequate temperature equilibration, the experimenter must allow an equilibration time $t_{\text{equilib}} \gg 100$ s, to be certain temperature gradients will be negligible.

A similar constraint applies to the thermal response *during* testing. The evolution of heat s will clearly, in general, lead to a change in temperature during the test - i.e. there is a further feedback loop linking s to T, not shown in *Figure 8*. For the temperature change to be restricted to a tolerance ΔT_{tol}, the heat source must be limited to

$$s < \frac{c\Delta T_{\text{tol}}}{\tau_{\text{conv}}}. \tag{17}$$

In the case of the specimen shown in *Figure 9*, the maximum heat source available is the power input (per unit mass) from the testing machine $\sigma_{\text{n}}\dot{\lambda}/\rho$ (in the absence of any latent heat effect from strain-induced structural change). Hence, to ensure isothermal conditions we require

$$\dot{\lambda} < \frac{\rho c \Delta T_{\text{tol}}}{\sigma_{\text{n}}\tau_{\text{conv}}}. \tag{18}$$

In large deformation tests on rubberlike solids, the nominal stress reaches values of order 10 MPa (see *Figures 3, 5,* and *7*), and the tightest tolerance normally imposed on temperature in standard test protocols is ±1 K. These figures indicate that to ensure testing under isothermal conditions as conventionally defined requires $\dot{\lambda} < 10^{-3}$ s^{-1}.

Alternatively, to ensure deformation proceeds uniformly under *adiabatic* conditions, requires that there is insufficient time for temperature equilibration within the specimen, and hence the test duration is restricted to $t_{\text{test}} \ll \tau_{\text{cond}}$. For a tensile test under constant nominal strain rate $t_{\text{test}} = (\lambda_{\text{max}} - 1)/\dot{\lambda}$. Hence the condition for uniform adiabatic conditions is

$$\dot{\lambda} \gg \frac{\lambda_{\text{max}} - 1}{\tau_{\text{cond}}}. \tag{19}$$

For a large deformation test on a rubbery material, for example with $\lambda_{\text{max}} \approx 6$, equation (19) gives the condition for uniform adiabatic conditions as $\dot{\lambda} \gg 0.5$ s^{-1}. Combining all these constraints produces the bounds on acceptable ranges of strain-rate shown in *Figure 10* below.

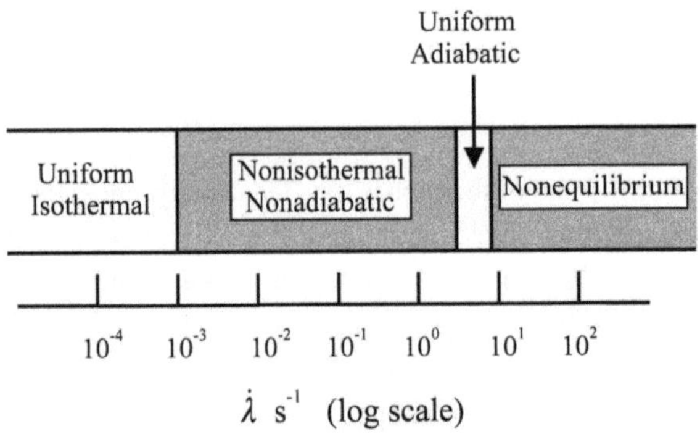

Figure 10. Typical ranges of acceptable and unacceptable strain-rate in the testing of a sample as shown in *Figure 9* (light and dark shading respectively).

2.4 Practical implementation

While the above notes give a framework for design of experiments on rubberlike solids, details of their practical implementation are largely dictated by the context of the experimental programme and its purpose. In particular, in many cases (especially where the data are to be used in a commercial context) good practice dictates that experiments are carried out according to a standardised protocol, laid down by a national or international standards organisation – e.g. ASTM (in the USA), BS (in the UK), or ISO (international). Brown (1996) gives a useful discussion of this aspect of testing. In other situations, the experimenter is free to design experiments to suit a particular research goal. The execution of any mechanical or thermomechanical test, however, always involves three steps A-D as follows.

A - Preparation of test specimens

In the case of crosslinked rubberlike solids, specimen manufacture is most likely to consist of the steps: thorough mixing of ingredients; moulding and curing of a shape such as a sheet; cutting of specimens from the shape. In the case of thermoplastic rubberlike solids, the procedure is more likely to consist merely of melt-forming a shape (by moulding or extrusion) and cooling; cutting of specimens from the shape. In both cases, there is the alternative possibility of moulding directly into the shape of test specimens. Recall, however, that if the aim is to determine the properties of the material continuum, the specimen must be structurally uniform (on length scales larger than Λ) *throughout* the assumed "uniform" region under test. This is usually difficult to achieve by direct moulding, because of adventitious inhomogeneity introduced by flow patterns or nonuniform cooling.

Clearly, the process of material and specimen manufacture is likely to have a major impact on the structural state of the test material, as expressed by S. In fact the diagram in *Figure 8* applies equally to the process history of the material. The purposes of the specimen preparation procedure, therefore, are twofold: to form a specimen of the required shape for testing *and* to bring the material into the initial (constant) structural state S_0 intended for testing. Precautions must be taken to ensure any slow post-mould structural evolution has become negligible, and also to prevent any unintended chemical degradation (for example from exposure to ultra-violet light) prior to testing. For this reason it is usual to store specimens after moulding for at least a period of 16 hours, in the dark, before testing.

A problematic, and unresolved, question for the experimenter is what to do about the effects of strain history seen in *Figure 7* (see also Section 4). These are especially pronounced in particle-filled rubberlike solids. They clearly reflect some form of structural breakdown caused by straining, as recognized in *Figure 8*. Since the stress response on first loading to a particular stretch is so unrepresentative of subsequent stretching (in showing a stiffer response), it is common practice to condition specimens with one or more load/unload cycles to a stretch equal to or greater than the maximum stretch to be encountered, before beginning a series of mechanical tests. Unfortunately, stress and strain data from the conditioning sequence are usually not included in published results. This introduces ambiguity into what constitutes the initial structural state of the material. For example, the structural changes caused by a tensile conditioning sequence on an initially isotropic material must be expected to introduce anisotropy into its response to subsequent straining, although careful experimenters such as Jones and Treloar (1975) have taken care to check this remains within acceptable bounds. The most appropriate course for the experimenter is to be guided by what the experimental data are required for. Are they intended to characterize the constitutive response of the material in its as-moulded structural state? Or are they intended to characterize the structural state existing after moulding, followed by a particular conditioning straining sequence?

B – Application of independent variables

As we saw above, the independent variables will consist of boundary: forces, displacements and temperature. Some of these will be *passive* (free surfaces providing zero load, or rigid grips providing zero lateral displacement). The common *active* means of applying them in mechanical or thermomechanical tests are as follows:

forces

- i. dead weights
- ii. servo-hydraulically driven pistons
- iii. internal pressure

displacements

- i. motor-driven screws
- ii. servo-hydraulic pistons

temperature

- i. convective heat transfer from circulated air
- ii. radiative heat transfer from infra-red heaters.

C – Measurement of dependent variables

Similarly, the measurable dependent variables will comprise boundary forces, displacements and temperature. In addition, it may be necessary to check values of the independent variables. Current measurement techniques available, and potentially suitable for rubberlike solids are as follows.

forces

- i. strain-gauged spring-element force transducers

ii. strain-gauged diaphragm pressure transducers

displacements

i. linear variable differential transformer (LVDT) transducers
ii. strain-gauged spring element extensometers
iii. potentiometric extensometers
iv. video extensometers
v. laser scanning extensometers
vi. photography or video recording with image analysis
vii. geometric moire interferometry with image analysis
viii. speckle pattern interferometry

Of the above, i.-v. measure relative displacements of two points and are useful for the uniform zone of *material* characterisation tests (Objectives 1-3 in Section 2.1), whereas vi.-viii. are full-field methods useful for situations of nonuniform strain, as needed for continuum model *validation* tests (Objective 4 in Section 2.1).

temperature

i. thermocouples
ii. platinum resistance sensors
iii. radiometers
iv. thermography

Methods i.-iii. provide single point measurements, while iv. is a full field method suited to Objective 4.

3 Exploration of Strain-States

3.1 Analytical framework

Hyperelasticity

To meet Objectives 1-3 of Section 2.1, experiments must explore systematically the functional relationships expressed symbolically in equations (8) and (9). Since strain $\boldsymbol{B}$ and stress $\boldsymbol{\sigma}$ are both second rank tensor functions of time, there are two aspects to the relationships concerning them. There is: (a) dependence on stress or strain *state* respectively (by which we mean variations in the relative values of their components), and (b) dependence on stress and strain *history* (by which we mean variations in the time-sequences of their components).

In exploring the dependence on strain-state, it greatly simplifies the interpretation of results if we exploit the experimental fact that rubberlike solids approximate to being *hyperelastic*: having a unique stored strain energy density function $W(\boldsymbol{B}(t))$ that depends only on current strain. Since the following relation exists between stress, strain and W – see Holzapfel (2000) and Haupt (2002)

$$\boldsymbol{\sigma} = 2J^{-1}\frac{\partial W(\boldsymbol{B})}{\partial \boldsymbol{B}}\boldsymbol{B} \tag{20}$$

it follows that in this approximation there is a unique relation between the current values of strain and stress $\boldsymbol{B}(t),\boldsymbol{\sigma}(t)$ independent of their previous history[3]. This simplifies enormously the relation given in equation (8), as the role of time disappears: the instantaneous value of $\boldsymbol{\sigma}$ depends only on the current $\boldsymbol{B}$, in addition to T and $\mathcal{S}_0$. We invoke the hyperelastic approximation here and postpone to Section 4 the investigation of deviations from it.

Isotropy

A sub-set of the initial structure $\mathcal{S}_0$ is the initial symmetry group $\mathcal{G}_0$ of the material microstructure. This usually provides further simplication in the anticipated response, as the symmetry of the structure must be reflected in the symmetry of the response. In fact, virtually all reported experimental work on rubberlike solids has been for materials in isotropic form. $\mathcal{G}_0$ is then the set of all orthogonal transformations – the "full orthogonal group". We shall therefore restrict attention here to *isotropic* materials. In this case, W may be expressed in terms of any three independent isotropic invariants of $\boldsymbol{B}$, for example the usually employed coefficients in its characteristic equation:

$$\begin{aligned} I_1 &= \operatorname{tr}\boldsymbol{B} = \lambda_1^2 + \lambda_2^2 + \lambda_3^2 \\ I_2 &= \tfrac{1}{2}\left[\left(\operatorname{tr}\boldsymbol{B}\right)^2 - \operatorname{tr}\boldsymbol{B}^2\right] = \lambda_1^2\lambda_2^2 + \lambda_2^2\lambda_3^2 + \lambda_3^2\lambda_1^2 \\ I_3 &= \det\boldsymbol{B} = J^2 = \lambda_1^2\lambda_2^2\lambda_3^2 \ , \end{aligned} \tag{21}$$

where we recall λ_i are the principal stretches. W also depends, of course, on T and the remaining structural variables $\mathcal{S}_0 / \mathcal{G}_0$. Holtzapfel (2000) and Haupt (2002) show that, in terms of I_i, equation (20) may be re-written

$$\boldsymbol{\sigma} = \alpha_0\mathbf{I} + \alpha_1\boldsymbol{B} + \alpha_2\boldsymbol{B}^2 \ , \tag{22}$$

where the coefficients are functions of I_i, T and $\mathcal{S}_0 / \mathcal{G}_0$.

The linear limit

From the definition of deformation gradient $\boldsymbol{F}$, it may also be written in terms of the gradient[4] of the displacement field $\boldsymbol{u} = \boldsymbol{x} - \boldsymbol{X}$:

$$\boldsymbol{F} = \mathbf{I} + \nabla\boldsymbol{u} \ . \tag{23}$$

Consequently, $\boldsymbol{B}$ may be related to the usual infinitesimal strain $\boldsymbol{\varepsilon}$ in the linear limit of small displacement gradients:

$$\tfrac{1}{2}\left(\boldsymbol{B} - \mathbf{I}\right) \rightarrow \tfrac{1}{2}\left(\nabla\boldsymbol{u} + \left(\nabla\boldsymbol{u}\right)^{\mathrm{T}}\right) = \boldsymbol{\varepsilon} \ . \tag{24}$$

Substitution from equation (23) into equation (22), and taking the linear limit in $\nabla\boldsymbol{u}$ leads to the most general expression for stress in an isotropic linear elastic solid under a small isothermal deformation:

$$\boldsymbol{\sigma} = 2\mu\boldsymbol{\varepsilon} + \eta \operatorname{tr}\boldsymbol{\varepsilon}\,\mathbf{I} \ , \tag{25}$$

[3] In Section 1 we saw that this is a good approximation for modest deformations of unfilled natural rubber, but is less accurate for the largest deformations and filled rubber.

[4] Note that the gradient here is computed with respect to the initial co-ordinates.

where μ, η are the elastic constants known as Lamé parameters. They depend on T and the remaining internal variables $\mathcal{S}_0 / \mathcal{G}_0$.

Incompressibility

A special feature of rubberlike solids is that the volume is relatively constant. This is revealed by separating equation (25) into the parts associated with shape change and volume change. Define the mean stress σ_{m}, dilation Δ, and deviatoric stress $\boldsymbol{s}$ and strain $\boldsymbol{e}$, in the usual way:

$$\begin{aligned} \sigma_{\mathrm{m}} &= \tfrac{1}{3}\mathrm{tr}\boldsymbol{\sigma}, \quad \Delta = \mathrm{tr}\boldsymbol{\varepsilon}, \\ \boldsymbol{s} &= \boldsymbol{\sigma} - \sigma_{\mathrm{m}}\mathbf{I}, \quad \boldsymbol{e} = \boldsymbol{\varepsilon} - \tfrac{1}{3}\Delta\mathbf{I}. \end{aligned} \tag{26}$$

Then equation (25) may be re-written as two separate equations governing the deviatoric and hydrostatic response of the material:

$$\boldsymbol{s} = 2G\boldsymbol{e}, \qquad \sigma_{\mathrm{m}} = K\Delta, \tag{27}$$

where G and K are the shear and bulk moduli. It is generally found that rubberlike solids have remarkably different stiffnesses under deviatoric and hydrostatic deformation. Typical orders of magnitude are (see Section 1)

$$G \sim 1\ \mathrm{MPa},\ K \sim 10^3\ \mathrm{MPa}.$$

It follows from equations (27) that rubberlike solids approximate[5] to being *incompressible* in the sense that

$$\|\boldsymbol{e}\| \gg \Delta \quad \text{for} \quad \|\boldsymbol{s}\| \approx |\sigma_{\mathrm{m}}|. \tag{28}$$

The situation at finite strains is less well documented. However, available data suggest that here, too, the approximation of incompressibility holds true. This approximation is frequently invoked in the interpretation of experimental results, and we shall exploit it below.

In the incompressible limit we have

$$J = I_3 = 1 \tag{29}$$

and the stress becomes indeterminate to within a hydrostatic stress $-p$ (to be found from boundary conditions). Thus equation (22) is modified to

$$\boldsymbol{\sigma} = -p\mathbf{I} + \alpha_1\boldsymbol{B} + \alpha_2\boldsymbol{B}^2, \tag{30}$$

where the coefficients are in this case

$$\alpha_1 = 2\left(\frac{\partial W}{\partial I_1} + I_1\frac{\partial W}{\partial I_2}\right), \quad \alpha_2 = -2\frac{\partial W}{\partial I_2}. \tag{31}$$

[5] We emphasise the word *approximate*, since these materials have finite bulk moduli.

3.2 Experimental exploration of strain-states

Experimental studies have employed two approaches: either using results to deduce empirically the form of the function W, or using results to check adherence to predictions made by assuming a particular form of the function. From equations (30) and (31), the principal stresses are readily obtained as

$$\sigma_i = -p + 2\left(\lambda_i^2 \frac{\partial W}{\partial I_1} - \frac{1}{\lambda_i^2}\frac{\partial W}{\partial I_2}\right). \tag{32}$$

An alternative approach is to employ the principal stretches as invariants, giving $W = W(\lambda_1, \lambda_2, \lambda_3)$, subject to the constraint $\lambda_1\lambda_2\lambda_3 = 1$ in the incompressible case. Principal stresses may then be found from equation (20). Invoking incompressibility they become:

$$\sigma_i = -p + \lambda_i \frac{\partial W}{\partial \lambda_i}. \tag{33}$$

Measurements of stress are employed therefore to obtain information on W, via its derivatives with respect to either the invariants or the principal stretches.

Experimental studies of the constitutive response of rubberlike solids are conducted usually under conditions of *plane stress*. There is a practical reason for this: it ensures the test specimen has at least one free surface where deformations can be measured. We here define axis $\mathbf{e}_3$ as the normal to the stress-free plane. It follows that

$$\sigma_{13} = \sigma_{23} = \sigma_{33} = 0. \tag{34}$$

Hence $\mathbf{e}_3$ is a principal direction of stress σ, and (in view of equation (22)) a principal direction of strain $\boldsymbol{B}$. Moreover, in view of the incompressibility assumption $I_3 = 1$, the free stretch is always given by

$$\lambda_3 = \frac{1}{\lambda_1\lambda_2}. \tag{35}$$

Exploring the effects of varying strain-state reduces therefore to the task of exploring the response throughout the two-dimensional space of λ_1, λ_2 or equivalently of I_1, I_2. The mapping between these two spaces, and their relation to different experiment types, is demonstrated conveniently by the diagrams given below in *Figures 11* and *12*.

The regions of these spaces accessible to each experiment type may be expressed in terms of *biaxiality ratio* ϕ, defined by:

$$\phi = \frac{\lambda_2 - 1}{\lambda_1 - 1}. \tag{36}$$

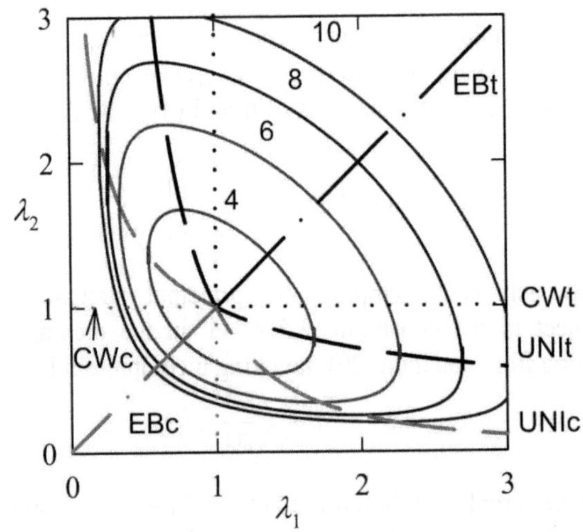

Figure 11. Contours of strain invariant I_1 and loci of various experiment types, for the case $\sigma_3 = 0$, $I_3 = 1$.

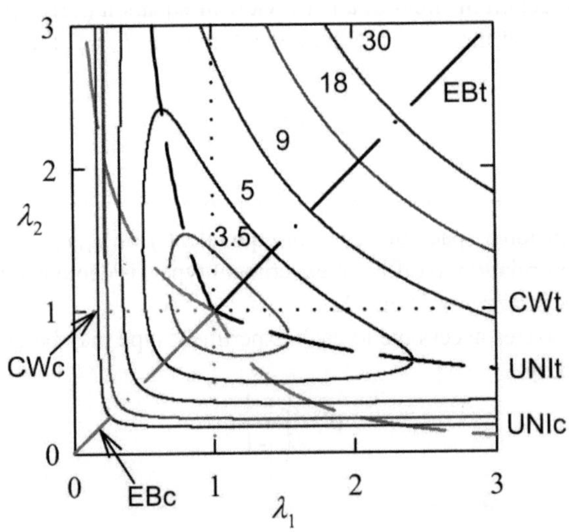

Figure 12. Contours of strain invariant I_2 and loci of various experiment types, for the case $\sigma_3 = 0$, $I_3 = 1$.

An important point arises from the combination of: the assumptions of hyperelasticity and isotropy (W is a symmetric function of the λ_i) and incompressibility (since $\lambda_3 = 1/\lambda_1\lambda_2$, strain-states are specified fully by their position in the above diagrams). The diagrams contain sets of equivalent strain-states with identical W, through permutation of the λ_i. Specifically, for a general point on the diagram there are five other equivalent points, except in the degenerate case when the point lies on the EB line, when there are only two others. Thus, for example, the EBt locus is equivalent to the two branches of the UNIc locus. Notes on the range of experiment types and their practical implementation are given below. We divide them into two classes, as follows.

A - Test methods giving access to two-dimensional regions of principal stretch space

1. Biaxial tension tests on sheet specimens

This is the most versatile type of test, allowing access to the largest region of principal stretch space, and with the most flexibility over the strain history. However, it requires either a biaxial testing machine with considerable complexity (see (1970), James and Green (1975; 1975) and Buckley and Turner (1999)) or the tedious manual biaxial adjustment of dimensions of a specimen, as used in some pioneering early work (see Rivlin and Saunders (1951) and Jones and Treloar (1975)). The accessible region is

$$\frac{\lambda_1^{-1/2} - 1}{\lambda_1 - 1} \le \phi \le 1 \ . \tag{37}$$

The use of a thin sheet specimen prevents the application of compressive stresses, since buckling would intervene.

The primary difficulty of this type of test is to avoid unwanted interference between the responses to the two axes of straining. Two axes of independent displacement are required, say along $\mathbf{e}_1$ and $\mathbf{e}_2$. Clearly, it is not possible to employ a single rigid grip on each axis, as this would restrict displacement in the transverse direction. The solution is to employ a set of discrete grips on each axis, as shown in *Figure 13* below.

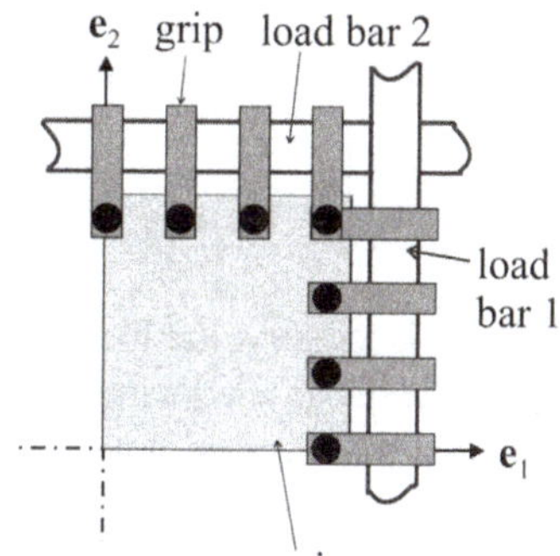

Figure 13. Schematic diagram of one quarter of a square specimen in a biaxial testing machine for materials in sheet form.

Each row of grips runs freely on a crosshead or "load bar", as shown, with the corner grip running on both load bars. Load bars on the same axis but opposing sides of the specimen move apart to stretch the specimen. Thus in the diagram above, relative to the origin shown, motion of load bar 1 imposes λ_1, while motion of load bar 2 imposes λ_2. Clearly, as in the uniaxial case in *Figure 9*, there is a zone of non-uniform strain in the region of the grips. Strain measurements must be made in the central uniform region. Stress on each axis is deduced from forces applied by the grips, sensed by load cells (usually strain-gauged spring elements) acted on by the load bar or one or more grips.

An early example of a machine of this type is the "Bistron" tester employed in the work of Obata et al. (1970), shown in *Figure 14*. In this test machine, one load bar on each axis is fixed and one moves. The effect is that the centre of the specimen moves as it stretches. A similar machine was used in the work of James et al. (1975; 1975) in the Dunlop laboratories. Such an arrangement minimises complexity of the machine mechanism, but is inconvenient if it is required to view the specimen during the experiment.

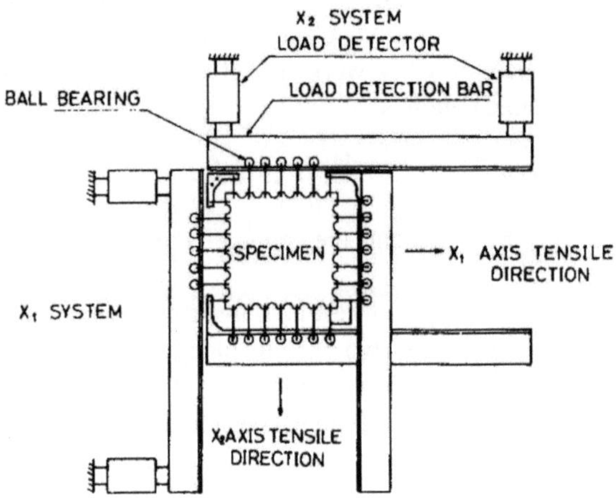

Figure 14. Diagram of the moving parts of the "Bistron" biaxial tester for rubberlike solids in sheet form, used in the work of Obata et al. (1970).

In more recent work in the author's laboratory, the Flexible Biaxial Film Tester (FBFT) was developed, with the added features of variable temperature and real-time video imaging of the specimen. To enable this, a mechanism was devised whereby both load bars on each axis are driven, in opposite directions, through use of backlash-free nuts on a lead screw with reversed thread – see *Figure 15*.

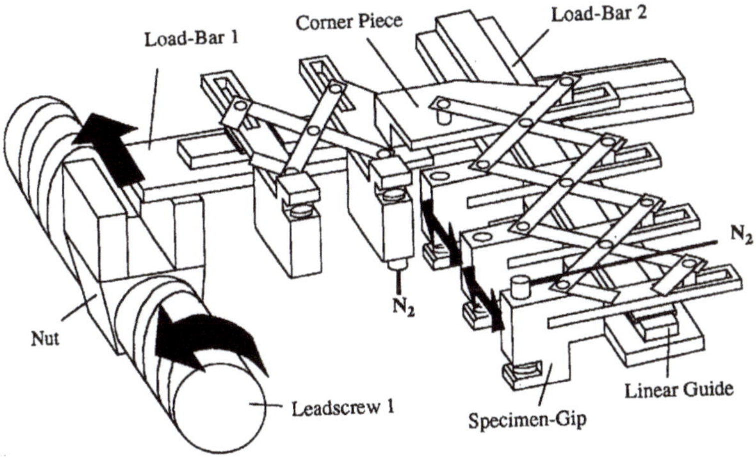

Figure 15. Cut-away diagram of part of the movement of the FBFT biaxial testing machine developed at Oxford. From the work of Gerlach (1998).

In this machine, the grips on each axis are interconnected by a scissors mechanism, to ensure they remain equi-spaced during straining. Another prominent feature is the use of pneumatic piston grips, to accommodate the large reductions in specimen thickness during biaxial straining. The FBFT illustrates how the various functions of providing movement on two axes and varying temperature, and the recording of measurements of force on each axis and of air temperature, can all be co-ordinated under computer control with current technology. A diagram of the overall system is shown in *Figure 16* below.

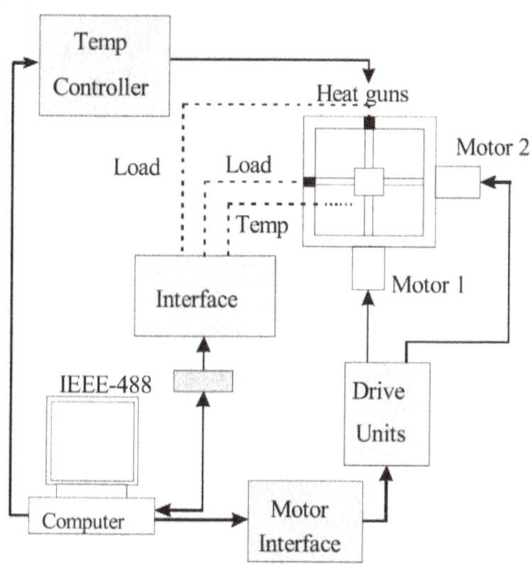

Figure 16. Schematic diagram of the total system of the FBFT biaxial testing machine, showing computer-intergration of the various drives and measurements.

The graphs in *Figure 17* illustrate data obtained with the FBFT with constant nominal strain rate on axis 1, and constant ϕ.

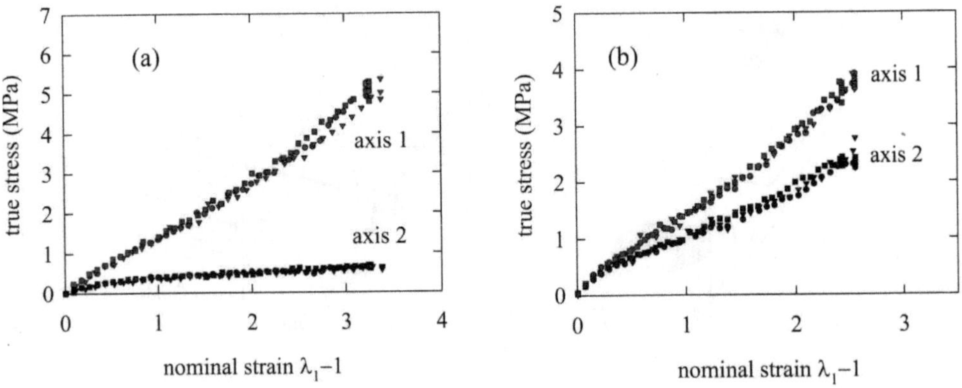

Figure 17. Sample data obtained at 20°C with the FBFT biaxial testing machine, for latex rubber films. The nominal stretch rates were $\dot{\lambda}_1 = 10^{-2}\ s^{-1}$, $\dot{\lambda}_2 = \phi\dot{\lambda}_1$: (a) $\phi = 0$; (b) $\phi = 2/3$. In each case, results are shown for first loading of three nominally identical specimens (indicated by different symbols).

2. Extension and inflation of a thin-walled tube

The test material is moulded into the form of a long, thin-walled tube with initial radius *R* and wall thickness *D* (<<*R*), inflated with a pressure *p* and subjected to an externally applied axial tensile force *F*. The need for a tubular specimen, however, is a potential limitation. The most extensive study using this method seems to be that of Vangerko and Treloar (1978). If we take the axis of the specimen as $\mathbf{e}_1$, the in-plane stresses are given by

$$\sigma_1 = \frac{\lambda_1\lambda_2^2 pR}{2D} + \frac{\lambda_1 F}{2\pi RD}, \qquad \sigma_2 = \frac{\lambda_1\lambda_2^2 pR}{D}. \tag{38}$$

The entire region of principal stretch space bounded from below by the UNIt line is then accessible, as for *A.1* above.

B – Methods giving access to only a one-dimensional path in principal stretch space

1. Uniaxial tensile test (UNIt)

This is the most widely used test method, because of the widespread availability of automated uniaxial testing machines. The literature records many studies of rubberlike solids using this approach. Since automated testing is readily available, this method has the advantage of allowing easy variation of the strain time-sequence – although (in the absence of feed-back from the "uniform" test section) the end-effect problem identified in Section 2 may cause the time-sequence of axial strain to differ from that intended. Clearly, with axes as defined in *Figure 9*, the locus of straining is

$$\phi = \frac{\lambda_1^{-1/2} - 1}{\lambda_1 - 1}. \tag{39}$$

2. Constant width tensile test (CWt)

If a tensile test is carried out with a specimen much wider than it is long ($B \gg L$ in *Figure 9*), the strain state approximates to pure shear, where $\lambda_2 = 1$, $\lambda_3 = 1/\lambda_1$. Now the non-uniform zone is close to the edges $X_2 = \pm B/2$, where the boundary condition is instead $\sigma_2 = 0$. Several authors have used this approach (Rivlin and Saunders (1951), Jones and Treloar (1975) and Gough et al. (1999) , to obtain another locus in principal stretch space:

$$\phi = 0. \tag{40}$$

In a recent study, a special instrumented specimen grip was used to obtain the stress σ_2 in the transverse direction - see *Figure 18* below.

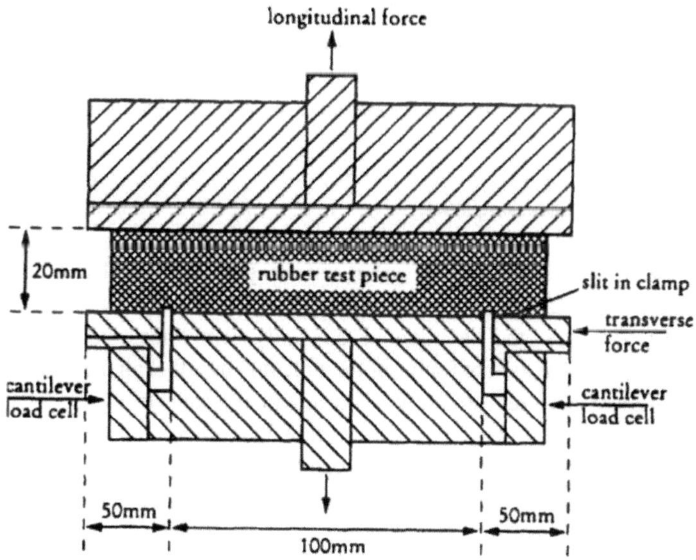

Figure 18. Special specimen grip used by Gough et al (1999), to determine the transverse stress σ_2 in constant width tensile tests on elastomers.

3. Uniaxial compression test (UNIc)

Referring to the diagrams of principal stretch space in *Figures 11* and *12*, only regions of the diagram to the right of the line UNIt represent purely tensile stress states. Experiments in compression allow exploration of the remaining portion of the diagrams. Uniaxial compression was the method employed in a study of constitutive response of elastomers by Bergstrom and Boyce (1998). In such tests, specimens take the form of squat solid cylinders, initial length L and diameter D, and are compressed between rigid anvils. The experiment is intended to apply a uniaxial compressive stress σ_{11}, but friction at the anvils usually gives rise to some shear stress ($\sigma_{12},\sigma_{13} \neq 0$) and hence a nonuniform zone near the anvils ("barrelling"). Unfortunately, the possibility of buckling prevents the use of specimens with L>>D and hence this can be a severe problem in interpreting results from compression tests. Normal practice is to minimise the friction by use of a lubricant such as a thin film of PTFE. Provided frictional shear stresses at the anvils are made negligible compared to σ_{11} uniaxial conditions prevail and the locus of straining in principal stretch space is the continuation of that for uniaxial extension (UNIt), extended into the compression region of the diagram.

4. Constant width compression test (CWc)

Another path in the compression region of principal stretch space is accessible with plane strain compression, i.e. constant width compression: $\phi = 0$. This is achieved with a rectangular specimen, which is constrained in direction $\mathbf{e}_2$ within a rigid die, while compressed along $\mathbf{e}_1$. The approach was used by Bergstrom and Boyce (1998) to provide additional strain states. Clearly, this test presents the specimen with four friction surfaces, and it is even more difficult to avoid the effects of friction than in the previous case. Provided this is achieved, the locus of straining is the continuation of that for constant width extension (CWt), extended into the compression region of the diagram.

5. Equal biaxial tension test (EBt)

In the absence of a biaxial tester, this strain state may be achieved by inflating a specimen in the form of a circular membrane, clamped around its perimeter – see for examples the work of James and Green (1975), Rivlin and Saunders (1951) and Johannknect and Jerrams (1999)). The sheet adopts a spheroidal shape, where (only) at the pole it is exactly spherical. The main difficulty with this test is that there is no uniform zone. The spherical shape strictly extends over only an infinitesimal area of the specimen. In practice, however, there exists an area large enough for measurements to be made, where the shape can be treated as spherical to within the resolution of the measurement methods. The experiment requires the measurement of: the inflation pressure p, the radius of curvature r at the pole, and the in-plane stretch $\lambda_2 = \lambda_1$ at the pole. If the initial thickness of the sheet was D, it becomes $d = \lambda_3 D = D / \lambda_1^2$ on stretching, and hence the in-plane Cauchy stress is obtained from

$$\sigma_1 = \sigma_2 = \frac{\lambda_1^2 pr}{2D}. \tag{41}$$

In early work such as that of Rivlin and Saunders (1951), the geometrical measurements were made manually with a cathetometer, but this experiment is now greatly facilitated by the use of video image capture and analysis (Johannknecht (1999)). Equation (41) however reveals an inherent limitation of the approach. If the sheet specimen is initially flat, the stress is indeterminate for small deformations, since

$$r \to \infty \quad \text{as} \quad p \to 0\,. \tag{42}$$

Consequently there is always a high degree of experimental uncertainty in results obtained at small stretches. Indeed, the method is more prone to experimental scatter than most other methods (compare *Figure 19* below with *Figure 17*).

The corresponding experiment in compression – the equal biaxial compression test (EBc) – is theoretically possible, by radial compression of a specimen in the form of a rod. However, it has not been employed as far as the author is aware, presumably because the difficulty of its implementation, compared to the ease of uniaxial tension tests, which give access to the same locus of straining.

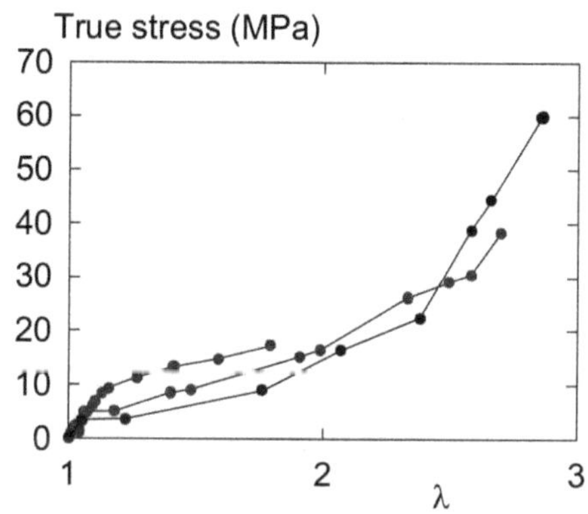

Figure 19. Sample equal biaxial stress-stretch curves for three polyurethane elastomers, with differing hard blocks: data were obtained by the inflated membrane method.

3.3 Results from exploration of strain-states

Interpretation in terms of strain invariants. Measurement of stresses for a range of strain-states provides information on the dependence of W on $\boldsymbol{B}$. If this dependence is expressed via the invariants I_i, equation (32) provides the means of determining the gradients. From tests where only one stress is measured, only a combination of the gradients may be obtained. When two stresses are available, however, the two gradients may be found separately. For all plane stress experiments with $\sigma_3 = 0$, equation (32) provides the unknown pressure term p:

$$p = 2\left(\lambda_1^2\lambda_2^2 \frac{\partial W}{\partial I_1} - \frac{1}{\lambda_1^2\lambda_2^2}\frac{\partial W}{\partial I_2}\right), \tag{43}$$

and substituting into the corresponding equations for σ_1 and σ_2 yields

$$\sigma_1 = 2\left(\lambda_1^2 - \frac{1}{\lambda_1^2\lambda_2^2}\right)\left(\frac{\partial W}{\partial I_1} + \lambda_2^2\frac{\partial W}{\partial I_2}\right),$$
$$\sigma_2 = 2\left(\lambda_2^2 - \frac{1}{\lambda_2^2\lambda_1^2}\right)\left(\frac{\partial W}{\partial I_1} + \lambda_1^2\frac{\partial W}{\partial I_2}\right). \tag{44}$$

These form a pair of simultaneous equations that can be solved for the gradients of W:

$$\frac{\partial W}{\partial I_1} = \frac{\dfrac{\lambda_1^2\sigma_1}{\lambda_1^2 - 1/\lambda_1^2\lambda_2^2} - \dfrac{\lambda_2^2\sigma_2}{\lambda_2^2 - 1/\lambda_2^2\lambda_1^2}}{2\left(\lambda_1^2 - \lambda_2^2\right)},$$
$$\frac{\partial W}{\partial I_2} = \frac{\dfrac{\sigma_1}{\lambda_1^2 - 1/\lambda_1^2\lambda_2^2} - \dfrac{\sigma_2}{\lambda_2^2 - 1/\lambda_2^2\lambda_1^2}}{2\left(\lambda_2^2 - \lambda_1^2\right)}. \tag{45}$$

It may be seen from these equations that there are regions of principal stretch space where the method fails: close to the origin where $\lambda_1 \approx \lambda_2 \approx 1,\ \sigma_1 \approx \sigma_2 \approx 0$ and near to the EB path where $\lambda_1 \approx \lambda_2,\ \sigma_1 \approx \sigma_2$.

This approach was pioneered by Rivlin and Saunders (1951) and was applied by others subsequently to unfilled, crosslinked natural rubber (see Obata (1970), James et al. (1975), and Jones and Treloar (1975)) and to carbon black filled, crosslinked natural rubber (see James and Green (1975)). All these studies employed biaxial extension experiments, to obtain the derivatives in equations (45) over a region of principal stretch space, while Gough et al (1999) used constant width tension tests (with the instrumented grip shown in *Figure 18*) to determine the derivatives along the CWt path, for crosslinked natural rubber with and without carbon black filler.

A general conclusion from published results is that the I_1 derivative dominates the I_2 derivative, exceeding it by a factor of at least 5-10. Beyond this, there is some conflict between results published so far. The original conclusion by Rivlin and Saunders (1951) was that the I_1 derivative is constant, while the I_2 derivative is independent of I_1 but decreases with increasing I_2. Subsequent work by Obata (1970) and Jones and Treloar (1975) has challenged this as something of an over-simplification. However, all studies concur in showing that the I_1 derivative is *relatively* more constant than the I_2 derivative. Moreover, the relatively small magnitude of the latter compared to the former has led to many suggestions (e.g. Gough et al (1999)) that the variation of W with I_2 can be neglected for practical purposes. The original results of Rivlin and Saunders are reproduced in *Figure 20* below.

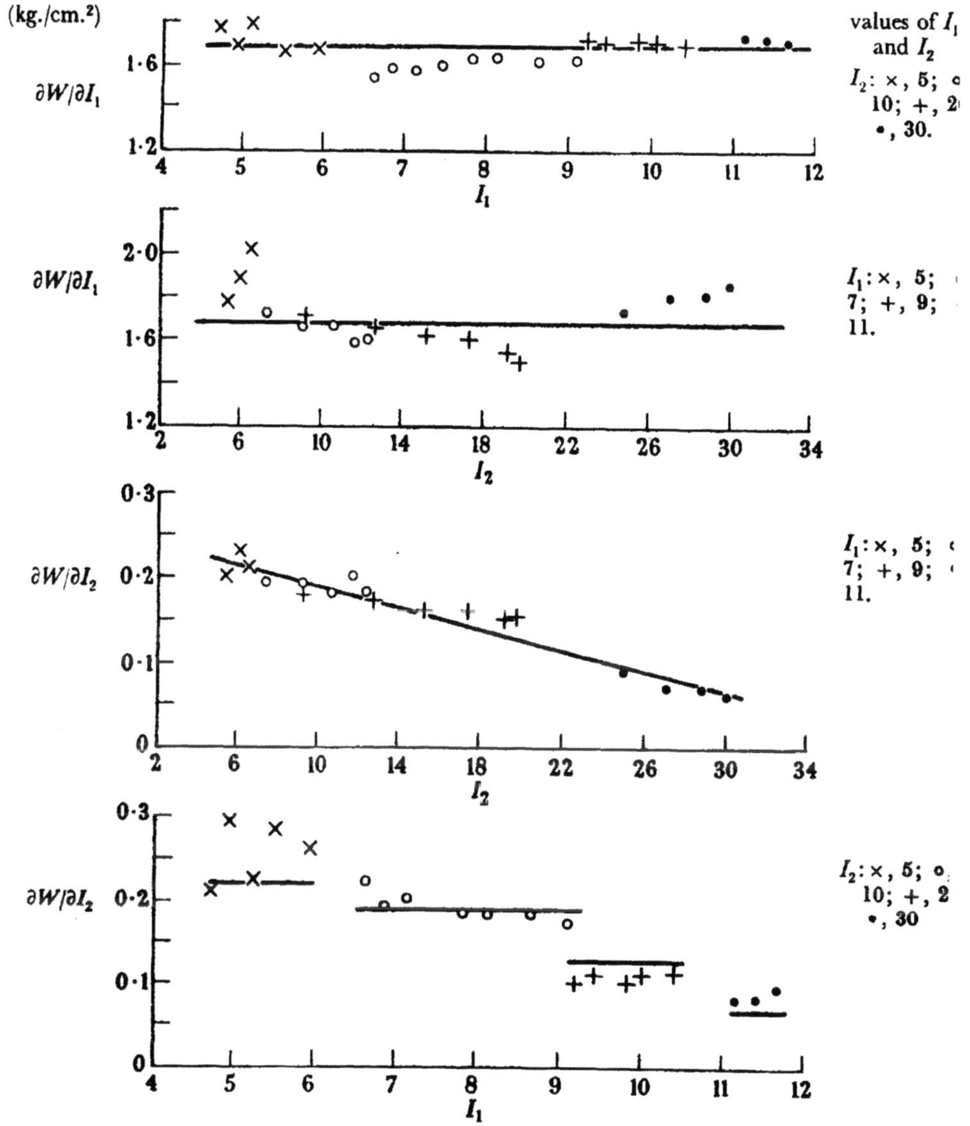

Figure 20. Measurements of $\partial W / \partial I_1$ and $\partial W / \partial I_2$ obtained by Rivlin and Saunders (1951) from the results of biaxial tensile tests on unfilled crosslinked natural rubber at room temperature.

Results expressed in terms of the variation of W with I_i are also employed to test the accuracy of specific mathematical expressions for this dependence. The most widely used form for this has been the polynomial:

$$W = \sum_{i=0}^{m} \sum_{j=0}^{n} C_{ij} (I_1 - 3)^i (I_2 - 3)^j, \quad \text{with } C_{00} = 0. \tag{46}$$

So-called neo-Hookean behaviour, corresponding to the molecular theory based on Gaussian chain statistics, is the simplest case: $m = 1$, $n = 0$, giving $\partial W / \partial I_1 = \text{constant}$, $\partial W / \partial I_2 = 0$. The data clearly do not support this over a significant range of the invariants (see *Figure 20*). A better fit to data for uniaxial extension is obtained with the expansion to first order in both invariants, also known as the Mooney form: $m = n = 1$, with $C_{11} = 0$, causing both derivatives to be constant. Substituting the Mooney form into the first of equations (44) for uniaxial extension or compression ($\lambda_2 = \lambda_1^{-1/2}$) leads to the prediction:

$$\frac{\sigma_1}{2\left(\lambda_1^2 - \frac{1}{\lambda}\right)} = \left(C_{10} + \frac{1}{\lambda} C_{01}\right). \tag{47}$$

Equation (47) is readily tested by plotting the left side versus $1/\lambda$. Reasonable agreement is obtained usually, with the predicted linearity (discrepancy is less than or comparable to experimental uncertainty) for the approximate range $1 \le \lambda_1 \le 1.5$. For higher extensions, however, this equation under-predicts stress, as it fails to capture strain-stiffening (e.g. see Mullins (1959)). Also, in the compression region this equation dramatically *over*-predicts the stress (e.g. see Rivlin and Saunders (1951)).

Clearly, increasing the number of terms in equation (46) can be made to improve the accuracy of fit to experimental data, although this must be done with caution, as the values of the coefficients determined become increasingly sensitive to random measuring error. James et al. (1975) investigated this point systematically in relation to their biaxial extension study, over the range $\lambda_1 = 1.3 - 3.5$, $\lambda_2 = 0.84 - 3.4$: see the *Table 1* below.

Maximum order of invariants	Number of coefficients C_{ij}	Residual error between calculated and measured stresses (MPa)
1st	2	0.087
2nd	5	0.078
3rd	9	0.051

Table 1 Residual error between measured and calculated stresses from fitting biaxial extension data to polynomials in the strain invariants: results from James et al. (1975).

Results expressed directly in terms of principal stretches. The alternative approach of expressing experimental results directly in terms of the principal stretches has also been widely used: $W = W(\lambda_1, \lambda_2, \lambda_3)$. This is largely because of the success of the hypothesis suggested by Valanis and Landel (V-L) that the function may be expressed in separated form

$$W = w(\lambda_1) + w(\lambda_2) + w(\lambda_3). \tag{48}$$

Perhaps surprisingly, in view of its simplicity, this form for the function W is remarkably successful in fitting experimental data. In particular, it was systematically tested against data from biaxial extension tests on unfilled natural rubber in the work of Jones and Treloar (1975) and Vangerko and Treloar (1978). If the V-L hypothesis were correct, equation (33) would allow the principal stress differences in a biaxial test to be written:

$$\sigma_1 - \sigma_2 = \lambda_1 w'(\lambda_1) - \lambda_2 w'(\lambda_2) , \tag{49}$$

$$\sigma_3 - \sigma_2 = \lambda_3 w'(\lambda_3) - \lambda_2 w'(\lambda_2) , \tag{50}$$

where $\sigma_3 = 0$, $\lambda_3 = 1/\lambda_1\lambda_2$, and a prime denotes derivative. Therefore the hypothesis is easily tested: for constant λ_2, $\sigma_1 - \sigma_2$ is plotted versus λ_1 (for $\lambda_1 > 1$) or $-\sigma_2$ is plotted versus λ_3 (for $\lambda_3 < 1$). If the plots superpose by vertical shifts to form a single "master curve" then equations (49) and (50), and hence the V-L hypothesis, are verified. The graph in *Figure 21* shows a master curve prepared in this manner by Treloar and Riding (1979), from the biaxial results of James et al. (1975).

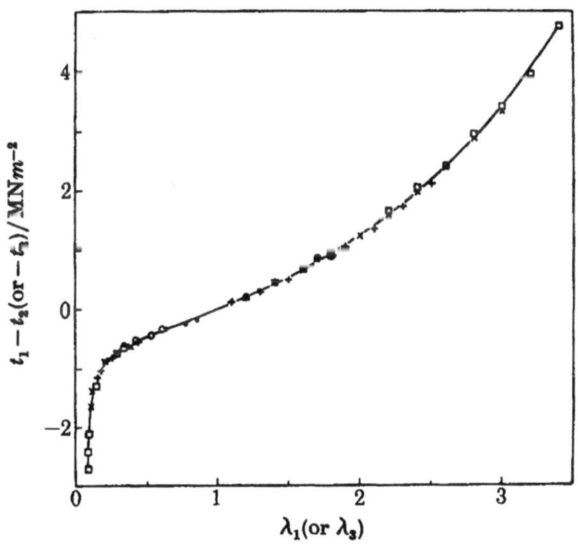

Figure 21. Master curve prepared by Treloar and Riding (1979) from the data of James et al. (1975) by vertical shifting of biaxial extension data for natural rubber, as described in the text. From Treloar and Riding (who use symbol t for true stress). Each symbol refers to a different strain state.

A particularly successful form of the function $w(\lambda)$ has proved to be the series expression proposed by Ogden (1972):

$$w = \sum_{i=1}^{n} \frac{\mu_i \lambda^{\alpha_i}}{\alpha_i} , \tag{51}$$

where the μ_i and α_i are treated as adjustable parameters. This appears to work well for natural rubber over a wide range of strain-states, provided at least three terms in the series are used – hence demanding at least six adjustable parameters. The first demonstration of the accuracy of the three-term expansion is shown in *Figure 22* below – later authors have obtained similar findings.

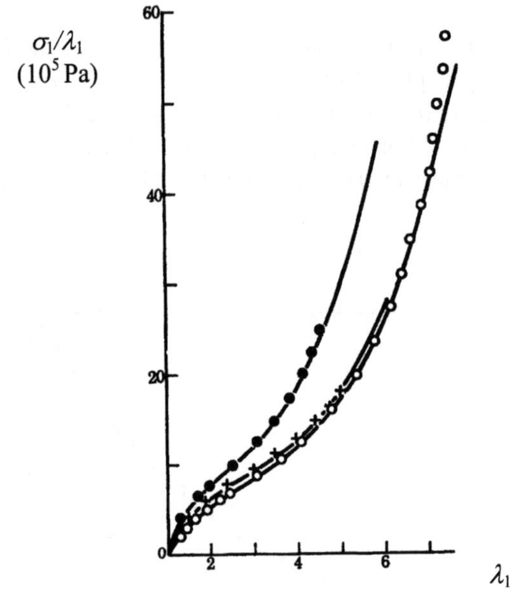

Figure 22. The demonstration given by Ogden (1972) that a three-term series of the form in equation (51) fits Treloar's data for nominal stress from tests of type UNIt (o), CWt(+) and EBt(•).

Comments on experimental data. Although much of the above published experimental data has been available for many years, there remain some puzzling inconsistencies and uncertainty concerning what is the "best" form to assume for the function *W*. The origin no doubt lies primarily in the inherent limitations of the hyperelastic approximation. Most experimental studies have included conditioning (or "scragging") of specimens before the recording of stress-stretch data (see Sections 2 and 4 of this Chapter). But the detailed procedures used have varied between experimenters. In addition, the stretch-time sequences used in different studies have varied: biaxial straining experiments provide a rich variety of possible sequences. Correspondingly, it cannot be assumed that results refer to specimens in identical structural states, even if they initially have the same structure.

At the present time, there is only incomplete experimental evidence on how material structure affects the stress-strain responses outlined above. However, it appears that the increases in stiffness associated with increasing crosslink density (Mullins (1959)) *or* with increasing carbon black content (Gough et al. (1999)) are both associated with an increase in $\partial W / \partial I_1$ with no significant increase in $\partial W / \partial I_2$. They are also both associated with strain-stiffening occurring at lower stretches, as predicted by physical theories that incorporate finite chain extensibility.

3.4 Measurement of compressibility

It is clear that the assumption of incompressibility has played a crucial role in the interpretation of results from measurements of the apparent hyperelasticity of rubberlike solids. To improve the accuracy of the material description, especially under stress states that are close to hydrostatic, requires experiments to measure dilation and the formulation of modifications to the strain energy function to include it. Currently, the usual approach is to *assume* additivity of strain energies arising from deviatoric strain and dilation, as is rigourously true in the linear elastic limit for an isotropic material (following from equations (27)). However, there is a need for more accurate information on the compressibility and the accuracy of this assumption at finite strains. To measure volume change for finite strains involves difficult experiments combining stretching with some form of dilatometry. A pioneering study of this sort was executed by Penn (1970).

In the small strain region, however, the measurement of bulk modulus K may be carried out conveniently as follows, as applied by Arridge (1974) to epoxy resins. The specimen has the form of a thin-walled tube: we let $\mathbf{e}_1$ represent the tube axis. It is subjected to internal pressure p, and the infinitesimal axial strain ε_1 is measured with a displacement transducer, for an example an LVDT. From equations (26), (27) and (38), for $\varepsilon_1 \ll 1$

$$\sigma_{\mathrm{m}} = \sigma_1 = \frac{pR}{2D}, \quad s_1 = 0, \quad \varepsilon_1 = \frac{\Delta}{3}, \quad K = \frac{\sigma_{\mathrm{m}}}{\Delta} = \frac{R}{6D}\frac{p}{\varepsilon_1}. \tag{52}$$

3.5 Stress–strain-temperature experiments

There is considerable theoretical interest in the thermoelastic response of rubberlike solids, as it reveals the predominantly entropic nature of W. The classical experiment to study this, first published in 1935, consists of two steps:

1. Apply a uniaxial stretch λ.
2. Change temperature from T_0 to T, with λ constant.

Assuming incompressible, isotropic elasticity in the linear limit, with $\varepsilon = \lambda - 1$:

$$\sigma = 3G\left[\varepsilon - \frac{\beta}{3}(T - T_0)\right], \tag{53}$$

where β is the coefficient of volumetric thermal expansion. Since the physical origin of G is primarily entropic, it rises with temperature. The experiment yields graphs of σ versus T of varying slope, depending on the stretch applied:

$$\frac{\mathrm{d}\sigma}{\mathrm{d}T} = 3\frac{\mathrm{d}G}{\mathrm{d}T}\left[\varepsilon - \frac{\beta}{3}(T - T_0)\right] - \beta G. \tag{54}$$

Since $\mathrm{d}G/\mathrm{d}T > 0$, the usual negative gradient found for most solids exists only for the smallest strains, to be replaced by a positive gradient at large strains. The transition from negative to positive stress-temperature gradient is known as *thermoelastic inversion* and occurs at a strain

$$\varepsilon^* \simeq \frac{\beta / 3}{\mathrm{d}\ln G / \mathrm{d}T}. \tag{55}$$

Since to a first approximation G is entropic and hence $G \propto T$, and typically $\beta \approx 6\times10^{-4}\,\mathrm{K}^{-1}$, we find in the region of room temperature

$$\varepsilon^* \simeq \beta T \approx 0.06 \tag{56}$$

in accord with experimental observations.

In view of the interest of polymer physicists in determining as precisely as possible the entropic and energetic contributions to G, there has been much discussion of how to determine its temperature derivative accurately. It can be shown (see Shen (1969)) that the energetic contribution to stress in uniaxial extension σ_e is obtainable from

$$\frac{\sigma_e}{\sigma} = 1 - \frac{\mathrm{d}\ln G}{\mathrm{d}\ln T} - \frac{\beta T}{3}. \tag{57}$$

Problems arise, however, from the deviations of rubberlike solids from purely elastic response. With increase in temperature, there is acceleration of stress relaxation (see next Section), easily causing significant errors if $\mathrm{dln}G/\mathrm{dln}T$ is obtained from a stress-temperature plot. Strategies have been proposed for minimizing the error. In particular, the specimen is allowed to stress relax at the highest temperature, and then stress measurements made during cooling. Although this approach certainly reduces the problem of stress relaxation, compared to the use of a rising temperature, McCrum (1984) pointed out the lack of rigour in design of such experiments. He proposed an improved method for determining the temperature-dependence of G, from (a) the induced shear creep compliance $\Delta J(t)$ when a temperature down-jump $T_0 \downarrow T$ is superposed on a shear stress and (b) the usual shear creep compliance in an isothermal creep test at temperature T: $J(t)$

$$\frac{\mathrm{d}\ln G}{\mathrm{dln}T} = \frac{T\Delta J(t)/\Delta T}{\left(\Delta J(t) - J(t)\right)}. \tag{58}$$

When suitable precautions are taken, typical measured values for σ_e/σ lie in the range 0.1-0.3.

Another interesting manifestation of the entropic nature of rubber elasticity is the quasi-reversible heating that occurs on adiabatic extension. Experiments to study thus have been conducted since Joule(1859). The usual experiment is conducted in uniaxial extension, and consists of applying a stretch rapidly, in the window of time-scale where uniform adiabatic conditions prevail (see Section 2). Temperature is measured with a device, for example a thermocouple or radiometer, that clearly must respond on a much faster time-scale. If a thermocouple is used, the experimenter must ensure

$$\tau_{\mathrm{cond}}\left(\text{thermocouple}\right) \ll \tau_{\mathrm{cond}}(\text{rubber specimen}). \tag{59}$$

We saw in Section 2 that for the rubber specimen of thickness 2 mm, $\tau_{\mathrm{cond}} \approx 10$ s. The requirement in equation (59) is met by means of a very thin thermocouple junction. For example, commercially available 50 μm thermocouple wire has $\tau_{\mathrm{cond}} < 1$ ms. The classic results of Joule (1859) and James and Guth (1943) are shown in *Figure 23*. The temperature rises are less than 1K, for the moderate stretch levels shown. Provided strain-induced crystallization does not intervene, the adiabatic heating does not normally exceed 3K even at the highest stretches attainable, but is almost completely reversible. In certain rubberlike solids (e.g. natural rubber and neoprene rubber), however, crystallization occurs on stretching. This is manifest as a considerable increase in adiabatic heating, to ca. 15 K, and in the hysteresis associated with it.

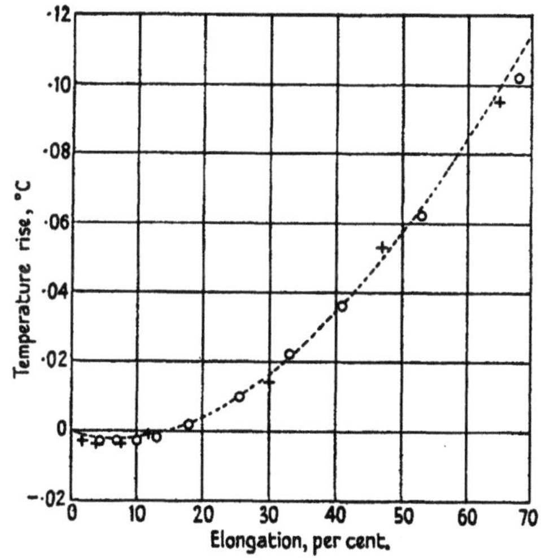

Figure 23. The classic data of Joule (+) and James and Guth (o) showing reversible adiabatic heating during stretching of natural rubber, as plotted by Treloar (1958).

4 Exploration of strain histories

4.1 Introduction

To complement knowledge of material response to varying strain-*state*, it is necessary to determine the response to varying strain *history*. Since rubberlike solids are not perfectly elastic, at any instant t the stress in equation (8) and the strain in equation (9) depend not just on the current strain and stress respectively, but also on their history. Thus, a constant uniaxial stress σ gives rise to a strain ε increasing with time in the same direction – the phenomenon of *creep*. A corollary is that a constant strain ε gives rise to a stress σ decreasing with time – the phenomenon of *stress relaxation*. Another corollary is that a cyclic stress σ produces a cyclic strain ε with a phase lag relative to the stress – the phenomenon of *hysteresis*.

For sufficiently modest strain/temperature histories these effects are reproducible. If the stress is returned to zero and the temperature to T_0, and the material is allowed sufficient time to recover, a subsequent application of the same stress/temperature history will produce precisely the same response. This can be interpreted as indicating: (a) the response of the material of given structure has a perfectly fading memory, and (b) there is no persistent change in any of the variables within S. If they changed under the influence of stress, they have returned to their original values S_0 during the recovery period. That is, the response of S also exhibits a perfectly fading memory.

For larger excursions of strain and/or temperature, the material response is generally *not* reproducible. The response to a subsequent loading sequence differs from the original. This may be interpreted in terms of a persisting strain-induced change in internal variables S. The distinction is of crucial importance in the design of experimental programmes. If the response is not reproducible, and the aim is to observe the response to different straining paths all starting from the same virgin state, a new specimen must be used for each straining path.

4.2 Linear viscoelastic characterization

Creep tests. Within the category of reproducible responses mentioned above is the special case of linear viscoelastic response. Since this often applies only in the case of small strains (especially in filled elastomers), we shall express deformation in terms of the infinitesimal strain $\boldsymbol{\varepsilon}$ instead of $\boldsymbol{B}$. Isothermal linear viscoelasticity corresponds to the case where T is constant and functionals $\mathcal{F}$ and $\mathcal{F}'$ (equations (8) and (9)) are *linear* functionals of the stress and strain history respectively. The phenomenological theory that results from this restriction has been discussed widely in the literature: Tschoegl (1989) has provided a particularly thorough treatment.

Consider the relation between one tensile component of stress σ and the corresponding tensile component of strain ε. In an ideal tensile creep test the stress undergoes a step and then remains constant:

$$\sigma(u) = \sigma_0 H(u) \ , \tag{60}$$

where $H(u)$ is the Heaviside unit step function and σ_0 is a constant. It follows from the linearity of $\mathcal{F}$ that the creep strain $\varepsilon(t)$ at any time $t \geq 0$ varies linearly with σ_0:

$$\varepsilon(t) = \sigma_0 D(t) \qquad (0 \leq t) \ , \tag{61}$$

where $D(t)$ is the tensile creep compliance. The usual aim of a tensile creep test in the linear viscoelastic region is to measure $D(t)$. The advantage of a tensile creep test is its simplicity. If data are to be gathered over a long period of time (days or weeks are common durations for creep tests), it is highly attractive for the equipment to be inexpensive, and it is imperative that it is free of time-dependent drift in either the loading system or the measurement instrumentation. The simplest procedure is for the specimen to be clamped at one end with a dead-weight load suspended from the other end, as shown in *Figure 24*. Note the use of an extensometer for measuring extension on a gauge length in the uniform region of the specimen, the use of flexible couplings to avoid any unwanted bending moments arising from slight misalignment of the load, and the use of a thermostated environmental chamber to provide the constant temperature T. Care must be taken to use an extensometer that does not subject the specimen to a stress that is significant compared to σ_0. If the material is a soft unfilled elastomer, this probably means that either a balanced potentiometric extensometer, or one of the non-contacting forms of extensometer (e.g. a video extensometer) is necessary.

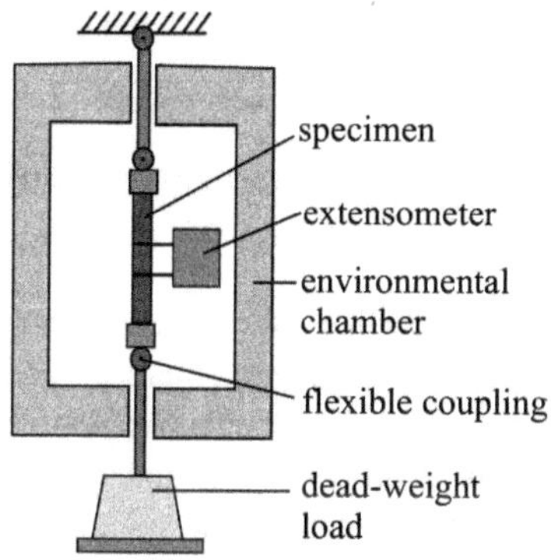

Figure 24. Schematic cross-section through a tensile creep apparatus.

After the required creep time t_c the load is removed, ideally instantaneously, and measurement of strain is usually continued, to follow the recovery of strain that follows – see *Figure 25*.

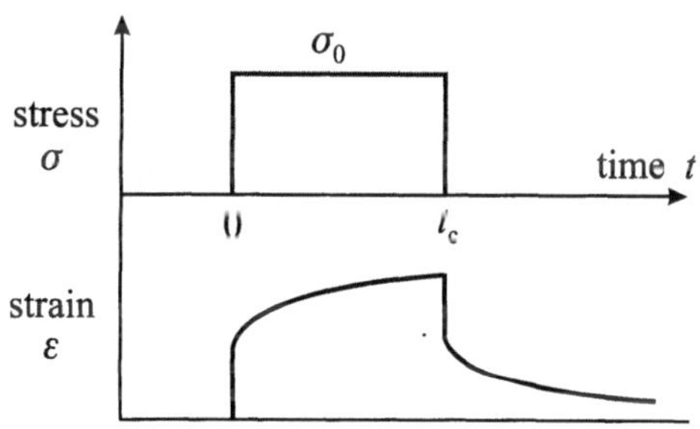

Figure 25. Schematic diagram of stress and strain time-sequences in a creep test on a crosslinked elastomer.

A second consequence of linear viscoelasticity is that the strain response to a sequence of stresses is the sum of the responses to each of them applied separately. Since we can represent the creep experiment stress time-sequence (*Figure 25*) as the superposition of two steps – one positive and one negative

$$\sigma(u) = \sigma_0 H(u) - \sigma_0 H(u - t_c)\,, \tag{62}$$

the strain response is, employing equation (61) and invoking superposition,

$$\begin{aligned} \varepsilon(t) &= \sigma_0 D(t) & (0 \le t \le t_c)\,, \\ \varepsilon(t) &= \sigma_0 \left[D(t) - D(t - t_c)\right] & (t_c \le t)\,. \end{aligned} \tag{63}$$

A crosslinked elastomer in the linear viscoelastic region exhibits a decaying rate of creep towards a limiting "relaxed" compliance D_R :

$$D \to D_R, \quad \dot{D} \to 0 \quad \text{as} \quad t \to \infty. \tag{64}$$

It follows by Taylor series expansion of the second of equations (63) that

$$\varepsilon(t) \to \dot{D}t_c \to 0 \quad \text{as} \quad t/t_c \to \infty. \tag{65}$$

This is an important result for a rubberlike solid. In the linear viscoelastic limit, provided there is sufficient crosslinking to ensure adherence to equation (64), it will show no *permanent set* (unrecovered strain) following a creep and recovery experiment.

Good practice dictates that, if results are to be interpreted in terms of linear viscoelasticity, linearity (to within an acceptable tolerance) should be checked as part of the experimental programme. This is readily achieved, for example, by checking that creep data show the linear relation between stress and strain required by the first of equations (63), and that recovery data agree with the superposition of creep strains required by the second of these equations.

An important operational issue is the practical impossibility of achieving a perfect step-wise increase or decrease in stress. It takes a time t_l to load and unload the specimen (typically $t_l \approx 1$ s). Whatever the detail of the loading sequence, however, because of the fading memory of the material, the response approaches the response to a pure step as $t_l/t \to 0$. If the creep compliance $D(t)$ is required, therefore, a simple (but unfortunately somewhat wasteful) solution is to obtain it from $\varepsilon(t)/\sigma_0$ ($t < t_c$), but ignoring the data for $t \le 10t_l$.

Stress relaxation tests Linear viscoelasticity may be characterized also by means of stress relaxation tests, where (ideally) a step in strain is applied and the stress is measured as a function of time. Following a parallel logic to that for linear viscoelastic creep above, a tensile stress relaxation experiment with removal of strain after a time t_s can be described

$$\varepsilon(u) = \varepsilon_0 \left[H(u) - H(u - t_s) \right] \tag{66}$$

and the stress response is given by

$$\begin{aligned} \sigma(t) &= \varepsilon_0 E(t) && (0 \le t \le t_s), \\ \sigma(t) &= \varepsilon_0 \left[E(t) - E(t - t_s) \right] && (t_s \le t). \end{aligned} \tag{67}$$

Here, $E(t)$ is the tensile stress relaxation modulus. $E(t)$ is a decaying function, but for a sufficiently crosslinked rubber it decays to a finite "relaxed" value E_R. Hence

$$E \to E_R, \quad \dot{E} \to 0 \quad \text{as} \quad t \to \infty. \tag{68}$$

Since E decays with increasing values of its argument, it follows from the second of equations (67) that the stress becomes negative after removal of strain. But parallel reasoning to that leading to equation (65) tells us that no stress will remain in the limit $t/t_s \to \infty$.

The practical implementation of stress relaxation tests in tension is in principle less straightforward than for tensile creep tests. The reason is as given in Section 2.2. A tensile test machine of the usual form may be used, as shown in *Figure 26*.

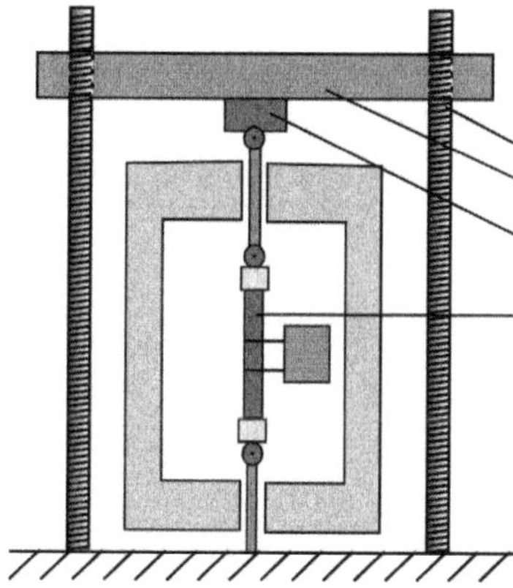

Figure 26. Schematic cross-section through a tensile test machine as might be employed for tensile stress relaxation experiments. A motor in the base of the machine drives the crosshead via leadscrews as shown.

The machine controls the motion of the ends of the specimen (as in *Figure 9*), but not the displacement in the uniform zone of the specimen. For precise definition of the strain time-sequence in the uniform zone, it is necessary to add a feedback loop from the measured strain in the zone to the motor drive. Commercial systems are now available to do this, but clearly the complexity (and hence cost) of the equipment is increased accordingly. The relative complexity of the test machine means that stress relaxation tests are less attractive than creep tests for characterizing linear viscoelastic response over long periods of time.

Another practical problem with *tensile* stress relaxation tests is that it is not possible to design a specimen that has a significant uniform zone (requiring $L \gg B, D$) *and* is resistant to buckling when the stress becomes compressive (imposing an upper limit on length $L \sim D$). Consequently, stress data may not be available after removal of strain.

To obtain stress relaxation data under both positive and negative stresses the simple shear mode of deformation (*Figure 27*) may be used instead – normally a balanced arrangement is used, with two specimens as shown. To minimize the nonuniform zones near the upper and lower free surfaces requires $L \gg D$. The measured quantity is the shear stress relaxation modulus $G(t)$. Given that rubberlike solids are incompressible to a close approximation (see Section 3), in this case the tensile stress relaxation modulus may be obtained from $E(t) = 3G(t)$ to within a tolerance acceptable for most purposes.

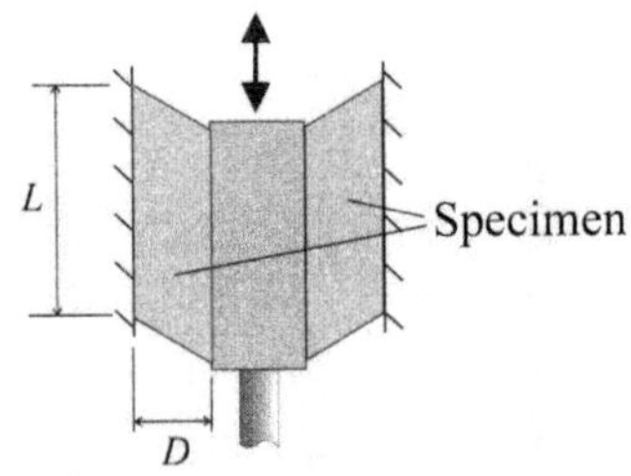

Figure 27. Schematic diagram of the simple shear test, that may be used to obtain stress relaxation data for both positive and negative stresses.

Stress relaxation in a rubberlike solid is most pronounced at low temperatures, as it approaches the glass transition – see *Figure 28*.

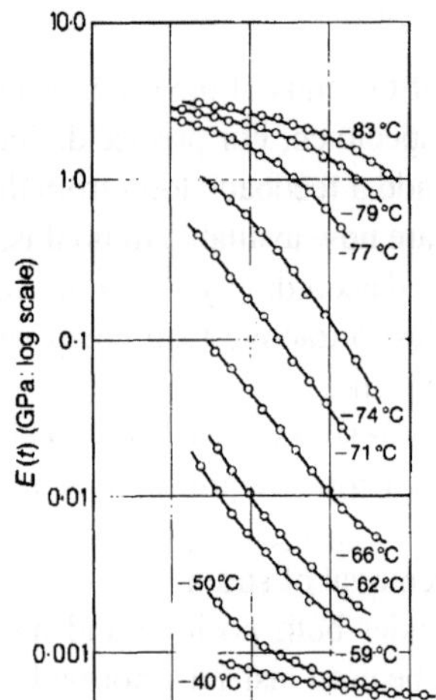

Figure 28. Tensile stress relaxation data for uncrosslinked butyl rubber at low temperatures, approaching the glass transition temperature of ca –80°C. From McCrum et al. (1997).

Dynamic mechanical tests For many applications of rubberlike solids, linear viscoelastic response in the frequency domain is required. Dynamic Mechanical Tests (DMA) are used to measure this. There are several test machines available commercially. Deformation modes include tension/compression, bending and simple shear. Alternatively, a general purpose test machine (e.g. as in *Figure 26*) may be employed, provided the drive and software have the flexibility to apply sinusoidal loading.

A sinusoidal time-sequence of strain is applied, for example in tension/compression $\varepsilon(u)=\varepsilon_0\cos\omega u$. In the linear viscoelastic regime, in the steady state, this causes a sinusoidal stress, with amplitude proportional to ε_0 :

$$\sigma(t)=E_{\mathrm{d}}\varepsilon_0\cos\left(\omega t+\delta\right)\ , \tag{69}$$

where E_{d} is the dynamic modulus and δ the phase lead of stress over strain. In a DMA test apparatus, a force transducer measures the oscillating force applied to the specimen, and from this and knowledge of specimen geometry the in-phase and π/2 out-of-phase components of stress are computed. Re-writing equation (69) in terms of these:

$$\sigma(t)=\left[E'\cos\omega t+E''\cos\left(\omega t+\pi/2\right)\right]\varepsilon_0\ , \tag{70}$$

where the *storage modulus* $E'=E_{\mathrm{d}}\cos\delta$ and the *loss modulus* $E''=E_{\mathrm{d}}\sin\delta$. Results are normally expressed in terms of $E'(\omega)$, $E''(\omega)$ and $\tan\delta(\omega)$, or their equivalents for shear. Some typical results are shown in *Figure 29*.

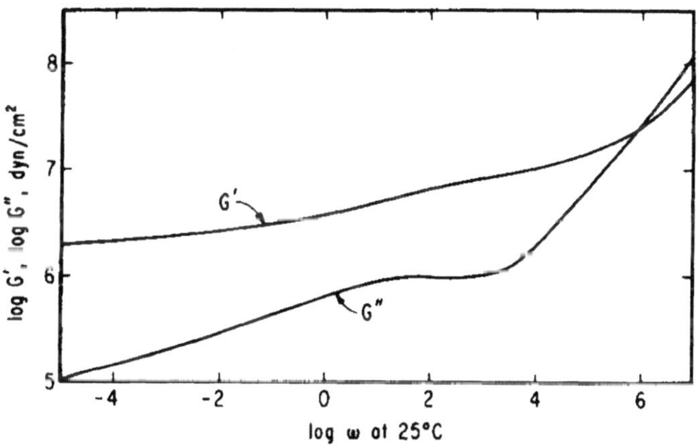

Figure 29. DMA measurements of storage and loss shear moduli, plotted versus excitation frequency, for an SBR rubber at room temperature. Data from Kramer et al. (1994).

Relations between results of tests in the linear viscoelastic regime In the linear viscoelastic regime, linearity of $\mathcal{F}$ and $\mathcal{F}'$ means that knowledge of the response to any one time-sequence of stress or strain over the whole domain of time or frequency can be used to predict the response to any other, through linear superposition. Thus the results of different experiments are related. This is extremely useful to the experimenter, as it enables material response functions to be determined over a wider range of time or frequency than is accessible directly in a single experiment, by combining results from different techniques.

In view of linearity, responses to any uniaxial stress history or strain history are predictable from the following two convolutions respectively:

$$\varepsilon(t) = \int_0^t D(t-u)\frac{\mathrm{d}\sigma}{\mathrm{d}u}\mathrm{d}u; \qquad \sigma(t) = \int_0^t E(t-u)\frac{\mathrm{d}\varepsilon}{\mathrm{d}u}\mathrm{d}u\,. \tag{71}$$

Substituting the result of a stress relaxation experiment ($\varepsilon = \varepsilon_0$, $\sigma(u) = E(u)\varepsilon_0$) into the first of equations (71), gives an expression relating $D(t)$ and $E(t)$:

$$\int_0^t D(t-u)\frac{\mathrm{d}E(u)}{\mathrm{d}u}\mathrm{d}u = 1\,. \tag{72}$$

In principle this integral equation may be solved numerically, to obtain $E(t)$ from $D(t)$ and vice versa. Similarly, consider stress response to a sinusoidal strain, as in a DMA experiment. Expressing the strain input in complex form $\varepsilon(u) = \varepsilon_0 \exp i\omega u$, we find from the second of equations (71)

$$\sigma(t) = \varepsilon_0 \int_{-\infty}^t i\omega E(t-u)\exp i\omega u\,\mathrm{d}u\,. \tag{73}$$

Changing the integration variable to $v = t-u$ gives

$$E^* = \frac{\sigma(t)}{\varepsilon_0 \exp i\omega t} = E' + iE'' = i\omega\int_0^\infty E(v)\exp(-i\omega v)\,\mathrm{d}v \tag{74}$$

and the complex modulus $E^*(\omega)$ is revealed to be simply the Fourier transform of $\dot{E}(t)$. Similarly, the complex compliance $D^* = 1/E^*$ is the Fourier transform of $\dot{D}(t)$. Thus, to convert between the transient responses (creep and stress relaxation) and dynamic responses, any of the techniques for finding Fourier transforms and their inverses may be deployed.

Modelling of linear viscoelastic responses In order to use experimental results in predictive simulations, it is convenient for them to be modelled by expressions in closed form. In the linear viscoelastic case there are several established means of doing this. One is to express the results of transient experiments in terms of a Prony series of the form

$$G(t) = G_{\mathrm{R}} + (G_{\mathrm{U}} - G_{\mathrm{R}})\sum_{i=1}^{n}\varphi_i \exp\left(-\frac{t}{\tau_i}\right), \tag{75}$$

and hence the results of dynamic experiments in terms of the corresponding series (from equation (74):

$$G^*(\omega) = G_{\mathrm{R}} + (G_{\mathrm{U}} - G_{\mathrm{R}})\sum_{i=1}^{n}\frac{\varphi_i\left(\omega^2\tau_i^2 + i\omega\tau_i\right)}{\left(1+\omega^2\tau_i^2\right)}. \tag{76}$$

In equations (75) and (76), G_{U} is the "unrelaxed" shear modulus corresponding to the instantaneous response to strain at $t \to 0$, and the set of *n* parameters φ_i form a discrete normalized spectrum over the relaxation times τ_i . It is a straightforward matter to find these in principle, for example by least-squares curve-fitting. Techniques have been suggested for constraining the fitting to avoid unphysical negative values. If a wide span of time (or frequency) is involved, however (e.g. 5 decades), the method leads to an awkwardly large number of adjustable parameters, since at least two points are required per decade of τ, if unphysical inflections in the response functions are to be avoided. A convenient means of reducing the number of parameters is to assume $\varphi_i(\tau_i)$ lie on a smooth curve, for example a q-th order polynomial in some base function $f(\tau)$ (for example $f = \ln\tau$), such that

$$\varphi(\tau) = \sum_{j=0}^{q-1} a_j [f(\tau)]^j , \tag{77}$$

thereby reducing the number of parameters to q. For experiments covering the ranges of time and frequency usually accessible, a good fit can be obtained with a small value for q – e.g. 2-4. Alternatively, other empirical functions have been suggested, that are able to provide a reasonable fit to data spanning many decades of time or frequency with only a small number (e.g. 4) of adjustable parameters, for example the Kohlrausch relation

$$G(t) = G_R + (G_U - G_R)\exp\left[-\left(\frac{t}{\tau_0}\right)^k\right] \tag{78}$$

and the Cole-Cole type of relation

$$G(t) = G_R + \frac{G_U - G_R}{1+(t/\tau_0)^c} . \tag{79}$$

Tschoegl (1989) has discussed the relative merits of the different approaches. *Figure 30* below shows an example of stress relaxation data over eight decades of time, fitted successfully to equation (79).

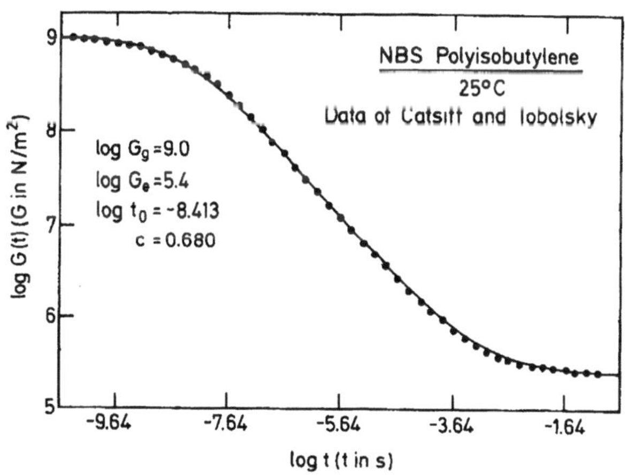

Figure 30. Stress relaxation data for uncrosslinked butyl rubber, fitted to equation (79) by Tschoegl (1989), who uses G_g, G_e and t_0 for G_U, G_R and τ_0 respectively.

Figure 28 illustrates the extreme sensitivity of viscoelastic responses to varying temperature. Fortunately, however, the dependence on temperature is readily modelled. As can be seen, the dominant effect of varying temperature is to shift stress relaxation curves along the $\log t$ axis.

This is the expected result if all the relaxation times $\tau_i (i = 1..n)$ in equations (75) and (76), and τ_0 in equations (78) and (79) vary with temperature thus

$$\tau(T) = a_T \tau(T_0), \tag{80}$$

where T_0 is a reference temperature, but the remaining parameters in the equations remain constant. All these expressions for $G(t,T)$ then take the form

$$G(t,T) = G(t / a_T, T_0). \tag{81}$$

Consequently, the effect of varying temperature is represented through a single function a_T, obtained from the shift $-\log a_T$ along the $\log t$ scale required to bring the data for different temperatures into coincidence with the data for T_0. In practice, the temperature-dependences of the limiting elastic responses G_U and G_R are not zero, but they are small compared to the effects on the time-scale. In the case of polymers displaying rubbery behaviour, the time-temperature shift factor a_T is found to be represented well by the Williams-Landel-Ferry (WLF) equation (see Ferry (1980)):

$$a_T = \exp\left(-\frac{C_1 (T - T_0)}{C_2 + T - T_0} \right), \tag{82}$$

where C_1 and C_2 are material constants. *Figure 31* presents data for a polyurethane elastomer, showing the closeness of fit.

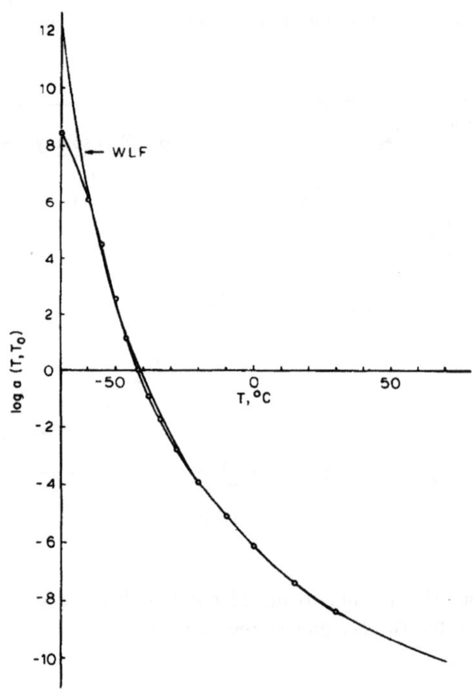

Figure 31. Time-temperature shift factor for a polyurethane elastomer (points) and the fitted WLF equation for a reference temperature $T_0 = -42°C$, with $C_1 = 38.5$, $C_2 = 68$ K, from Kramer et al (1994).

4.3 Measurement of reproducible nonlinear viscoelasticity

Payne effect For levels of strain exceeding some critical value, the viscoelasticity is measurably nonlinear. Provided the sample has been previously "conditioned" however (see Sections 2.4 and 4.4), the response may nevertheless be reproducible to within an acceptable tolerance. In the case of unfilled elastomers, a strain of at least 0.1 is required to achieve measurable nonlinearity. In filled systems, however, discernible nonlinearity appears much sooner.

Since many applications of elastomers involve oscillatory loading, the response in DMA experiments is of particular interest. Recall that the imposed strain is usually a sinusoidal time-sequence: $\varepsilon(u) = \varepsilon_0 \cos \omega u$. In steady-state after many cycles, the stress response must also be of constant amplitude and repeat with a frequency ω. When nonlinearity intervenes, however, the stress will in general deviate from a pure sinusoid. If the stress time-sequence is analysed as a Fourier series, there will be not only the fundamental with frequency ω but also harmonics with frequencies 2ω, 3ω, 4ω.... In the case of simple shear oscillations, however, symmetry of the deformation dictates that the shear stress τ obeys $\tau(t+\pi/\omega) = -\tau(t)$ and hence the even harmonics must be missing.

From, for example, the fundamental frequency component of the stress, an *apparent* complex modulus may be obtained[6]. In such dynamic experiments in which the strain amplitude is increased into the nonlinear regime, filled elastomers show a characteristic feature known as the *Payne effect* after the seminal study of it made by him (e.g. see Payne (1962; 1963)). The apparent storage and loss shear moduli G' and G'' at a frequency of 0.1 Hz are shown in *Figure 32*, plotted by Kramer et al (1994) from the original work of Payne.

The apparent storage modulus G' falls dramatically with increasing strain amplitude γ_0 over the range 0.001-0.1 for the filled samples, with the fall largest for the greatest volume fraction of carbon black. The apparent loss modulus G'' passes through a peak between 0.001 and 0.1. This is the pattern for samples reinforced with different proportions of the *same* black. When samples containing *different* blacks are compared, the Payne effect is seen to be most pronounced for samples containing black with the highest specific surface area. Such evidence suggests that the Payne effect arises from strain-induced structural breakdown associated with the elastomer-filler interfaces. In terms of the present notation: increasing strain amplitude in a dynamic test modifies one or more of the set of internal variables S , thereby changing the stress response to strain.

When the large strain amplitude is removed, however, the dynamic response recovers on a time-scale τ_s . This may be interpreted in terms of a recovery of the structure on this time-scale. The recovery may be followed by means of experiments with variable strain amplitude. Following oscillations in the nonlinear region, the strain amplitude is reduced to a low value in the linear region. The apparent complex shear modulus G^* is found to return gradually to its linear viscoelastic values over a time τ_s from the change in amplitude. This time is typically 10^3-10^4 s, decreasing with increasing temperature. It is noteworthy that usually $\tau_s \gg 2\pi/\omega$. Hence there is clearly insufficient time available for significant structural recovery *within* each cycle of strain. Consequently the steady-state high amplitude response corresponds to a constant structure

---

[6] In the earliest work the fundamental was not separated out from the total stress signal, hence strictly this approach lacked rigour. In practice, however, since the higher frequencies make such a small contribution, the error is small.

$\mathcal{S} \neq \mathcal{S}_0$. This explains why it displays *pseudo*-linear viscoelasticity[7], and the higher harmonics expected in general in the nonlinear region are found to be essentially absent – the "harmonic paradox" observed, for example by Chazeau et al (2000).

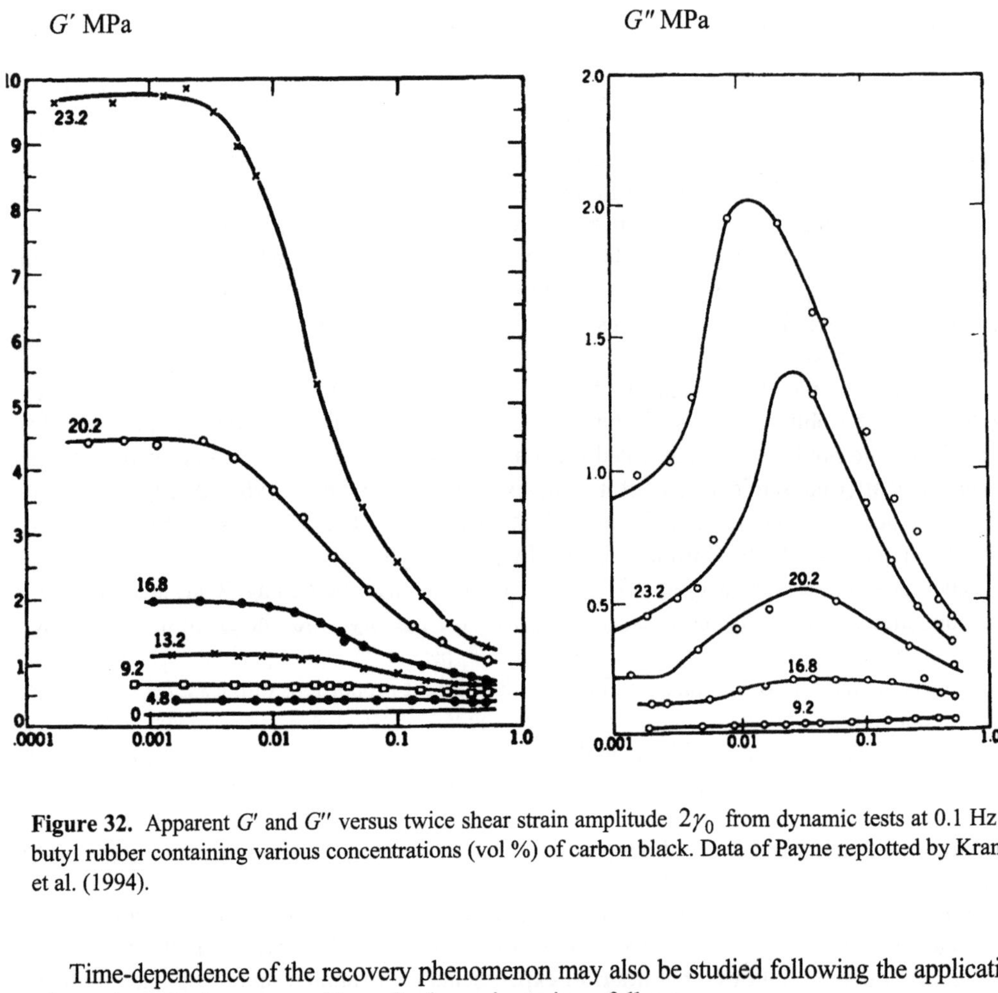

Figure 32. Apparent G' and G'' versus twice shear strain amplitude $2\gamma_0$ from dynamic tests at 0.1 Hz on butyl rubber containing various concentrations (vol %) of carbon black. Data of Payne replotted by Kramer et al. (1994).

Time-dependence of the recovery phenomenon may also be studied following the application of a static strain, by superposition of a dynamic strain as follows:

$$\gamma(u) = \gamma_1 H(u) + \gamma_2 \cos \omega u \quad , \tag{83}$$

where γ_2 lies in the linear region. The effect of a large static strain γ_1 is initially to cause a change in the apparent shear modulus G^* similar that caused by a large dynamic strain (i.e. $G' \downarrow, G'' \uparrow$), but it evolves back to its linear viscoelastic limit over the time-scale τ_s.

[7] "Linear" is used here in a restricted sense: stress is a linear functional of strain on a time-scale of one period of oscillation.

Special experiments for testing and fitting nonlinear constitutive models Several authors have proposed special experiments for testing specific constitutive models. We mention just two.

A convenient model framework for capturing time-dependent departures from hyperelasticity is to invoke two internal tensor variables – elastic and elasto-viscoplastic stresses σ_e, σ_v - such that at all times the Cauchy stress is given by

$$\sigma = \sigma_e + \sigma_v \ , \tag{84}$$

where the relation between σ_e and strain $\boldsymbol{B}$ is purely hyperelastic, while σ_v exhibits finite elasto-viscoplasticity in its dependence on $\boldsymbol{B}$. Thus, for constant $\boldsymbol{B}$ the stress contribution σ_v decays, while σ_e is constant. This proposition may be tested, and the magnitudes of the two contributions found, by means of a constant rate-of-strain test with interruptions. In uniaxial tension or compression, the magnitude of the axial nominal strain varies with time as in *Figure 33*, for example.

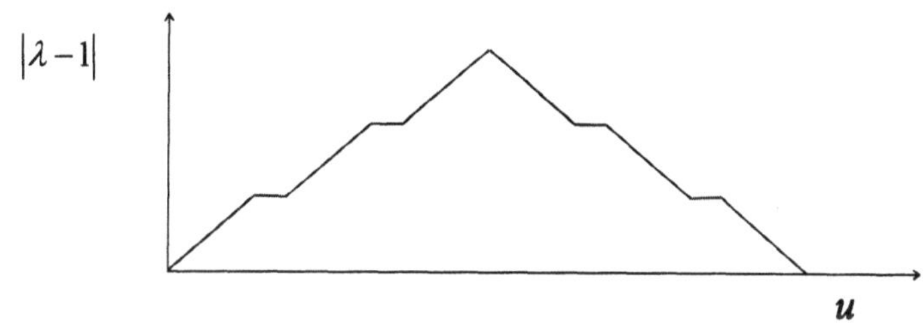

Figure 33. Interrupted constant rate-of-strain uniaxial test used to separate out hyperelastic and viscoelastic contributions to stress.

Such a constitutive model was proposed by Bergstrom and Boyce (1998), and this type of experiment was found useful for demonstrating qualitatively the validity of the model and also in determining model parameters. In the direction of straining, in steady state, $\text{sgn}(\sigma_v) = \text{sgn}(\dot{\lambda})$. Hence, on the positive ramp this contribution to stress is positive and relaxes downwards to zero during interruptions, while on the negative ramp it is negative and relaxes upwards to zero during interruptions. The observed relaxations downward and upward, obtained on interruptions during loading and unloading respectively, are consistent with expectations for the model *and* provide useful bounds on the magnitude of the elastic contribution σ_e, to assist with the fitting of model parameters. *Figure 34* shows results from Bergstrom and Boyce, obtained in compression.

Another variant has been found useful in examining closely the response to cyclic deformation in the nonlinear region. A small amplitude high frequency sinusoid is superposed on a large amplitude low frequency sinusoid – the "dual sine" experiment. Thus, in simple shear for example

$$\gamma(u) = \gamma_1 \cos \omega_1 u + \gamma_2 \cos \omega_2 u \quad \text{where} \quad \gamma_2 < \gamma_1, \ \omega_2 > \omega_1 \tag{85}$$

An experimental force-displacement curve is shown *Figure 35*, from the work of Coveney et al. on natural rubber filled with 70 phr carbon black, subjected to simple shear with

$$\gamma_1 = 0.1, \ \omega_1 = 0.5\pi \ \text{s}^{-1}, \ \gamma_2 : \gamma_1 = 1:10, \ \omega_2 : \omega_1 = 15:1 \ .$$

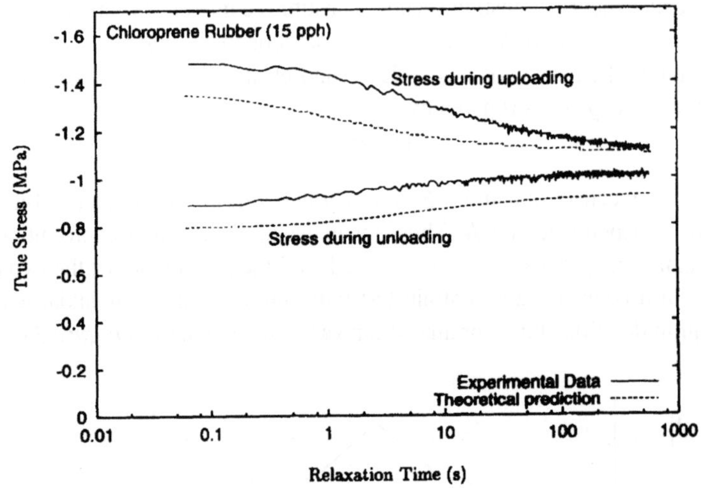

Figure 34. Stress relaxation during interrupts on loading and unloading of chloroprene rubber with 15 phr carbon black, from the work of Bergstrom and Boyce (2000).

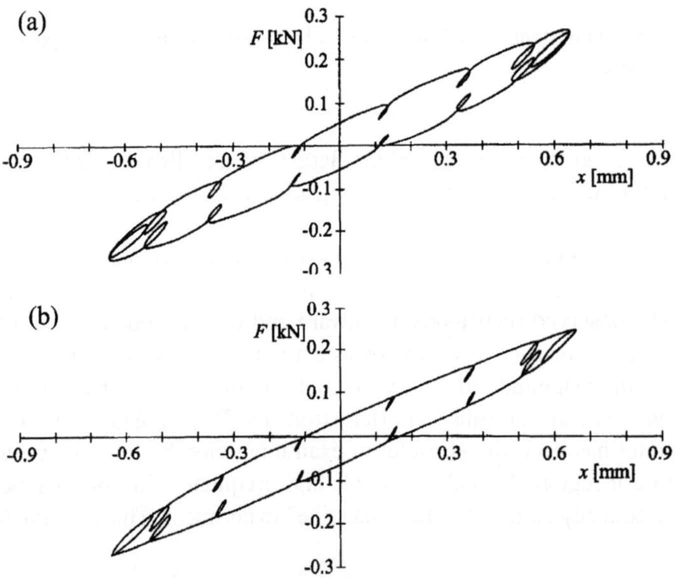

Figure 35. (a) Force versus displacement from a simple shear dual-sine experiment on natural rubber with 70 phr carbon black (parameters given in the text); (b) simulation of the results in (a) using the "triboelastic" constitutive model with three adjustable constants. From Coveney et al. (1999).

The measured loop as in *Figure 35a* has a characteristic shape, that seems to be general for filled elastomers. Note the localization and relatively high gradient of the low amplitude stress reversals compared to the high amplitude component. Coveney, Johnson and Turner (1995) had previously suggested a "standard triboelastic solid" constitutive model, with the same general form as equation (84), but where σ_v responds to stress as a series-coupled linear spring and *friction* element combination. It can be seen to reproduce these features well. The corresponding three-parameter *visco*-elastic model was found to be woefully inadequate in this respect.

4.4 Measurement of *non*-reproducible nonlinear viscoelasticity

All the results quoted above were obtained on pre-conditioned specimens (see Section 2.4) and were apparently reproducible for each specimen tested. Following the *first* deformation of an elastomer to a given strain[8] level outside the linear region, however, there is an effectively permanent change in the constitutive response – the well-known *Mullins effect.* This is studied by repeated strain cycles to a particular strain level. It is found that the stiffness and hysteresis reduce asymptotically to limiting values that remain apparently indefinitetly[9] during further cycles, provided this strain level is not exceeded. Thus the material may be considered to have suffered a permanent change in constitutive response, and a new equilibrium state S_0 to have been established.

The Mullins effect involves a measurable but only minor perturbation from hyperelasticity in the response of unfilled elastomers, except at the highest extensions - see *Figure 36a* below. In elastomers filled with reinforcing particles, however, it greatly exceeds the usual experimental uncertainty and must be accommodated in the design of experimental programmes. *Figure 36b* shows results from successive tensile stress-strain cycles of increasing amplitude applied to a single virgin specimen of natural rubber filled with 45 phr of carbon black. The dotted lines in *Figure 36* show the corresponding curves for continuous straining of separate specimens. A manifestation of the Mullins effect is clearly apparent in *Figure 36b*. During loading on the 2nd-4th cycles, the stress lies below the previous loading curve, but as the stretch approaches the previously highest stretch, the stress rises rapidly, essentially rejoining the dotted line.

Excessive temperatures also cause permanent changes in mechanical response of elastomers, as a result of temperature-accelerated changes to the molecular network. Such effects are studied by means, for example, of stress relaxation tests at elevated temperatures. In *Figure 37*, stresses are shown normalised to their initial values. The time-dependent response is clearly quite different from the reproducible viscoelastic behaviour discussed in Section 4.1. Firstly the stress is seen to relax fully. Secondly the relaxation is seen to be highly localised in the time domain – essentially confined to 2 decades of time, c.f. *Figures 28* and *30*. The cause in this case is molecular scission.

[8] At the present time, it is unclear what measure of *three-dimensional* strain determines the subsequent response.

[9] Gradual recovery is observed, but over a longer time-scale than the recovery discussed in Section 4.3.

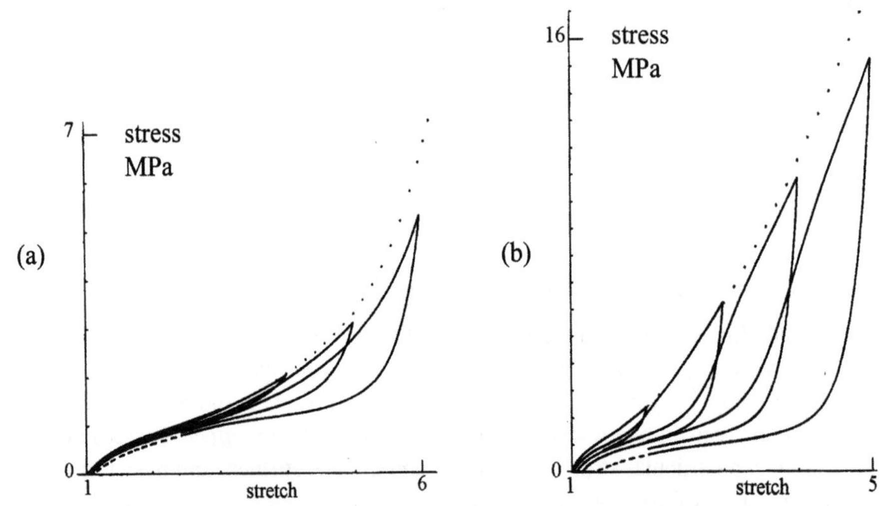

Figure 36. Successive load-unload cycles at a nominal strain-rate of 0.04 s^{-1}, with increasing strain amplitude, on single specimens of (a) natural rubber, and (b) natural rubber reinforced with 45 phr carbon black. Dotted lines show the corresponding curves for specimens strained continuously. From the work of Muhr et al. (1999).

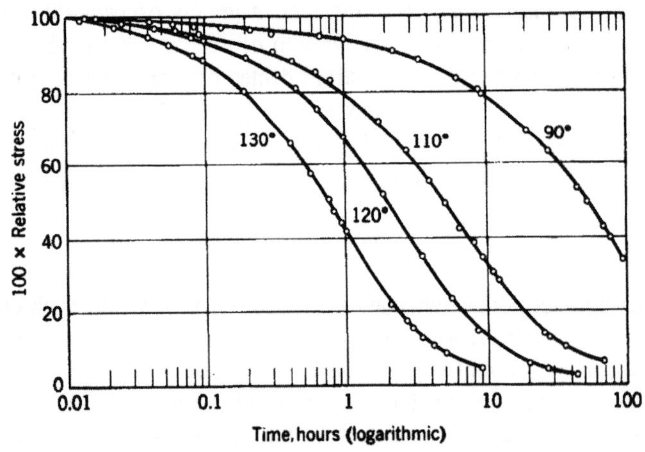

Figure 37. Stress relaxation at elevated temperatures (°C) in a crosslinked polyester elastomer, revealing the effect of chain scission. Data of Offenbach and Tobolsky quoted by Ferry (1980).

Thus, after stress has relaxed, the specimen has different molecular connectivity, and a permanent change in structure and properties has occurred. The response to a subsequent strain cannot be predicted *a priori*. During relaxation there may have been further crosslinks forming, but since they form with the strained state as the stress-free state, they make no contribution to stress during the relaxation process. Their formation is revealed only by further straining of the relaxed specimen, whereupon the incremental tensile modulus may have increased if the rate of new crosslinking exceeded the rate of chain scission. Such effects correspond to formation of a genuinely new equilibrium structural state S_0. The original structure and properties cannot be recovered.

5 Validation experiments

5.1 Introduction

The aim of many studies of mechanical and thermomechanical properties of rubberlike solids is to generate better constitutive models, for embedding in continuum models of components of engineering interest. The purpose of the continuum model is to solve for inhomogeneous distributions of stress and strain. In the isothermal case it consists of three sets of equations applying to the solution domain - equilibrium, compatability and constitutive - and it is solved subject to the known history of boundary forces and displacements. There are various possible solution methods: analytical methods, finite difference numerical methods and finite element numerical methods. Whatever method is used, the accuracy of the model must be checked against experiment, before it can be used with confidence for a problem of practical importance. Thus there is a need for *validation experiments*, to meet this need – Objective 4 of Section 2.1.

Desirable features of a validation experiment are as follows.

- *Problem match*: a close match between geometry and boundary conditions of the experiment and of the model.
- *Material match*: a close match between the state of the material in the validation experiment and in experiments from which model parameters were determined.
- *Benchmarks*: existence of known benchmark solutions for special cases of the experiment, for checking assumptions made.
- *Relevance* of the experimental stress/strain-states to the problems of practical interest.

5.2 Stretch of a perforated sheet

The classical problem of a thin sheet with a central circular hole provides a useful validation test for large deformation plane stress problems. It is most readily realisable with a biaxial testing machine for sheet specimens, as described in Section 3. The sheet is of initial thickness H and length L, with a central circular hole of diameter D (radius A=D/2) – see *Figure 38*, with $H \ll D \ll L$. The advantage of this test is the existence of the well-known analytical solution in the limit of infinitesimal displacement gradients and linear elasticity (e.g. see Timoshenko and Goodier), providing a useful benchmark in the small strain limit, that must be obeyed if the boundary conditions are as assumed, and if the material is isotropic and homogeneous as assumed[10]. For example, in the uniaxial tension case $\sigma_1 = \sigma,\ \sigma_2 = 0$, the tangential component of stress at initial position R, θ is

[10] This remains true even if the material is viscoelastic, by virtue of the correspondence between linear elastic and linear viscoelastic stress states.

$$\sigma_{\theta\theta} = \frac{\sigma}{2}\left[1+\left(\frac{A}{R}\right)^2\right] - \frac{\sigma}{2}\left[1+3\left(\frac{A}{R}\right)^2\right]\cos 2\theta \,. \tag{86}$$

The solution for an equal-biaxial tensile test (EBt) on the sheet is obtained by superposing the above on the corresponding solution for the case $\sigma_1 = 0,\ \sigma_2 = \sigma$. The result is

$$\sigma_{\theta\theta} = \sigma\left[1+\left(\frac{A}{R}\right)^2\right] \tag{87}$$

and at the edge of the hole $R = A$ we see $\sigma_{\theta\theta} = 2\sigma$. The far-field strain in the EBt case is simply (with the incompressibility assumption) $\varepsilon = \sigma / 2E$ and the tangential strain at the hole $\varepsilon_{\theta\theta}^{\mathrm{h}} = 2\sigma / E$. In this axisymmetric case, the hole stretch $\lambda^{\mathrm{h}} = d / D = 1 + \varepsilon_{\theta\theta}^{\mathrm{h}}$ is hence related to the far-field stretch $\lambda = 1 + \varepsilon$ by

$$\lambda^{\mathrm{h}} - 1 = 4(\lambda - 1)\,. \tag{88}$$

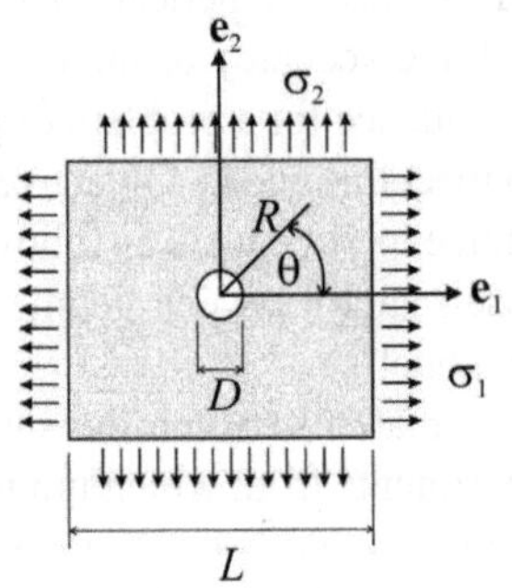

Figure 38. Initial configuration of a square sheet specimen, perforated with a central circular hole: a classic validation experiment.

An example is shown below, from the work of Buckley and Turner (1999). Latex rubber sheet had been characterised using the FBFT biaxial testing machine (see Section 3) and the data fitted to an extension of the Turner-Brennan constitutive model. A specimen 70 mm square, with thickness 150 μm was provided with a sharply-cut central circular hole of 7 mm diameter, and then stretched equal biaxially at a nominal extension rate 0.01 s^{-1} at 22°C. A video camera mounted above the sheet recorded the deformation. The problem was also solved numerically by the finite difference method. Results are shown in *Figure 39*, including as a benchmark the limiting (linear theory) case of equation (88). As expected, both the data and numerical model are in accord with this equation for small strains, but deviate at higher strains. It is interesting to note the manner of deviation, which reflects the form of the tensile stress-strain relation. The prominent hole stretch deficit at large far-field stretches reflects the pronounced strain-stiffening in this material, associated with the strain-induced crystallisation of natural rubber.

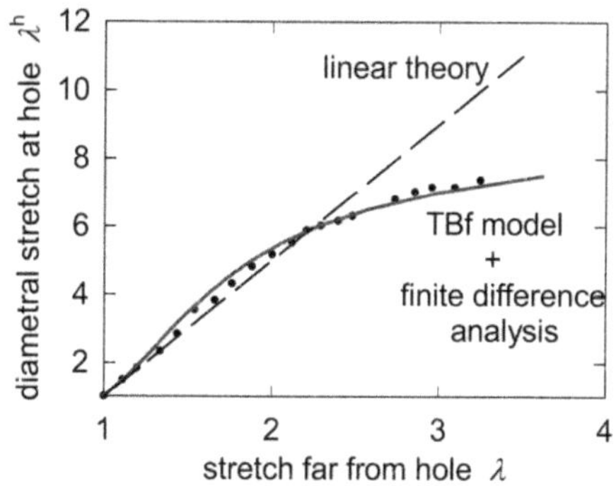

Figure 39. Equal biaxial stretch of a perforated latex rubber sheet (data points), together with calculated results from small strain linear elasticity theory (dashed line) and finite difference analysis using the Turner-Brennan 'filament' (TBf) constitutive model (full line). Data from Buckley and Turner (1999).

Further information obtainable from EB stretching of a perforated sheet is the radial distribution of the stretch. This was measured in the classic experiments of Rivlin and Thomas (1951), who subjected to EB stretching a circular sheet of natural rubber, diameter 127 mm, thickness 1.6 mm, with a central hole of diameter 25 mm. By manual measurement of concentric rings drawn on the sheet, the radial stretch was determined, for various far field stretches. Results were compared with finite difference calculations employing the Mooney strain energy function – see Section 3.4 – with $C_{01} = 0.1C_{10}$. Good agreement was obtained, as may be seen in *Figure 40*. This is a little surprising, since Rivlin and Saunders had previously highlighted the deficiency of the Mooney form for W. However, the explanation is that the strain field in an EBt test on a perforated sheet is characterized by rather constant I_2, and hence any variation of $\partial W / \partial I_2$ with I_2 has little effect.

It is interesting to compare the Rivlin-Thomas results with the benchmark linear elastic solution for the radial stretch:

$$\frac{r}{R} = 1 + (\lambda - 1)\left[1 + 3\left(\frac{A}{R}\right)^2\right]. \tag{89}$$

Comparison with the data in *Figure 40* again reveals a similar pattern to *Figure 39*: λ^h / λ deviates initially slightly above and then finally below the linear elastic solution.

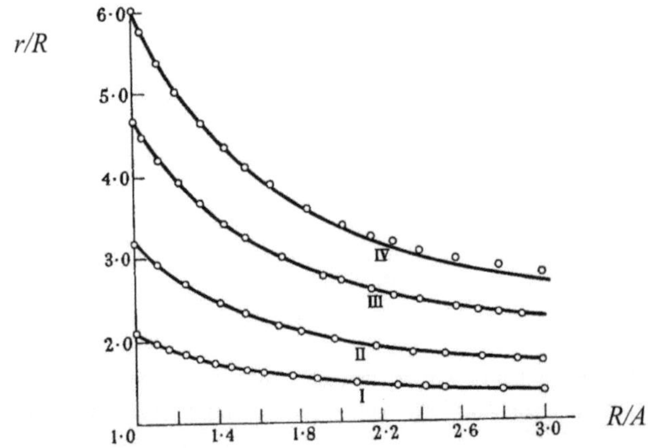

Figure 40. Radial stretch versus initial radial position, during equal biaxial stretching of a sheet of natural rubber containing a central circular hole (data points), compared with a numerical solution employing the Mooney hyperelastic constitutive model (lines). From Rivlin and Thomas (1951). Curves I-IV refer to different values of far-field stretch.

If the far-field strain-state is *non*-equal biaxial, with biaxiality ratio

$$\phi = \frac{\lambda_2 - 1}{\lambda_1 - 1} \neq 1 \tag{90}$$

the circular hole becomes elongated during stretching. The experiment then provides another measure that may be used for validating continuum models: the variation of tangential stretch around the hole, i.e. $\lambda_\theta(\theta)$ at R=A. For example, this was used by Lindley (1971) to validate a very early Finite Element analysis of large inhomogeneous deformation in an elastomer. An 80 mm square sheet of natural rubber, with thickness 1.5 mm and a central circular hole of diameter 17.5 mm, was subjected to uniaxial extension from rigid grips. Ink marks were used to measure displacements, for comparison with the FE calculations. Lindley employed the Gent-Thomas strain energy function

$$W = C_1\left(I_1 - 1\right) + C_2 \ln(I_2 / 3)\,. \tag{91}$$

Generally good agreement was obtained between the measured and calculated displacements. Specifically, the measured and calculated tangential nominal strains $\varepsilon_{\theta\theta} = \lambda_\theta - 1$ around the hole are shown in *Figure 41*. Also included is $\varepsilon_{\theta\theta}$ as calculated from the linear elastic solution for tangential tensile stress at R=A , given in equation (86).

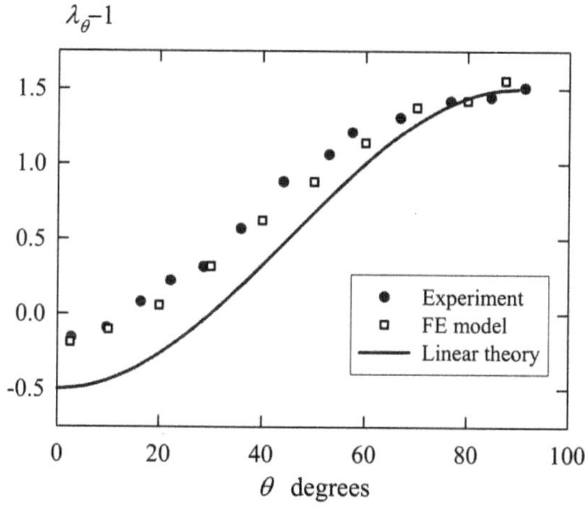

Figure 41. Extension of a square sheet of natural rubber with a central circular hole to a far-field nominal strain of 0.5: tangential nominal strain around the hole, as obtained by experiment, as calculated by FE analysis, and as predicted by linear elasticity theory. Data points replotted from Lindley (1971).

Although the use of rigid grips provides a boundary condition at $X_1 = \pm L/2$ different from that for which equation (86) applies strictly, this linear elastic benchmark solution again provides useful insight. Note that the measured and predicted total curvilinear extensions of the perimeter of the hole (the mean values of $\lambda_\theta - 1$) clearly exceed the linear theory prediction. Mean values of $\varepsilon_{\theta\theta}$ from experiment and FE analysis are 0.77 and 0.69 respectively, while the linear theory predicts 0.5 as may be shown from equation (86). This difference reflects the initial strain-softening of the material in tension, causing the more highly stressed hole region to accept relatively larger strains than elsewhere, as compared to the linear solution .

An important practical point in all such experiments on specimens in the form of thin sheets, is that buckling instability must be avoided. Thus, even when the far-field stress is a positive tension, inhomogeneity of stress adjacent to the hole may cause local buckling, as a result of a compressive stress. Thus, equation (86) shows that for a far-field uniaxial tension one third of the hole perimeter will be subject to a tangential compressive stress. If the sheet is thin ($H \ll D$ – as is required for the plane stress assumption to be valid) this is likely to cause buckling. In fact, it can be shown by applying equation (86) to the case of a far-field *biaxial* tension, that the occurrence of a compressive stress and hence the possibility of buckling is prevented by restricting far-field strain biaxialities to the range $\phi >$ -0.2. Even though this argument is based on results from infinitesimal linear elasticity, it provides a useful guide for design of experiments on perforated sheets at finite strains.

5.3 Inflation experiments

As we saw in Section 3, a convenient means of creating biaxial stress states in the laboratory is to employ inflation of a specimen in the form of a membrane (e.g. a sheet, a sphere or a tube) by application of a pressure differential between the two surfaces. Unless the boundary displaces to ensure uniformity, only in the case of a perfect sphere can the strain induced be uniform. Therefore, careful measurements on inflated *non*-spherical membranes provide a further readily achievable means of checking predictions of inhomogeneous strain.

One check is provided by comparing the three dimensional shape of the deformed membrane with the calculation. Consider the case of a flat circular sheet with initial radius A and thickness H, clamped around its circumference and subject to compressed air of increasing relative pressure p on one face, with atmospheric pressure on the other face. The sheet bulges away from the pressurized face and inflates. At all times the strain-state varies from $\phi = 1$ (EBt) at the pole, to $\phi = 0$ (CWt) at the perimeter where it is clamped. It thus forms a non-spherical shape. Such an experiment was used by James et al. (1975) to check the predictions of finite difference solutions, employing different numbers of terms in the strain energy function W expressed in terms of invariants (equation (46)). A biaxial testing machine had been used already for biaxial extension tests, to obtain the data for fitting to the constitutive model. *Figure 42* shows their results.

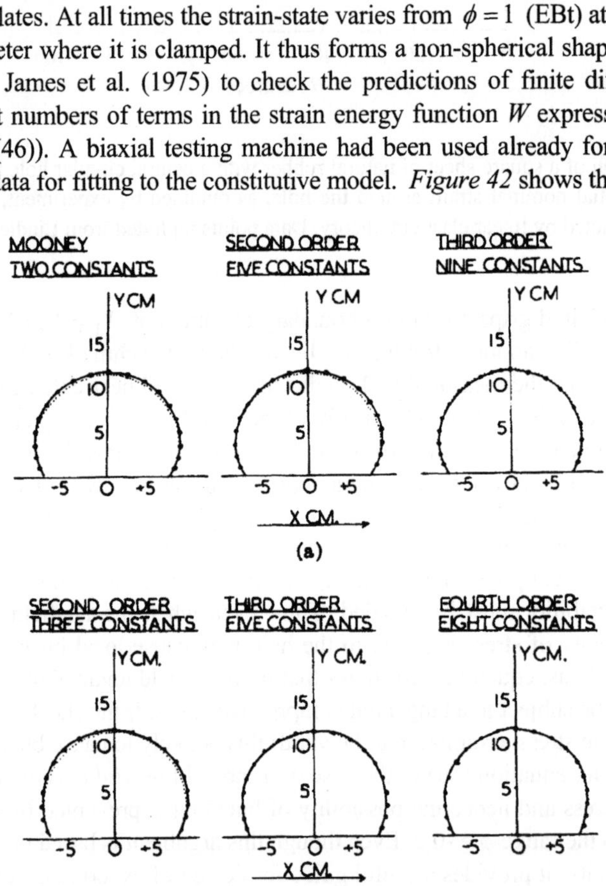

Figure 42. Cross-section profiles of a flat circular sheet of natural rubber, inflated to a maximum stretch $\lambda \approx 3$ at the pole (points), and calculation using different numbers of terms in equation (46) for $W(I_1,I_2)$ (dotted lines). From James et al. (1975).

The same team of researchers also moulded their rubber into the form of a thin-walled tube (thickness 1 mm, diameter 25 mm, length 115 mm). It was clamped over a mandrel at each end and inflated with compressed air. The cross-sectional profile obtained at various pressures was then compared with the results of finite difference calculations. Results for three pressures are shown in *Figure 43*, where the calculations employed the expansion to third order in equation (46). Again the fit is excellent. Note again that the strain-states involved lie between the EBt and CWt bounds in principal stretch space.

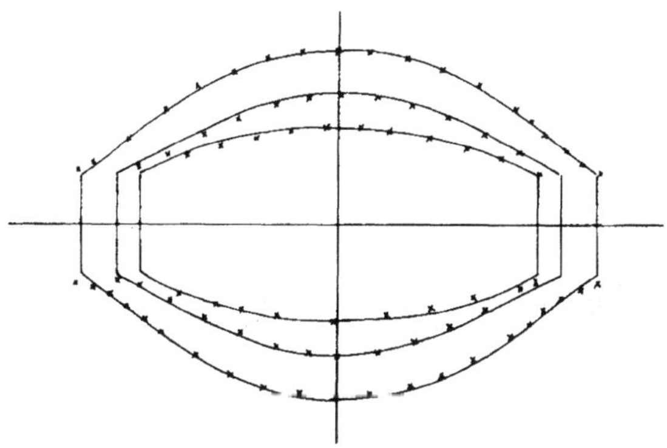

Figure 43. Cross-section profile of a tube of natural rubber inflated with ends clamped over mandrels: measurements (crosses) and results of finite difference calculations employing the expansion to third order in equation (46) (lines). From James et al. (1975).

Finally it is worth noting that, while such fits to the *shapes* of inflated membranes are successful, it is clear they do not provide a validation of the *stresses* calculated. For some validation purposes, therefore, they provide insufficient discrimination. In these cases, experiments are required that include measurement of at least the boundary forces in addition to the deformed shape. An example is the measurement of pressure p during inflation of an initially flat circular sheet (radius A, thickness H) clamped around its edge. The pressure is found to pass over a peak. The accuracy of prediction of this peak with a numerical model provides a good check on whether the constitutive model employed captures quantitatively the stress-stretch response of the material in the region of principal space bounded by the EBt and CWt loci. An example is shown in *Figure 44*, where the pressure (in dimensionless form) is plotted versus the bulge displacement z of the pole of the inflated sheet (also in dimensionless form), from the work of Feng (1992). In this case the specimen was a sheet of latex rubber (A = 25 mm, H = 0.762 mm), and the constitutive model employed was neo-Hookean. The fit is remarkably successful, given the known limitations of this constitutive model. Note that the pressure peak observed (and predicted) in experiments such as this is a potential experimental pitfall. The potential instability may lead to asymmetric deformation if there is significant nonuniformity of either specimen thickness or properties.

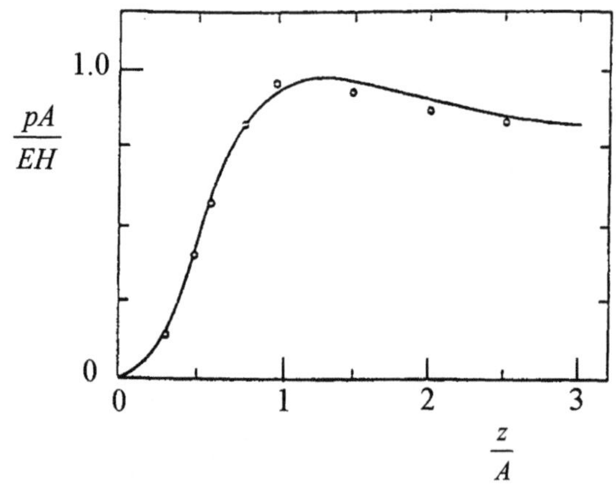

Figure 44. Pressure versus bulging displacement when an initially flat circular rubber sheet is inflated. Experimental data (points) and finite difference calculations (lines). From Feng (1992).

References

Arridge, R.G.C. (1974). A new method for measurement of bulk modulus of solid polymers. *Journal of Physics E: Scientific Instruments* **7**:399-401.

Bergstrom, J.S. and Boyce, M.C. (1998). Constitutive modelling of the large strain time-dependent behaviour of elastomers. *Journal of the Mechanics and Physics of Solids* **46**:931-954.

Bergstrom, J.S. and Boyce, M.C. (2000). Large strain time-dependent behaviour of filled elastomers. *Mechanics of Materials* **32**:627-644.

Brown, R.P. (1996). *Physical Testing of Rubber*. London: Chapman and Hall.

Buckley, C.P. and Turner, D.M. (1999). Application of flexible biaxial testing in the development of constitutive models for elastomers. In A. Dorfmann and A. H. Muhr, eds., *Constitutive Models for Rubber*. Rotterdam: A.A.Balkema. 59-64.

Chazeau, L., Brown, J.D., Yanyo, L.C. and Sternstein, S.S. (2000). Modulus recovery kinetics and other insights into the Payne effect for filled elastomers. *Polymer Composites* **21**:202-222.

Coveney, V.A., Jamil, S., Johnson, D.E., Keavey, M.A. and Menger, C. (1999). Implementation in FEA of a triboelastic law for dynamic behaviour of filled elastomers. In D. Boast and V. A. Coveney, eds., *Finite Element Analysis of Elastomers*. London: Professional Engineering Publishing.

Coveney, V.A., Johnson, D.E. and Turner, D.M. (1995). A triboelastic model for the cyclic mechanical behaviour of filled vulcanizates. *Rubber Chemistry and Technology* **68**:660-670.

Feng, W.W. (1992). *Transactions of the ASME Journal of Applied Mechanics* **59**:S29-S34.

Ferry, J.D. (1980). *Viscoelastic properties of polymers*. New York: John Wiley and Sons.

Gerlach, C., (1998). *Sequential biaxial drawing of polyester film*. DPhil, Oxford University.

Gough, J., Gregory, I.H. and Muhr, A.H. (1999). Determination of constitutive equations for vulcanized rubber. In D. Boast and V. A. Coveney, eds., *Finite Element Analysis of Elastomers*. London: Professional Engineering Publishing. 5-26.

Haupt, P. (2002). *Continuum Mechanics and Theory of Materials*. Berlin: Springer-Verlag.

Holzapfel, G.A. (2000). *Nonlinear Solid Mechanics*. Chichester: John Wiley and Sons.

James, A.G. and Green, A. (1975). Strain energy functions of rubber. II. The characterisation of filled vulcanizates. *Journal of Applied Polymer Science* **19**:2319-2330.

James, A.G., Green, A. and Simpson, G.M. (1975). Strain energy functions of rubber. I. Characterisation of gum vulcanizates. *Journal of Applied Polymer Science* **19**:2033-2058.

Johannknecht, R. and Jerrams, S.J. (1999). The need for equi-biaxial testing to determine elastomeric material properties. In A. Dorfmann and A. H. Muhr, eds., *Constitutive models for rubber*. Rotterdam: A.A.Balkema. 73-76.

Jones, D.F. and Treloar, L.R.G. (1975). The properties of rubber in pure homogeneous strain. *Journal of Physics D: Applied Physics* **8**:1285-1304.

Kramer, O., Hvidt, S. and Ferry, J.D. (1994). Dynamic Mechanical Properties. In J. E. Mark, B. Erman and F. R. Eirich, eds., *Science and Technology of Rubber*. San Diego: Academic Press.

Lindley, P.B. (1964). *Engineering Design with Natural Rubber*. Brickendonbury: Malaysian Rubber Producers' Research Association.

Lindley, P.B. (1971). Plane-stress analysis of rubber at high strains using finite elements. *Journal of Strain Analysis* **6**:45-52.

McCrum, N.G. (1984). The determination of dlnr^2/dT by the the method of temperature induced creep. *Polymer Communications* **25**:213-217.

McCrum, N.G., Buckley, C.P. and Bucknall, C.B. (1997). *Principles of Polymer Engineering*. Oxford: Oxford University Press.

Medalia, A.I. and Krauss, G. (1994). Reinforcement of elastomers by particulate fillers. In J. E. Mark, B. Erman and F. R. Eirich, eds., *Science and Technology of Rubber*. 2, San Diego: Academic Press. 387-418.

Muhr, A.H., Gough, J. and Gregory, I.H. (1999). Experimental determination of model for liquid silicone rubber: hyperelasticity and Mullins' effect. In A. Dorfmann and A. H. Muhr, eds., *Constitutive Models for Rubber*. Rotterdam: A.A.Balkema.

Mullins, L. (1959). Determination of degree of crosslinking in natural rubber vulcanizates. Part IV. Stress-strain behaviour at large extensions. *Journal of Applied Polymer Science* **2**:257-263.

Obata, Y., Kawabata, H. and Kawai, H. (1970). Mechanical properties of natural rubber vulcanizates in finite deformation. *Journal of Polymer Science: Part A-2* **8**:903-919.

Oertel, G., Ed. (1994). Polyurethane Handbook. Munich, Carl Hanser Verlag.

Ogden, R.W. (1972). Large deformation isotropic elasticity - on the correlation of theory and experiment for incompressible rubberlike solids. *Proceedings of the Royal Society* **A 326**:565-584.

Payne, A.R. (1962). The dynamic properties of carbon black-loaded natural rubber vulcanizates. Part I. *Journal of Applied Polymer Science* **6**:57-63.

Payne, A.R. (1963). Dynamic properties of heat-treated butyl vulcanizates. *Journal of Applied Polymer Science* **7**:873-885.

Penn, R.W. (1970). Volume changes accompanying the extension of rubber. *Transactions of the Society of Rheology* **14**:590-617.

Rivlin, R.S. and Saunders, D.W. (1951). Large elastic deformations of isotropic materials. VII Experiments on the deformation of rubber. *Philosophical Transactions of the Royal Society* **A243**:251-288.

Rivlin, R.S. and Thomas, A.G. (1951). Large elastic deformation of isotropic materials VIII. Strain distribution around a hole in a sheet. *Philosophical Transactions of the Royal Society* **A 243**:289-298.

Shen, M. (1969). Internal energy contribution to the elasticity of natural rubber. *Macromolecules* **2**:358-364.

Timoshenko, S. and Goodier, J.N. (1951). *Theory of Elasticity*. New York: McGraw-Hill.

Treloar, L.R.G. (1958). *The Physics of Rubber Elasticity*. Oxford: Clarendon Press.

Treloar, L.R.G. and Riding, G. (1979). A non-Gaussian theory for rubber in biaxial strain. *Proceedings of the Royal Society* **A 369**:261-280.

Tschoegl, N.W. (1989). *The Phenomenological Theory of Linear Viscoelastic Behaviour*. Berlin: Springer-Verlag.

Vangerko, H. and Treloar, L.R.G. (1978). The inflation and extension of rubber tube for biaxial strain studies. *Journal of Physics D: Applied Physics* **11**:1969-1978.

Molecular Aspects of Rubber Elasticity

Burak Erman

Department of Chemical and Biological Engineering, Koc University, Rumelifeneri Yolu, Sariyer 34450, Istanbul Turkey

Abstract. In this paper, the molecular aspects of rubber-like solids are reviewed. The elasticity of the single polymer chain, which is the basic unit of the elastomeric solid, is discussed first. The structure of the network obtained by joining the single chains together is described. The elementary and more advanced theories of rubber elasticity are discussed. Swelling of networks, thermoelastic behavior, strain-induced crystallization and bioelastomers are reviewed.

1 Elasticity of the Single Polymer Chain

The basic unit of a network is the single polymer chain. Before seeing the properties of networks, we will see the important features of the single polymer chain (Flory, 1953). In Figure 1, part of a polymer chain is shown lying in the plane of the paper. Atoms (or larger units) along the backbone are shown with the black dots in the figure and numbered. The lengths of bonds of a polymcr chain usually do not change much. Similarly, the

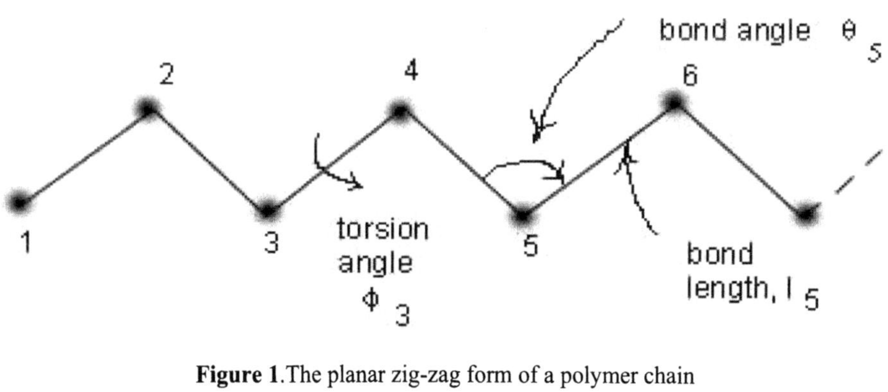

Figure 1.The planar zig-zag form of a polymer chain

bond angles, one of which is shown as θ_5, for example, do not change much. In the first approximation, they may be taken as fixed. These may change if a very large force stretches the chain, but this will not be considered here. Rotations about torsion angles, as shown with ϕ_3, for example, take place easily due to thermal energy. The value of the torsion angle in the planar form shown in the figure is zero.

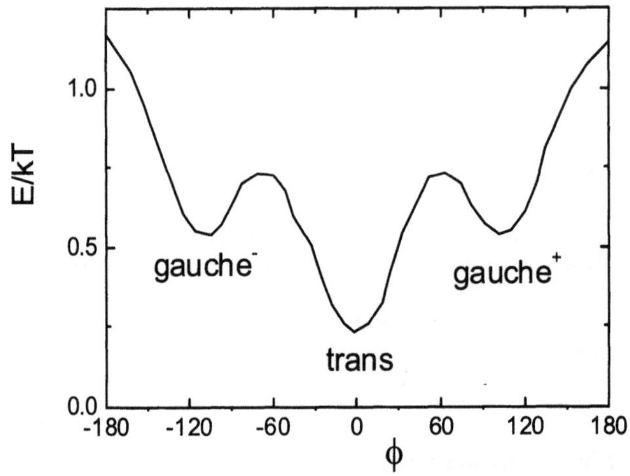

Figure 2. An example of the energy of the torsion angle.

As the torsion angle of the bond changes, the torsional energy of the bond changes, as shown in Figure 2. Here, E is the energy plotted as a function of the torsion angle. k is the Boltzmann constant, and T is the absolute temperature. kT has dimensions of energy, so the ordinate is dimensionless. Three minima are identified as trans, gauche+, and gauche$^-$. The probability of having a bond at a given angle ϕ depends on its torsional energy, and is given by the expression

$$p(\phi) = \frac{e^{-E(\phi)/kT}}{\int_{-180}^{180} e^{-E(\phi)/kT} d\phi} \tag{1}$$

According to eq 1, the angles for which the energy is minimum, has the maximum probability of occurrence. In order to simplify the study of the single chain, we assume that the bond may exist in only the three minima shown in Figure 2. These three minima, corresponding to the trans, gauche+ and gauche$^-$ states correspond to the three isomeric states of the bond. Accordingly, the probabilities of observing each of the three states become:

$$p_t = \frac{e^{-E_t/kT}}{e^{-E_t/kT} + e^{-E_{g+}/kT} + e^{-E_{g-}/kT}}$$

$$p_{g+} = \frac{e^{-E_{g+}/kT}}{e^{-E_t/kT} + e^{-E_{g+}/kT} + e^{-E_{g-}/kT}} \qquad (2)$$

$$p_{g-} = \frac{e^{-E_{g-}/kT}}{e^{-E_t/kT} + e^{-E_{g+}/kT} + e^{-E_{g-}/kT}}$$

If all the bonds are in the trans state, the chain will have a rod-like zig-zag shape. The presence of the gauche states changes the shape of the chain into a disordered, three-dimensional structure, an example of which is shown in Figure 3.

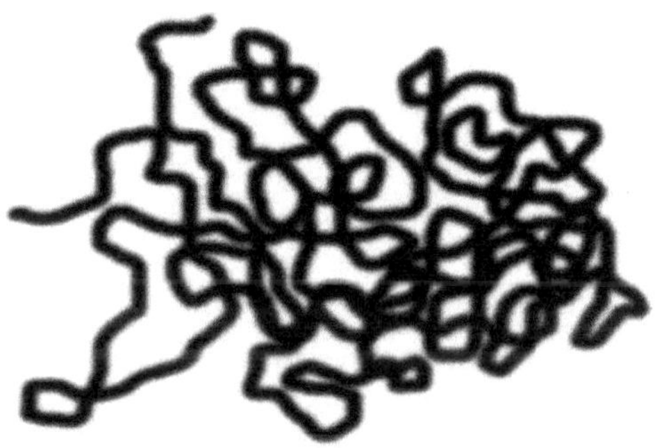

Figure 3. A random configuration of a long polymer chain

The disordered spatial configuration of the chain results from the disorientation of the individual bonds because of the gauche+ and gauche$^-$ states. However, this picture of the chain is not stationary. The state of each bond changes continually between the three isomeric states. The rate of this change depends on the environment of the chain. If the chain is in a suitable solvent, the rate of change between two states is about 10^{10} times per second at room temperature. In the bulk state, i.e., when several chains are together, this rate decreases, because the chains are all entangled together, and each chain becomes a highly viscous environment to its neighbors. The rate depends on temperature strongly. When the temperature is decreased below the glass temperature, this rate drops to zero. Although the state of the individual bonds changes between the three states, the populations of bonds of the entire chain in the trans, gauche+ and gauche- states are proportional to p_t, p_{g+} and p_{g-} given by eq 2, respectively.

If we have a chain of 100 bonds, and if each bond may exist in three states, then the total number of configurations a chain can have is 3^{100}, an enormous number. We can only consider the average properties of the chain. For example, the average squared end-to-end distance, shown as $\langle r^2 \rangle$, is a quantity that gives us an idea about the size of the chain. There are two types of averages, the time and the ensemble average. In the time average, we wait long enough so that the chain can take all the possible configurations, and we measure **r** for each configuration, and average over r^2. In the ensemble average, we consider an instantaneous collection of many identical chains, measure the **r** of each of them, and average over r^2. We will not differentiate between time and ensemble averages. We will assume that they are identical. The assumption that these two are identical is based on a hypothesis, called the ergodic hypothesis.

Let us now generate a random configuration of a chain of 10 bonds on a square lattice, as shown in Figure 4.

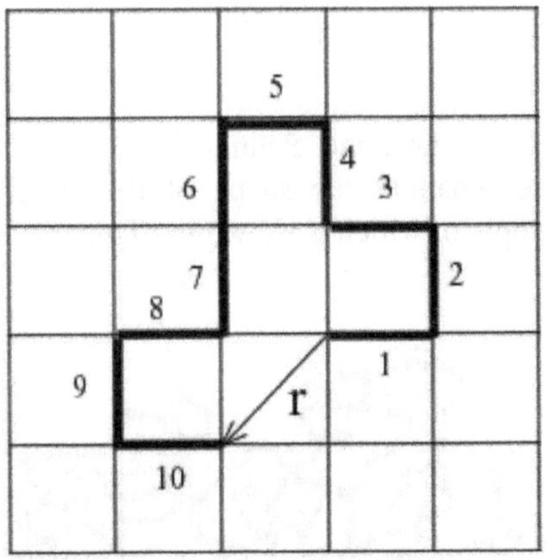

Figure 4. A random configuration of a 10-bond chain

We start from any lattice point. We have four possible directions to place the first bond. We randomly choose one direction. In the figure it is to the right. For the second bond, we again have four possible directions to choose. We randomly choose one. In the figure, it is up. So for every bond, we have a choice of right, up, left and down. We generate one configuration as shown in the figure. If we are interested in the end-to-end vector, we measure the distance from the beginning to the end and record it. We generate several of such configurations, each time randomly, and form the average:

$$< r^2 >= \frac{\sum_{i=1}^{N} r_i^2}{N} \tag{3}$$

where $<r^2>$ is the mean-square end-to-end distance, and N is the number of chains generated. This way of calculating averages is called the Monte Carlo (MC) method. Since we generated the chain on the lattice, it is called the on-lattice Monte Carlo Method. Here, we gave the four directions to each bond with equal probability. For realistic cases, a bond can lie along a direction if the atom at the end of it does not end at a lattice point that is already occupied by another atom. This is the 'excluded volume' effect, according to which two atoms cannot occupy the same point in space. When chain dimensions are not perturbed by excluded volume effects, or by other effects such as external forces, the dimensions are referred to as unperturbed dimensions. The unperturbed state is indicated with a subscript zero. The mean-squared dimensions of an unperturbed chain, for example, is shown as $<r^2>_0$.

How do we generate the direction of a bond at each step: Assume that a bond can be directed along the four directions with the following probabilities: Right, 0.2; Up, 0.3; Left, 0.4; Down, 0.1. Then we make an interval between 0 and 1, and divide this interval according to the four given probabilities, as shown in Figure 5. We then generate a random number, x. If, $0 < x < 0.2$, then we are in the first interval and the bond directs to the right, if, $0.2 < x < 0.5$, then we are in the second interval and the bond directs up, if, $0.5 < x < 0.9$, then the bond directs left, if, $0.9 < x < 1.0$, then the bond directs down. One can thus build any desired statistics into a theoretical model by using the Monte Carlo method. It is a very efficient

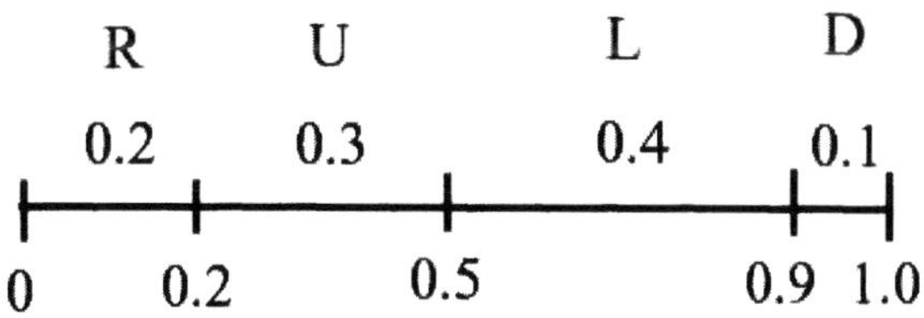

Figure 5. Interval of probabilities of different orientations

simulation method, and helps the newcomer visualize how averages of polymers are obtained. It is straightforward to generalize the method to three dimensions. If, in three dimensions, a bond can take any direction with equal probability, then the bond acts as if it is freely jointed to the previous bond, and that chain is called a freely jointed chain. For the freely jointed chain, $<r^2> = nl^2$, where n is the number of bonds and l is the length of a bond.

The end-to-end vector distribution. Let us place one end of a chain at the origin of a cartesian coordinate system, Oxyz. Let us assume that we perform a Monte Carlo simulation in three dimensions and record the end-to-end vector for all of the chains generated. The distribution of the end-to-end distance will appear as the one shown in Figure 6. There will be many chains with small end-to-end distances, and fewer chains with large r. For a sufficiently long chain, the probability of finding the other end at a distance r from the origin is given by the Gaussian function (Treloar, 1975)

$$W(r) = \left(\frac{a}{\pi}\right)^{3/2} \exp\left\{-a(x^2 + y^2 + z^2)\right\} \tag{4}$$

where, x, y and z are the cartesian coordinates of the end-to-end vector **r**, and $a = \dfrac{3}{2 < r^2 >_0}$. Equation 4 may be written in the more familiar form

$$W(r) = \left(\frac{3}{2 < r^2 >_0 \pi}\right)^{3/2} \exp\left\{-\frac{3}{2 < r^2 >_0} r^2\right\} \tag{5}$$

The form of W(r) will look like the one given in Figure 6:

The elasticity of the single chain. The thermodynamic expression (from the Helmholtz free energy, A=U-TS, and the Boltzmann relation S = k ln W for the entropy) for the elastic free energy, A_{el}, of a chain having the end to end vector r is

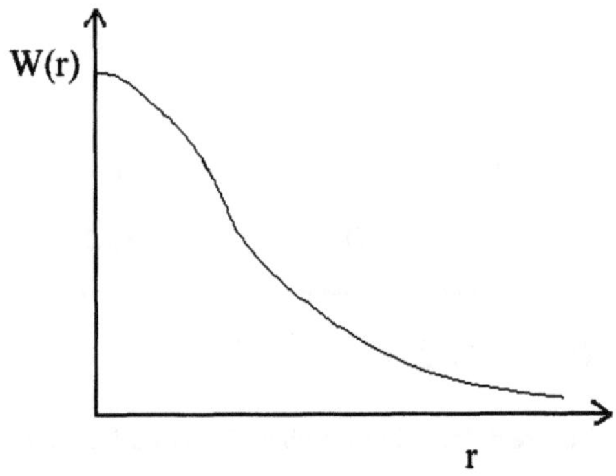

Figure 6. The distribution of the end-to-end vector.

$$A_{el} = U - kT \ln W(r) \tag{6}$$

where, U is the internal energy, a function of temperature only. Substituting eq 4 into 6 leads to the elastic free energy of the single chain at fixed end-to-end separation

$$A_{el} = A^*(T) + \left(\frac{3kT}{2 < r^2 >_0}\right) r^2 \tag{7}$$

where A*(T) is a function of T only.

The force to keep the chain with its two ends at a distance of r is obtained from the elastic free energy according to the thermodynamic relation:

$$f = \left. \frac{\partial A_{el}}{\partial r} \right|_T \tag{8}$$

Substituting eq 7 into 8 yields

$$f = \left(\frac{3kT}{< r^2 >_0}\right) r \tag{9}$$

Equation 9 shows that the force to keep the two ends of a chain at a distance r varies linearly with r. If r is zero, no force is needed. This is exactly the linear spring law, f = kΔx. Therefore, the single chain behaves like a linear spring. The spring constant of the chain is $(\frac{3kT}{<r^2>_0})$.

The mean-squared end-to-end distance is proportional to the number of repeat units, n, of the chain,

$$<r^2>_0 = Cnl^2 \tag{10}$$

Here, C is the constant of proportionality, called the characteristic ratio, and l is the length of a repeat unit. C is unity for a freely jointed chain, and is in the order of 6-10 for real chains. Equation 9 shows that the spring constant of a chain varies inversely with chain length.

2 Network structure

The elastic free energy of the single chain is derived in the previous section. The total elastic free energy of the network is the sum of the elastic free energies of the individual chains (Flory, 1953). Before adding the free energies, however, we have to discuss the structure of a network.

A network is obtained by permanently linking chains together in the form of a three-dimensional structure. An example is shown in Figure 7, where the lines indicate the chains and the dots indicate the cross-link points at which different chains are chemically and permanently joined together. These points are also referred to as the junctions. The total number of junctions is denoted by μ.

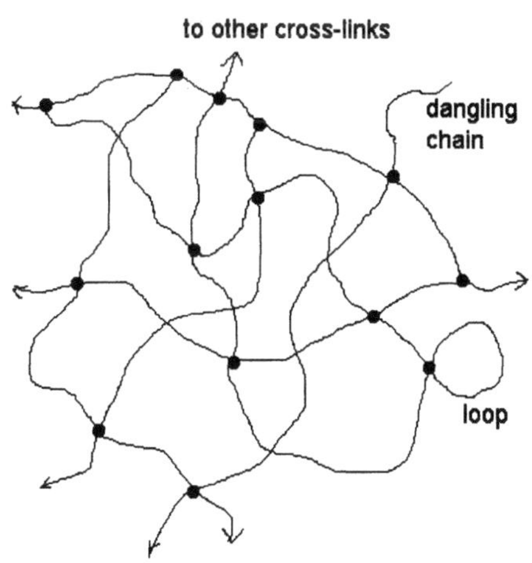

Figure 7. The structure of a network.

A chain that is joined at its two ends to the network is an elastically active chain and contributes to the elastic free energy of the network. A chain connected to the network only at its one end is called a dangling chain. A chain that is connected to the same junction at its two ends is called a loop. A dangling chain and a loop do not contribute to the elasticity of the network. A network with no loops and dangling chains is called a perfect network. The number of chains meeting at a junction is called the functionality, ϕ, of that junction. The junctions of a network may have different functionalities. It may then be characterized by the average functionality.

The number of chains, ν, that should be cut in order to reduce a network into a tree is called the cycle rank, ξ. A tree is a structure that contains no closed cycles.

The example network shown in Figure 8 contains 9 junctions, 18 chains, and the functionality of each junction is four. We need to cut 10 chains to reduce it to a tree. Therefore its cycle rank is 10.

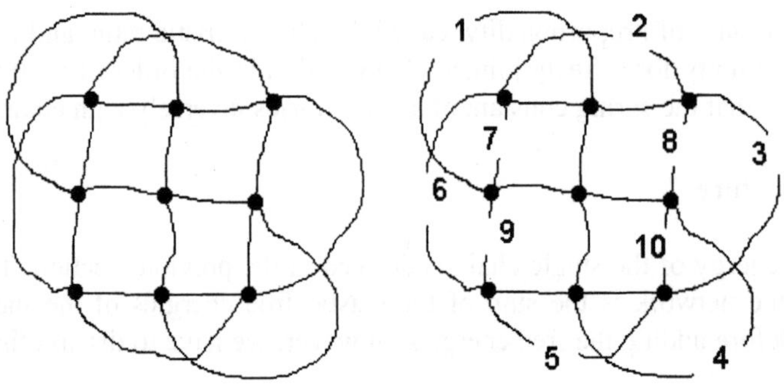

Figure 8. A perfect tetrafunctional network, and the number of cuts to reduce it to a tree.

The three variables, ν, μ, ξ are not free. There are two relations between them:

$$\mu = \frac{2\nu}{\phi}$$
$$\xi = (1 - \frac{2}{\phi})\nu + 1 \tag{11}$$

The second of eq 11 may be divided by the volume of the dry network, V_0, and cast into a useful form as

$$\frac{\xi}{V_0} = \frac{(1-2/\phi)\rho}{M_c/N_A} \quad \text{or} \quad \frac{\nu kT}{V_0} = \frac{\rho RT}{M_c} \tag{12}$$

where, ρ is the density of the network, M_c is the molecular weight of network chains, and N_A is the Avogadro number. Equation 12 gives the cycle rank density of the network. The parameters on the right-hand side are experimental quantities.

Networks are usually obtained by joining a repeat unit on a chain to another repeat unit on another chain with a suitable cross-linker. These cross-links are tetrafunctional, i.e., $\phi = 4$. Another way of joining chains together into a network is to link them to each other at their ends by a chemically active linker. The linker has arms that join the ends of the chains. This way, one can prepare a network of any desired functionality, depending on the type of the linker. Tri and tetra functional networks are the most common ones. However, it is possible to have functionalities up to 50 or 60.

The molecular weight M_c is not the same for each chain of the network and exhibits a distribution. Usually the average value of M_c is reported. In end-linked networks, the network chains may be very close to unimodal (same size), if the starting chains, called the precursor chains, are close to unimodal. Such well controlled networks are called 'model networks'.

Once the chains are permanently incorporated into the network, their entanglements which exist in the uncross-linked state are also permanently trapped. If the network chains are very long, these trapped entanglements may contribute to the elasticity of the network. However, if the chains between two cross-links are short, the effect of these trapped entanglements are small. Here, we will assume that the contributions from the trapped entanglements will be negligible. This is equivalent to assuming that the chains are phantom-like and do not see their neighbors along their contours. They interact with each other only at their ends. This picture forms the basis of the statistical theories of rubber elasticity as they have been introduced by Flory, Wall, Rehner and Treloar.(1,2)

3 Elementary statistical theory for idealized networks (Flory 1953, Treloar 1975)

In section 1 we saw that the single chain behaves like a linear spring. How does a network of these linear springs behave?

The elementary molecular theory of networks is based on the postulate that the elastic free energy of a network is equal to the sum of the elastic free energies of the individual chains. This follows from the phantom-like behavior of the chains. We assume that the chains are of the same size. We sum the elastic free energy of a single chain, given by eq 7, over ν chains of the network:

$$A_{el} = \nu A^*(T) + \left(\frac{3kT}{2 < r^2 >_0}\right)\sum_{i=1}^{\nu} r_i^2 \tag{13}$$

Dividing the summation in eq 13 by ν gives the mean-squared end-to-end distance of the network chains, $< r^2 > = \dfrac{\sum_{i=1}^{\nu} r_i^2}{\nu}$ in the deformed state. In the unperturbed state, this relation is written as $< r^2 >_0 = \dfrac{\sum_{i=1}^{\nu} r_{i,0}^2}{\nu}$, where the subscript zero indicates that the chain is unperturbed by external effects.

The total elastic free energy of the network relative to the undeformed network is obtained from eq 13 as

$$\Delta A_{el} = A_{el} - A_{el,0} = \frac{3kT}{2<r^2>_0}\left(\sum_{i=1}^{\nu} r_i^{\,2} - \sum_{i=1}^{\nu} r_{i,0}^{\,2}\right) \tag{14}$$

Using the definition of the mean-squared end-to-end distance of the perturbed and the unperturbed chains, eq 14 may be rearranged into

$$\begin{aligned}\Delta A_{el} &= \frac{3\nu kT}{2}\left(\frac{<r^2>}{<r^2>_0} - 1\right) \\ &= \frac{\nu kT}{2}\left(\frac{<x^2>}{<x^2>_0} + \frac{<y^2>}{<y^2>_0} + \frac{<z^2>}{<z^2>_0} - 3\right)\end{aligned} \tag{15}$$

where the second line follows from the entity

$$\frac{<r^2>}{<r^2>_0} = \frac{<x^2>}{3<x^2>_0} + \frac{<y^2>}{3<y^2>_0} + \frac{<z^2>}{3<z^2>_0} \tag{16}$$

The macroscopic state of deformation is assumed to be homogeneous. The extension ratios along the three principal coordinate directions x, y, and z are defined as

$$\lambda_x = \frac{L_x}{L_{x,0}}, \lambda_y = \frac{L_y}{L_{y,0}}, \lambda_z = \frac{L_z}{L_{z,0}} \tag{17}$$

where, λ_x is the extension ratio along the x-direction, and L_x and $L_{x,0}$ are the lengths along the x-direction in the deformed and undeformed states, respectively. The other two extension ratios are defined similarly.

In the elementary statistical theory of networks, the junction positions are assumed to deform affinely with macroscopic deformation. According to this assumption, the mean-squared dimensions of the chains along the x, y, and z-directions are written as

$$\begin{aligned}\frac{<x^2>}{<x^2>_0} &= \lambda_x^{\,2} \\ \frac{<y^2>}{<y^2>_0} &= \lambda_y^{\,2} \\ \frac{<z^2>}{<z^2>_0} &= \lambda_z^{\,2}\end{aligned} \tag{18}$$

Substituting eqs 18 into eq 16 leads to

$$\frac{<r^2>}{<r^2>_0}=\frac{1}{3}\left(\lambda_x^{\ 2}+\lambda_y^{\ 2}+\lambda_z^{\ 2}\right) \tag{19}$$

Equation 19 expresses the state of deformation of the chains of the network. Substituting this into eq 15 leads to the expression for the elastic free energy of the network in terms of macroscopic deformation:

$$\Delta A_{el}=\frac{\nu kT}{2}\left(\lambda_x^{\ 2}+\lambda_y^{\ 2}+\lambda_z^{\ 2}-3\right) \tag{20}$$

In deriving eq 20, we ignored the change in the distribution of the cross-links in space under deformation. When a network swells, as we will see later, the change of the distribution of the cross-links in space, like the atoms in a gas, introduces another contribution to the elastic free energy. The correct form is therefore:

$$\Delta A_{el}=\frac{\nu kT}{2}\left(\lambda_x^{\ 2}+\lambda_y^{\ 2}+\lambda_z^{\ 2}-3\right)-\mu kt\ln(\frac{V}{V_0}) \tag{20'}$$

where, μ is the number of cross-links, and V and V_0 are the volumes of the network in the deformed and the undeformed state, respectively.

The volume of a solid increases when it is stretched, and decreases when it is compressed. In the dry state, elastic networks exhibit very small volume change under external extension or compression. To a good approximation, we can assume a network to be incompressible. Therefore, the second term in eq 20' vanishes. The volume of the block in Figure 9 is $V_0 = L_{x0}L_{y0}L_{z0}$ in the undeformed state and
$V= L_xL_yL_z$ in the deformed state. For an incompressible material, the ratio V/V_0 is unity:

$$\frac{V}{V_0}=\frac{L_xL_yL_z}{L_{x0}L_{y0}L_{z0}}=\lambda_x\lambda_y\lambda_z=1 \tag{21}$$

So, for incompressible materials, one can choose only two of the extension ratios independently, and the third is obtained from eq 21.

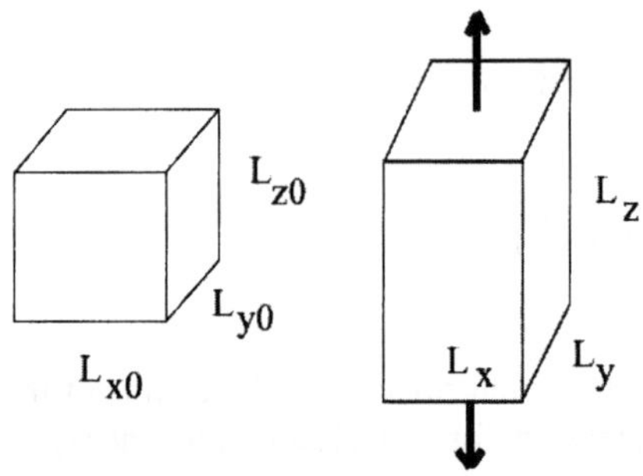

Figure 9. Dimensions of an undeformed and deformed network.

Simple tension: In this type of loading, forces act only in one direction, as shown in Figure 9 (Although we say tension, the force may be compressive. So a more appropriate term would be 'uniaxial loading'). For the case of simple tension shown in Figure 9, the deformations along the x and y directions will be equal to each other. Letting $\lambda_z = \lambda$, we have from eq 21,

$$\lambda_z = \lambda$$
$$\lambda_x = \lambda_y = \frac{1}{\lambda^{1/2}} \tag{22}$$

The elastic free energy, eq 20', for incompressible materials takes the form

$$\Delta A_{el} = \frac{\nu kT}{2}\left(\lambda^2 + \frac{2}{\lambda} - 3\right) \tag{23}$$

We will assume incompressible networks in this section.

The force-deformation relations in simple tension: The force f acting on a network is obtained from the elastic free energy expression according to the thermodynamic relation

$$f = \frac{\partial \Delta A_{el}}{\partial L}\Big|_T \tag{24}$$

This expression follows from the definition of the elastic free energy, $\Delta A_{el} = \int_{L_0}^{L} f dL$, i.e., the elastic free energy is the work done at constant temperature when a network is stretched (or compressed) from an initial length L_0 to a final length L.

Equation 24 may be written as

$$f = \frac{1}{L_0}\frac{\partial \Delta A_{el}}{\partial \lambda}\Big|_T \tag{25}$$

Substituting eq 23 into eq 25 leads to the force on a network as

$$f = \frac{\nu kT}{L_0}(\lambda - \lambda^{-2}) \tag{26}$$

The force extension relation given by eq 26 is based on the assumption of affine deformation of chain vectors. The theoretical model used in deriving this expression is therefore called the affine network model.

The nominal stress or the engineering stress, σ, is defined as the ratio of the force to the initial cross-sectional area, A_0. Dividing eq 26 by A_0, we obtain the stress as

$$\sigma = \frac{\nu kT}{V_0}(\lambda - \lambda^{-2}) \tag{27}$$

The deformation term in eq 27 may be expanded as $\lambda - \lambda^{-2} = \frac{(\lambda^2 + \lambda + 1)(\lambda - 1)}{\lambda^2}$. For small deformations, $\lambda \approx 1$, and this expression may be approximated as $\lambda - \lambda^{-2} = 3\varepsilon$, where $\varepsilon = \lambda - 1$ is the strain. With this approximation, eq 27 becomes

$$\sigma = 3(\frac{\nu kT}{V_0})\varepsilon = E\varepsilon \tag{28}$$

Equation 28 establishes the connection between the molecular quantity $\nu kT/V_0$ and the modulus of elasticity, E, of the network:

$$\frac{\nu kT}{V_0} = \frac{E}{3} = G \tag{29}$$

where G is the shear modulus, which is equal to the third of the modulus of elasticity for incompressible solids. Equation 27 may now be written as

$$\sigma = G(\lambda - \lambda^{-2}) \tag{30}$$

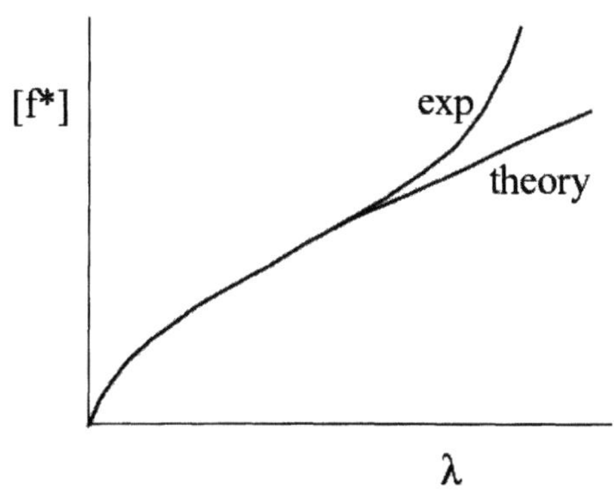

Figure 10. Stress-extension curve predicted by the simple theory, and comparison with experiment.

Equation 30 shows that the stress is not linear in λ. It should increase with a decreasing rate at low λ, and then increase at constant rate at higher values of λ. The predictions of the simple model is compared with theory in Figure 10. There is a general agreement between theory and experiment, as is apparent from the figure, except at high values of λ. Here, experiments show an upturn, which is due to either crystallization or limited chain extension which will be discussed later.

A closer look at the experimental and theoretical results shows, however, that there are stronger differences than seen at first sight. In order to elaborate on this, we define the reduced force, [f*]

$$[f^*] = \frac{\sigma}{\lambda - \lambda^{-2}} \tag{31}$$

Comparing equation 30 with eq 31 shows that the affine model predicts a constant reduced force, which is equal to the shear modulus, [f*] = G. The graph of the reduced force as a function of inverse λ is called the Mooney plot. The affine model and the experimental results look like the following on the Mooney plot:

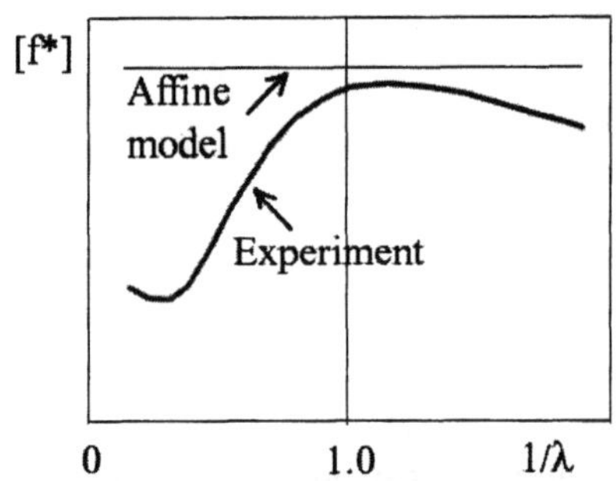

Figure 11. Mooney plot of the stress strain relations for rubber in simple tension.

The Mooney representation shows that experimentally the reduced force decreases with extension and compression, whereas the theoretical affine model gives a constant value. Also at high extensions, there is the upturn in the experimental data. The differences between the theory and experiment will be discussed in the next section.

4 More advanced models of rubber elasticity

The simple molecular theory of rubber elasticity described by the affine network model, gives an approximate estimate of the stress-strain behavior of rubbers. Experiments performed on well characterized networks show that the modulus of a network is predicted well with the affine network model. However, a closer inspection shows that there are certain shortcomings of the model. The first and the most important one is that it predicts a strain independent reduced force, whereas real networks exhibit a decrease, or 'softening' in the reduced force with increasing deformation, as shown in Figure 11. In this section, we discuss more advanced models of rubber elasticity which give better agreement with experiment (Mark, Erman 1988, Erman, Mark 1997, Erman, Flory 1982)

One obvious shortcoming of the model is that it assumes that the network junctions are rigidly embedded in the network and transform affinely with macroscopic deformation. A better and more realistic picture is one where the junctions fluctuate about their mean positions, rather than being fixed in the network. This picture leads to the 'Phantom Network' model, which we discuss here.

The instantaneous end-to-end vector of every chain i, may be represented as a sum of a mean $\bar{r}_i$ and a fluctuation Δr_i from this mean:

$$r_i = \bar{r}_i + \Delta r_i \tag{32}$$

The scalar product of both sides of eq 32 gives

$$r_i^2 = \bar{r}_i^2 + 2\bar{r}_i \cdot \Delta r_i + (\Delta r_i)^2 \tag{33}$$

Averaging both sides of eq 33 over all chains of the network and writing in terms of x, y, and z components, we obtain

$$<r^2>_0 = <\bar{r}^2>_0 + <(\Delta r)^2>_0$$

$$= <\bar{x}^2>_0 + <\bar{y}^2>_0 + <\bar{z}^2>_0 + <(\Delta x)^2>_0 + <(\Delta y)^2>_0 + <(\Delta z)^2>_0 \tag{34}$$

The cross-term in eq 33 has vanished because the means and random fluctuations from this mean are uncorrelated. The subscript zero indicates that the quantity is in the unperturbed state of the network. When the network is deformed, we remove the subscript zero. The phantom network model assumes that the mean vectors transform affinely with macroscopic deformation, but the fluctuations are not affected by macroscopic deformation. Thus, we can write the following expressions:

$$\begin{aligned} <\bar{x}^2> &= \lambda_x^2 <\bar{x}^2>_0 \\ <\bar{y}^2> &= \lambda_y^2 <\bar{y}^2>_0 \\ <\bar{z}^2> &= \lambda_z^2 <\bar{z}^2>_0 \end{aligned} \tag{35}$$

$$
\begin{aligned}
&<(\Delta x)^2>=<(\Delta x)^2>_0 \\
&<(\Delta y)^2>=<(\Delta y)^2>_0 \\
&<(\Delta z)^2>=<(\Delta z)^2>_0
\end{aligned}
\tag{36}
$$

In the undeformed state, the mean square values of chain vectors and of the fluctuations are related to the mean squared chain vectors by the relations:

$$
<\bar{r}^2>_0=(1-\frac{2}{\phi})<r^2>_0 \tag{37}
$$

$$
<(\Delta r)^2>_0=\frac{2}{\phi}<r^2>_0 \tag{38}
$$

Substituting eqs 35-38 into eq 15 leads to

$$
\begin{aligned}
\Delta A_{el} &= \frac{\nu kT}{2}\left[(1-\frac{2}{\phi})\left(\lambda_x{}^2+\lambda_y{}^2+\lambda_z{}^2\right)+\frac{2}{\phi}-1\right] \\
&=(1-\frac{2}{\phi})\frac{\nu kT}{2}\left[\lambda_x{}^2+\lambda_y{}^2+\lambda_z{}^2-1\right]
\end{aligned}
\tag{39}
$$

Following the same arguments to obtain the stress-strain relations, we obtain

$$
\sigma=(1-\frac{2}{\phi})\frac{\nu kT}{V_0}(\lambda-\lambda^{-2})=\frac{\xi kT}{V_0}(\lambda-\lambda^{-2}) \tag{40}
$$

For a tetrafunctional network, the front factor, hence the modulus is one half that of the affine model. Results for the phantom model are depicted in Figure 12. Like the affine model, the phantom network model is also unable to predict the decrease of the reduced force with increasing deformation. This strain dependence is predicted by more advanced models. Two of these models have been successful in predicting the stress-strain dependence of rubbers. These are (i) The constrained junction model, and (ii) The slip-link model. Here, we give only a qualitative description of these models.

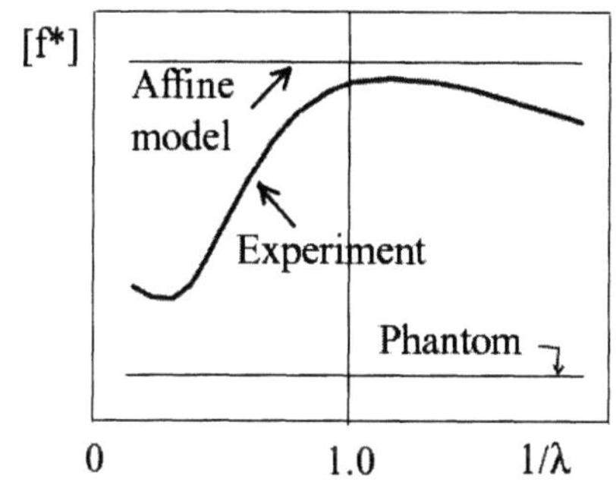

Figure 12. Mooney plot of the stress strain relations for rubber in simple tension. Comparison of experiment, affine and phantom model predictions.

The constrained junction model assumes that the fluctuations of junctions are not totally independent of macroscopic deformation. It assumes that at small deformations, the fluctuations are more strongly affected by their environments, and therefore transform closer to the affine assumption. But at higher deformations, it is easier for the junctions to fluctuate because the deformation decreases the density of the chains around a given junction, and therefore the fluctuations are closer to those of the phantom network model. These considerations lead to results that are consistent with the experimentally observed ones.

The slip-link model assumes that the chains are entangled, and this entanglement of one chain with others contributes to the elasticity of the network. In this model, entanglement effects are represented by a slip-link which joins two chains at a point along their contours. The slip-links may move along the network chain when a deformation is applied to the network. The predictions of the slip-link model are also in good agreement with experiment.

The phenomenological approach. The phenomenological theory of rubber elasticity assumes that the elastic free energy may be expanded in powers of the three invariants of strain. We write the three invariants as

$$I_1 = \lambda_x^2 + \lambda_y^2 + \lambda_z^2$$
$$I_2 = \lambda_x^2\lambda_y^2 + \lambda_x^2\lambda_z^2 + \lambda_y^2\lambda_z^2 \qquad (41)$$
$$I_3 = \lambda_x^2\lambda_y^2\lambda_z^2$$

Here, λ_x, λ_y and λ_z are the three extension ratios along the three principal directions defined previously. In view of the incompressibility assumption, the third invariant equates to unity. Expanding the elastic free energy of the network in terms of the remaining two invariants into Taylor series, and retaining the lowest order terms leads to

$$\Delta A_{el} = C_1 I_1 + C_2 I_2 + C_0 \qquad (42)$$

For simple tension, this expression simplifies into

$$\Delta A_{el} = C_1(\lambda^2 + \frac{2}{\lambda}) + C_2(2\lambda + \frac{1}{\lambda^2}) \tag{43}$$

The force is obtained by using eq 25:

$$f = \frac{2C_1}{L_0}(\lambda - \frac{1}{\lambda^2}) + \frac{2C_2}{L_0}(\lambda - \frac{1}{\lambda^2})\frac{1}{\lambda} \tag{44}$$

The reduced force is obtained, after redifining the C_1 and C_2, as

$$[f^*] = 2C_1 + \frac{2C_2}{\lambda} \tag{45}$$

This form of the reduced force is referred to as the Mooney-Rivlin form. It exhibits the dependence on the extension ratio, which the affine and the phantom network models fail to do. However, although the Mooney-Rivlin expression is satisfactory over a significant range of tensile deformations, it fails in compression, as shown in Figure 13. Nevertheless, since rubbers are used mostly under tension, the Mooney-Rivlin equation continues to be quite useful in applications.

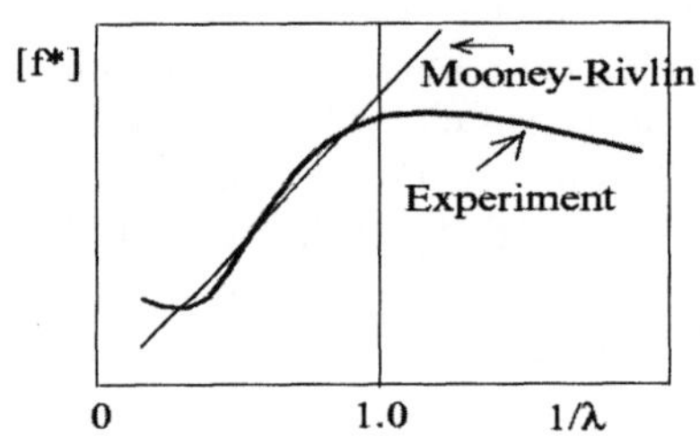

Figure 13. Comparison of experiment and the Mooney-Rivlin equation

5 Swelling of networks

The volume of a dry network does not change much when it deforms under external forces. The network is therefore considered incompressible. The volume of a network may change drastically, however, when it is exposed to the action of a suitable solvent. For example, the volume of a lightly cross-linked network may increase more than hundred-fold, when it is immersed into a good solvent. A good solvent is a solvent where the solvent molecules prefer to be together with the polymer chains, and therefore readily diffuse into the network. Thus mixing decreases the energy of the system relative to the state when the solvent and network are separate. When the solvent molecules enter the network, the network chains become stretched, and their elastic energy is increased. Therefore, the swelling of a network is a competition between a decrease of energy due to mixing, and an increase of energy due to the deformation of the network. The total free energy (Helmholtz) of the system, ΔA, is written as the sum of the mixing and elastic free energies

$$\Delta A = \Delta A_{el} + \Delta A_{mix} \tag{46}$$

where the elastic free energy ΔA_{el} is given for example by eq 20' for the affine network model and by eq 39 for the phantom network model. The mixing free energy ΔA_{mix} is given as

$$\Delta A_{mix} = kT\left[n_1 \ln(1 - v_2) + \ln v_2 + \chi n_1 v_2\right] \tag{47}$$

Here, n_1 is the number of solvent molecules in the network, χ is the interaction energy between the solvent and the polymer, and v_2 is the volume fraction of polymer in the swollen network.

At equilibrium swelling, at constant temperature and pressure, the change of the free energy by the entrance or exit of a solvent molecule into the system does not make a difference. This is given by the condition that the free energy is a minimum with respect to n_1:

$$\left(\frac{\partial \Delta A}{\partial n_1}\right)_{T,p} = \left(\frac{\partial \Delta A_{mix}}{\partial n_1}\right)_{T,p} + \left(\frac{\partial \Delta A_{el}}{\partial n_1}\right)_{T,p} = 0 \tag{48}$$

Here, the subscripts T and p indicate that the differentiations are made at constant temperature and pressure. The term on the left is the chemical potential of the solvent in the network relative to the surrounding solvent.

For isotropic swelling of a network, which is considered in this section, the linear extension ratio λ is the same in all three directions, and is related to volumes by

$$\lambda = \left(\frac{V}{V_0}\right)^{1/3} = \left(\frac{V_0 + \bar{v}_1 n_1}{V_0}\right)^{1/3} = v_2^{-1/3} \tag{49}$$

where $\bar{v}_1$ is the molar volume of solvent. Performing the differentiations indicated in eq 48, and equating to zero results in:

$$\ln(1 - v_2) + \chi v_2^2 + v_2 + \left(\frac{\bar{v}_1}{RT}\right)\left(\frac{\nu kT}{V_0}\right)\left[v_2^{1/3} - \frac{\mu}{\nu} v_2\right] = 0 \tag{50}$$

Equation 50 is an important equation. If one experimentally measures the degree of equilibrium swelling v_2, and if one knows the polymer solvent interaction parameter, then one can calculate the number of chains ν in the network from eq 50. The number of chains is related to the molecular weight M_c of a chain between two cross-links by eq 12. Substituting this in eq 50 and solving for M_c leads to:

$$M_c = -\frac{\rho \bar{v}_1 (v_2^{1/3} - \frac{\mu}{\nu} v_2)}{\ln(1 - v_2) + \chi v_2^2 + v_2} \tag{51}$$

Alternatively, if one knows the degree of cross-linking and the interaction parameter, one can predict how much the network will swell. As the third case, if one knows the degree of cross-linking, and measures the amount of swelling, one can calculate the interaction parameter χ between the polymer and the solvent.

Gels used in today's technological applications are formed not in the dry state, but in the presence of a solvent, or in the highly diluted state. When a gel is formed in this way, and for certain interaction parameters, it is possible for gels to exhibit a sudden collapse in volume as the temperature is decreased. This is a specific case where the polymer and the solvent system is said to undergo a first order phase transition. The change of temperature essentially changes the interaction energy of the system. Other parameters, such as an external electric or magnetic field or the pH of the solvent may also change the interaction parameter and result in sharp changes in the volume of the swollen network. Many biological systems undergo similar phase transitions upon the change of environmental parameters.

6 Molecular basis of thermoelasticity

When a rubber band is stretched, it gets warmer. This is because the internal energy of the system does not change much, and most of the work done in stretching the rubber goes into decreasing the entropy of the network chains. This in turn leads to heat leaving the system which is felt as an increase in temperature. Also, a rubber band that is stretched by hanging a load at one end becomes shorter upon heating. This is contrary to the behavior of a metal rod, which becomes longer upon heating. In this section, we will see the thermodynamics of deformation and try to answer why stretching the rubber band causes warming, and why a stretched rubber band gets shorter when heated (Treloar, 1975).

From the first law of thermodynamics, we have

$$dU = dQ + dW \tag{52}$$

where dU is the increase of the internal energy of the system (the rubber band in our case), dQ and dW are the heat absorbed by the system and the work done on it.

The heat absorbed by a system is related to the change in entropy, dS, in a reversible process by the second law:

$$dQ = TdS \tag{53}$$

and substituting this equation in eq 52 gives the change in internal energy as

$$dU = TdS + dW \tag{54}$$

The work done when the system is deformed under a force f is

$$dW = fdl - pdV \tag{55}$$

Here the second term is the work done by the hydrostatic pressure, which we will ignore because it will be small compared to the force term. So the change in internal energy from eqs 54 and 55 is

$$dU = TdS + fdl \tag{56}$$

The work done on the network is equal to the change in its elastic free energy, or its Helmholtz free energy at constant temperature. We can show this, by using the definition of the Helmholtz free energy, A=U – TS, as follows:

At constant temperature, the change in the Helmholtz free energy is written as

$$dA = dU - TdS \tag{57}$$

Substituting from eq 54 leads to:

$$dW = dA \tag{58}$$

At constant temperature, the force on the system is related to the Helmholtz free energy by

$$f = \left(\frac{\partial \mathrm{W}}{\partial 1}\right)_T = \left(\frac{\partial A}{\partial 1}\right)_T \tag{59}$$

and substituting for the Helmholtz free energy from eq 57, we have

$$f = \left(\frac{\partial A}{\partial l}\right)_T = \left(\frac{\partial U}{\partial l}\right)_T - T\left(\frac{\partial S}{\partial l}\right)_T \tag{60}$$

For any process, not necessarily at constant temperature, we have

$$dA = dU - TdS - SdT \tag{61}$$

Substituting from eq 56 for dU gives:

$$dA = fdl - SdT \tag{62}$$

which may be written as

$$dA = \left(\frac{\partial A}{\partial l}\right)_T dl + \left(\frac{\partial A}{\partial T}\right)_l dT \tag{63}$$

Comparing eq 62 and 63 term by term yields

$$f = \left(\frac{\partial A}{\partial l}\right)_T \qquad S = -\left(\frac{\partial A}{\partial T}\right)_l \tag{64}$$

Since $\frac{\partial}{\partial l}\left(\frac{\partial A}{\partial T}\right)_l = \frac{\partial}{\partial T}\left(\frac{\partial A}{\partial l}\right)_T$, eq. 64 gives

$$\left(\frac{\partial S}{\partial l}\right)_T = -\left(\frac{\partial f}{\partial T}\right)_l \tag{65}$$

Equation 65 is a convenient expression for estimating changes of entropy upon stretching a network, because the right-hand side is an experimentally accessible quantity, which may be obtained by measuring the change in the force to keep the network at fixed length when its temperature is changed. Inserting this equation into eq 60, we obtain

$$\left(\frac{\partial U}{\partial l}\right)_T = f - T\left(\frac{\partial f}{\partial T}\right)_l \tag{66}$$

Equation 66 gives the change in the internal energy of the system in terms of the applied force and the thermal coefficient of the applied force, both of which are experimentally accessible.

We can rearrange eq 66 and write as

$$f = \left(\frac{\partial U}{\partial l}\right)_T + T\left(\frac{\partial f}{\partial T}\right)_l$$

or,

$$f = f_e + f_{ent} \tag{67}$$

where f_e is the part of the force that goes into changing the internal energy of the system, and f_{ent} is the part that goes into changing the entropy.

What are the changes taking place in a deforming network? First, we assume that the work done on the system goes into the individual chains. This is intramolecular. Intermolecular effects such as work against friction, like the one in viscoelastic behavior are assumed negligible. When a chain is stretched, its entropy decreases. We know this from the Boltzmann relation,

$$S = k \ln W(r) \tag{68}$$

where the distribution function is given by eq 5. Writing eq 5 for end-to-end distance r_1 and r_2, substituting in eq 68 and subtracting the two entropies leads to

$$S_2 - S_1 = \Delta S = -\frac{3}{2 < r^2 >_0}(r_2^{\ 2} - r_1^{\ 2}) \tag{69}$$

Equation 69 shows that the change in entropy is negative as r is increased. This is a consequence of the fact that more configurations are available to chains in the undeformed state than when they are deformed. For a chain that is stretched to its full length, there is only one configuration.

The energy of a chain that is stretched changes if the bond lengths and bond angles change. Usually these changes are small if the chain is not stretched up to its full length. Therefore, the change in energy in deforming a network is usually small. For a network with freely jointed chains, $\left(\frac{\partial U}{\partial l}\right)_T$ =0. This means f_e=0, and all of the force goes into decreasing the entropy of the chains. In reality, the energy change is not zero, but small, so that we may safely assume that the deformation of the network is perfectly entropic. Careful experiments performed on networks show that the change of energy with deformation is small. Thus rubbers exhibit entropic elasticity, in contrast to the energetic elasticity of metals where internal energy changes are dominant and entropy changes are negligible.

For polyethylene, for example, as the temperature is increased, the chains of the network try to move into different configurations. It is only possible to do this if the chain decreases the distance between its two ends, because the number of configurations is large if the end-to-end vector is small. As a result, the overall length of the rubber sample decreases with increasing temperature.

If the change of internal energy of the system is zero, then from eq. 51, we have $dW = -TdS$ and from second law of thermodynamics, we have $dW = -dQ$, where dQ is the heat absorbed by the rubber band. Thus, when a force stretches the rubber band from length l_0 to l, the work done is positive, and therefore dQ is negative, which means heat is given off during stretching. This is felt as heating of the rubber band. In fact we can find the magnitude of this small increase in temperature, using the expression |dQ| = C_l |dT|, where C_l is the specific heat of rubber at constant length. Thus

$$\Delta T = \frac{1}{C_l}\int dW = \frac{1}{C_l}\int_{l_0}^{l} f\, dl \tag{70}$$

where, the integral on the right hand side is the area under the force vs length curve for the rubber.

Some polymeric systems show strain-induced crystallization upon stretching, which we will discuss in the next section. This happens when the network is highly stretched, so that the segments of the chain find a chance to align relative to each other along the direction of stretch. For some polymer systems, this alignment may induce the formation of crystallites, where segments belonging to different chains come together and therefore their energy is decreased, resulting in the decrease of the energy of the network.

7 Strain induced crystallization, finite chain extensibility, bioelastomers

The chemical structure of a polymer chain may be of three types shown in Figure 14. In this figure, the planar zig-zag form is shown. The side groups are indicated by A and B. These side groups may all lie on the same side of the chain. This chain has the isotactic structure. If the side groups alternate regularly as shown in the middle figure, the chain is called syndiotactic, and if the side groups are irregularly placed on each side of the chain, the structure is called atactic. Isotactic chains always crystallize if the temperature is decreased, or if they are stretched. Because, the side groups are regularly placed and the corresponding groups of two neighboring chains may match easily to form an ordered structure. An atactic chain never crystallizes, because the side groups, having different sizes and chemical properties may not match in space due to the irregularity of their positions relative to the chain backbone. Syndiotactic chains may or may not crystallize.

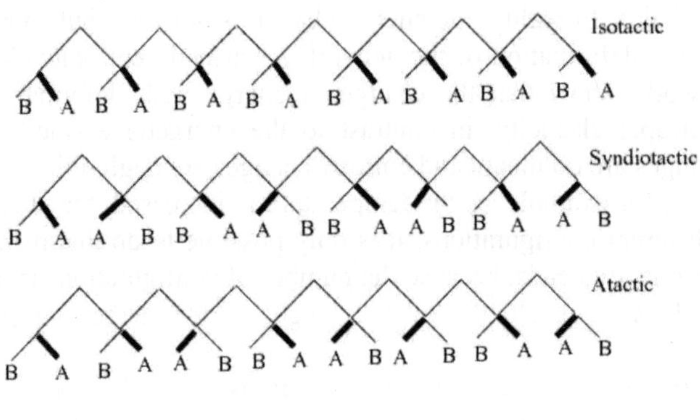

Figure 14. Isotactic, syndiotactic and atactic chains.

Now consider the interaction energy of group A and a group B. If they are widely separated in space, they do not interact with each other, and their interaction energy is zero. If they are very close together, they repel each other very strongly. There is an equilibrium distance where their interaction energy is a minimum, and they would like to stay that way. The interaction energy between a group A and a group B is shown in Figure 15 as a function of the distance between them.

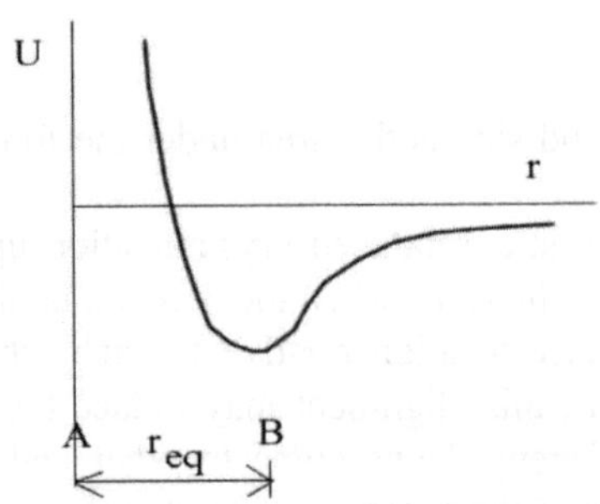

Figure 15. Interaction energy of two groups A and B, as a function of distance between them.

However, when the two groups A and B are connected into the chain structure, due to steric and excluded volume effects between the parts of the chain, their spatial positions deviate from that obtained in the minimum energy configuration. If the chains are isotactic, the groups will be able to come as close as possible to the minimum energy configuration. If several successive groups on one chain align with several others on the neighboring chains, a crystallite forms, which eventually grows as stretching increases. This is referred to as strain induced crystallization.

The crystallites act as huge junctions or cross-links. This has the effect of decreasing M_c. Therefore the modulus of the network increases. This may best be seen on a Mooney plot, as shown in Figure 16.

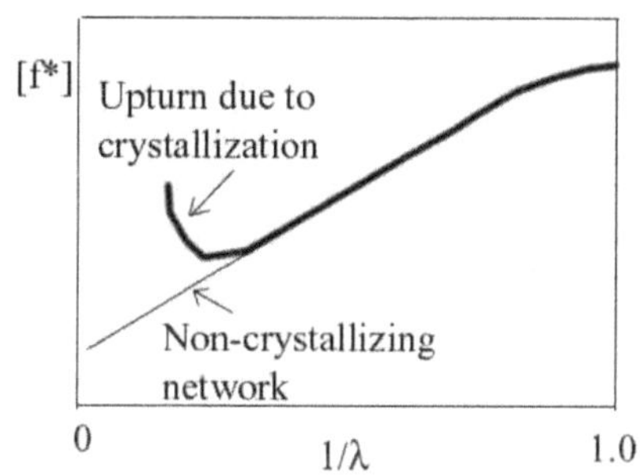

Figure 16. Upturn in the Mooney plot due to crystallization.

Crystallizing networks are in a sense desirable because the presence of crystallites strengthens the structure. In some other ways, it is undesirable because crystallization reduces the 'rubbery ness' and extensibility of the material.

The upturn in the Mooney plot may also arise from the effects of finite chain extensibility. We can have a rough estimate of the maximum extension ratio for a network that consists of freely jointed chains of n repeat units. On the average, the end-to-end distance is $<r^2>^{1/2} = n^{1/2} l$, where l is the length of a repeat unit. When the chains are maximally stretched, their end-to-end distance is nl. The ratio of the maximum stretch to mean value is a good measure of the maximum extensibility, i.e.,

$$\lambda_{max} = \frac{nl}{n^{1/2} l}$$

$$= n^{1/2} \quad (71)$$

Thus, a network having chains of about 400 repeat units may be stretched up to 20 times its original length. When the full extension is approached, the Mooney plot starts exhibiting an upturn, because most of the chain repeat units become aligned along the direction of stretch, and one has to start stretching the bonds, and changing bond angles, both of which require larger energies than that of changing the configurations of the chains.

Bioelastomers are among the most spectacular networks that exhibit both rubber-like elasticity and strength. The individual chains of the bioelastomers are protein molecules that have the peptide backbone, and side groups that are chosen from a repertoire of 20 different amino acids. Some examples of bioelastomers are, elastin in mammals, resilin in insects, abductin in mollusks, arterial elastomers in octopuses, locomotional proteins in cephalopods, and silk in spider webs.

Elastin is found in blood vessels, lungs and skin. It absorbs the energy at each pulse of the heart to prevent blood pressure to rupture delicate capillaries, and then releases the energy to maintain blood pressure between pulses. In elastin, the amino acids are placed along the protein backbone with sufficient irregularity to prevent crystallization with elongation and decrease in temperature. The chains are kept as flexible as possible in two ways. First, the amino acids are chosen as small as possible to prevent congestion. Larger side groups interact with other neighboring groups repulsively due to excluded volume effects. The side groups that predominate are glycine (-H), alanine ($-CH_3$), serine ($-CH_2OH$) and valine ($-CH(CH_3)_2$). Second, the network is highly swollen by body fluids so that the interactions of side groups belonging to neighboring chains are minimized. For example, the amount of stretching required for crystallizing the network increases significantly by swelling. Also, one must go to much lower temperatures to crystallize a swollen network. The cross-linking is of chemical nature and occurs at lysine groups that are positioned at specific sites along the protein chains.

The question of storage of elastic energy is important. A deforming elastomer absorbs energy. Part of this energy is released or recovered upon retraction, and part is stored. The stored energy is degraded into the surroundings by viscous effects. This part of the energy is wasted. More importantly, in cyclic deformations, the heat buildup is the main problem for the elastin in the arteries of mammals and resilin in the wings of insects. Nature has been very successful in designing these bioelastomers in which this heat buildup is minimized.

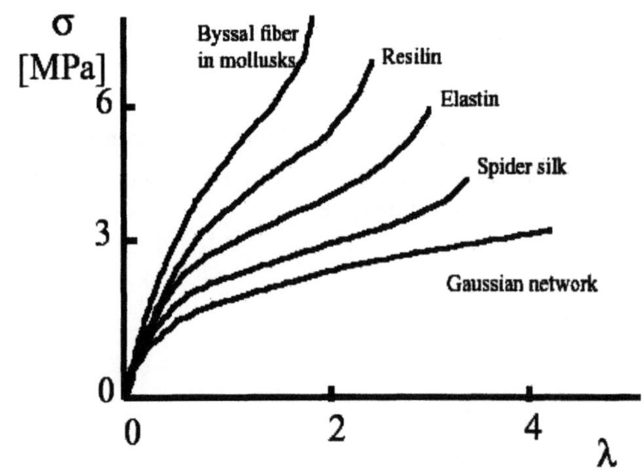

Figure 17. Stress-strain diagrams for some bioelastomers.

The upturns in the stress-strain diagrams of the bioelastomers shown in Figure 17 result from effects of both finite chain extensibility and crystallization. The byssal fiber shows the strongest upturn. It is designed as a spring-dashpot system, where the ends of the molecule exhibits crytallinity and the central portion is non-crystallizing. The mollusk anchors itself to the seabed by byssal fibers. A wave moves the mollusk and energy is stored in the fibers stretched between the anchor points. Most of this energy is non-recoverable and is dissipated as heat. If the energy were fully recoverable, the mollusk would snap back towards the seabed and result in a catastrophic impact. Similar behavior is observed in spider webs. The kinetic energy of a fly that flies into the web is absorbed by the fibers, and transformed into heat by the fibers. If the energy stored in the deforming fibers were fully recoverable, the fly would snap away from the web like a sling-shot.

References

Flory, P. J. (1953). Principles of Polymer Chemistry, Cornell University Press, Ithaca N. Y.

Erman, B and Mark, J. E. (1997). Structures and Properties of Rubberlike Networks, Oxford University Press.

Erman, B., Flory, P. J. (1982). Relationship Between Stress, Strain, and Molecular Constitution of Polymer Networks. Comparison of Theory with Experiments. Macromolecules, 15, 806-816.

Mark, J. E. and Erman, B. (1988). *Rubberlike elasticity: A molecular primer*, John Wiley and Sons.

Treloar, L. R. G. (1975). The Physics of Rubber Elasticity, 3rd ed. Clarendon Press: Oxford.

Phenomenology of Rubber-Like Materials *

Giuseppe Saccomandi
Dipartimento di Ingegneria dell'Innovazione,
Sezione di Ingegneria Industriale,
Università di Lecce
73100, Lecce, Italy
giuseppe.saccomandi@unile.it

Abstract. These lectures are devoted to an overview of the basic balance laws and constitutive requirements of continuum mechanics. Some solutions of simple boundary value problems are considered and a critical review of strain-energy density functions is proposed.

1 Introduction

These notes are about the possible application of the theory of finite elasticity to the understanding of the mechanics of rubber-like materials. There is a large literature on nonlinear elasticity and for this reason I begin with some bibliographic notices. This is also an opportunity to declare my sources of main inspiration.

First of all I have to mention the book by R. W. Ogden *Non-linear elastic deformations* (first published in 1984 by Ellis Horwood Ltd, Chichester and then reprinted in 1997 by Dover Publications Inc., Mineola). This is a complete and meticolous account of the theory of finite and nonlinear elasticity. Another interesting textbook is the basic treatise by R. J Atkin and N. Fox: *An Introduction to theory of Elasticity* (1980, Longman, London). A short but very interesting text is also the introductory survey by M. F. Beatty: *Introduction to Nonlinear Elasticity* (first published as *Topics in Finite Elasticity* in *Applied Mechanics Review*, 40, 1987, pp 1699-1734 and then updated and reprinted in *Nonlinear Effects in Fluids and Solids*, edited by M. M. Carroll and M. Hayes (1986) Plenum Press, New York). This is an excellent survey of some selected topics in finite elasticity with an updated list of references. For mathematically oriented readers, in my opinion, it is not possible to skip the monograph by J. E. Marsden and T. R. J. Hughes, *Mathematical Foundations of Elasticity* (first published in 1983 by Prentice-Hall, Englewood Cliffs and then reprinted in 1994 by Dover, Mineola) and the monograph by S.S. Antman, *Problems of Nonlinear Elasticity* (1995, Springer, New-York) where also theories of elastic structures are considered. Finally, all these books have been updated in very recent years by two collections of original papers: *Nonlinear Elasticity: Theory*

*These lecture notes are dedicated to the memory of Professor Pietro Benvenuti (1933-2002). I would never have been able to learn the little mechanics I know without the wonderful advice and the fatherly help of Pietro Benvenuti.

and Applications, edited by Y. B. Fu and R. W. Ogden, LMS Lectures Note Series 283 (2001) Cambridge University Press and *Topics in Finite Elasticity*, edited by M. Hayes and G. Saccomandi, CISM Lectures Note Series 429 (2001) Springer, Berlin.

Obviously this is only a partial list dictated by personal taste and for any aspect of continuum mechanics not covered by these books my standard references are C. Truesdell and C.A. Toupin, *The Classical Fields Theories* (in S. Flugge, ed., Encyclopedia of Physics, Volume III/1, Springer-Verlag Berlin, 226-793, 1960) and C. Truesdell and W. Noll, *The Non-Linear Field Theories of Mechanics*, 2nd edn., Springer-Verlag, 1992.

I am indebted for valuable suggestions and comments to Mike Hayes, Robin Knops, Edvige Pucci, Ray Ogden and Kumbakonam Rajagopal, moreover I must particularly thank Cornelius Horgan my dear friend and collaborator because a great part of these lectures are based on joint work over the past four years. For all this I owe Cornelius a profound debt of gratitude.

1.1 Basic equations

A continuum body, briefly a body $\mathcal{B}$, is a contiguous set of material points called particles. A reference frame is a set $\mathcal{R} = \{O; \mathbf{e}_i\}$ consisting of an origin O and an orthogonal vector basis $\mathbf{e}_i = \{\mathbf{e}_1, \mathbf{e}_2, \mathbf{e}_3\}$. The motion of a particle P relative to $\mathcal{R}$ is the time locus of its position vector $\mathbf{x}(P, t)$ in $\mathcal{R}$. A typical particle P may be identified by its position vector $\mathbf{X}(P)$ in $\mathcal{R}$ at some reference time t_0, say. The domain $\mathcal{C}_0$ of $\mathbf{X}$, the region occupied by $\mathcal{B}$ at the time t_0, is called a reference configuration of $\mathcal{B}$. Then, relative to $\mathcal{R}$, the motion of a typical particle P from $\mathcal{C}_0$ is described by the vector function $\mathbf{x} = \mathbf{x}(\mathbf{X}, t)$. The domain $\mathcal{C}_t$ of $\mathbf{x}$, the region occupied by $\mathcal{B}$ at the time t, is called the current configuration of $\mathcal{B}$. Hence, $\mathbf{x}$ denotes the place in the current configuration $\mathcal{C}_t$ at time t, that is occupied by the particle P whose place was at $\mathbf{X}$ in the reference configuration $\mathcal{C}_0$ initially.

The velocity and acceleration of a particle P relative to $\mathcal{R}$ are defined by $\mathbf{v}(\mathbf{X}, t) \equiv \dot{\mathbf{x}}(\mathbf{X}, t)$ and $\mathbf{a}(\mathbf{X}, t) \equiv \dot{\mathbf{v}}(\mathbf{X}, t) = \ddot{\mathbf{x}}(\mathbf{X}, t)$, respectively. As usual, $\cdot \equiv \partial/\partial t$ denotes the material time derivative, the time rate of change following the particle P.

The forces that act on a part[1] $\mathcal{P}$ of a body $\mathcal{B}$ of mass m are of two kinds: a distribution of contact force $\mathbf{t_n}$ per unit area a of the boundary $\partial\mathcal{P}$ of $\mathcal{P}$ in $\mathcal{C}_t$, and a distribution of body force $\mathbf{b}$ per unit volume v of $\mathcal{P}$ in $\mathcal{C}$. The global balance laws of momentum and moment of momentum can be written as

$$\int_{\partial\mathcal{P}} \mathbf{t_n} da + \int_{\mathcal{P}} \mathbf{b} dv = \frac{d}{dt} \int_{\mathcal{P}} \rho \mathbf{v} dv, \tag{1.1}$$

$$\int_{\partial\mathcal{P}} \mathbf{x} \times \mathbf{t_n} da + \int_{\mathcal{P}} \mathbf{x} \times \mathbf{b} dv = \frac{d}{dt} \int_{\mathcal{P}} \rho \mathbf{x} \times \mathbf{v} dv. \tag{1.2}$$

Herein we note that $dm = \rho dv$ is the material element of mass with density ρ per unit volume in $\mathcal{R}$. These equations must hold for any part $\mathcal{P}$ of a body $\mathcal{B}$.

[1] Here $\mathcal{P} \subset \mathcal{B}$ strictly, therefore $\partial\mathcal{P} \cap \partial\mathcal{B} = \emptyset$. If $\partial\mathcal{P} \cap \partial\mathcal{B} \neq \emptyset$ the discussion must take into account boundary conditions.

The principle of balance of mass requires also that $dm = \rho_0 dV$, where ρ_0 is the density of mass per unit volume V in $\mathcal{C}_0$, and hence the respective mass densities are related by the local equation of continuity:

$$\rho_0 = J\rho, \tag{1.3}$$

where $J \equiv \det \mathbf{F}$ and $\mathbf{F} \equiv \partial\mathbf{x}/\partial\mathbf{X}$ in $\mathcal{R}$.

Application of the first law (1.1) to an arbitrary tetrahedral element leads to Cauchy's stress principle:

$$\mathbf{t_n} = \mathbf{Tn}, \tag{1.4}$$

where $\mathbf{n}$ is the exterior unit normal vector to $\partial\mathcal{P}$ in $\mathcal{R}_t$. Hence, the stress vector $\mathbf{t_n}$ is a linear transformation of the unit normal $\mathbf{n}$ by the Cauchy stress tensor $\mathbf{T}$.

Use of this result and the divergence theorem because (1.1) for an arbitrary $\mathcal{P} \subset \mathcal{B}$ we have the local version of Cauchy's first law of motion:

$$\operatorname{div} \mathbf{T} + \mathbf{b} = \rho\mathbf{a}, \tag{1.5}$$

where div denotes the divergence with respect to $\mathbf{x}$ in $\mathcal{R}$. Then use of (1.1) in the second law (1.2) yields the local moment balance condition restricting the Cauchy stress $\mathbf{T}$ to the space of symmetric tensors:

$$\mathbf{T}^T = \mathbf{T}. \tag{1.6}$$

The Cauchy stress characterizes the contact force distribution $\mathbf{t_n}$ in $\mathcal{R}$ per unit current area in $\mathcal{R}$; but it may be often interesting to characterize the contact in the reference configuration, therefore, the engineering stress tensor $\mathbf{T}_R$, also known as the first Piola-Kirchhoff stress tensor, is introduced. The tensor $\mathbf{T}_R$ defines the contact force distribution $\mathbf{t_N}$ in $\mathcal{C}_t$ per unit reference area A in $\mathcal{C}_0$, where $\mathbf{N}$ is the exterior unit normal vector to $\partial\mathcal{P}$ in $\mathcal{C}_0$ whose image in $\mathcal{C}_t$ is $\mathbf{n}$. For the same contact force $d\boldsymbol{\mathcal{F}}_c(\mathcal{P}, t) = \mathbf{t_n}da = \mathbf{t_N}dA$. Let us consider two line elements, $d\mathbf{X}$ and $d\mathbf{Y}$, at a point $\mathbf{X}$ in $\mathcal{C}_0$ and the corresponding line elements, $d\mathbf{x}$ and $d\mathbf{y}$ in $\mathcal{C}$. Therefore

$$d\mathbf{X} \wedge d\mathbf{Y} = \mathbf{N}dA, \tag{1.7}$$

and

$$d\mathbf{x} \wedge dy = \mathbf{n}da. \tag{1.8}$$

From (1.8)

$$n_i da = \varepsilon_{ijk} dx_j dy_k = \varepsilon_{ijk} x_{j,L} y_{k,M} dX_L dY_M$$

and

$$n_i x_{i,K} da = \varepsilon_{ijk} x_{i,K} x_{j,L} y_{k,M} dX_L dY_M = J N_K dA$$

or

$$\mathbf{n}da = J\mathbf{F}^{-T}\mathbf{N}dA.$$

Considering

$$\mathbf{t_n}da = \mathbf{Tn}da = J\mathbf{TF}^{-T}\mathbf{N}dA,$$

we thus find the rule

$$\mathbf{T}_R = J\mathbf{TF}^{-T}, \tag{1.9}$$

relating the engineering and Cauchy stress tensors. Referred to $\mathcal{C}_0$, (1.4), (1.5) and (1.6) become

$$\mathbf{t_N} = \mathbf{T}_R\mathbf{N}, \ \operatorname{Div}\mathbf{T}_R + \mathbf{b}_R = \rho_R\mathbf{a}, \ \mathbf{T}_R\mathbf{F}^T = \mathbf{F}\mathbf{T}_R^T, \tag{1.10}$$

where $\mathbf{b}_R \equiv J\mathbf{b}$ denotes the body force per unit volume V in $\mathcal{C}_0$ and Div denotes the divergence operator with respect to $\mathbf{X}$ in $\mathcal{C}_0$. The last of (1.10) shows that the engineering stress $\mathbf{T}_R$ generally is not symmetric.

To derive a purely mechanical energy principle, we take the scalar product of equation (1.5) with the velocity $\mathbf{v}$ i.e.

$$(\operatorname{div}\mathbf{T})\cdot\mathbf{v} + \mathbf{b}\cdot\mathbf{v} = \rho\mathbf{a}\cdot\mathbf{v}, \tag{1.11}$$

which is possible to rewrite as

$$\operatorname{div}(\mathbf{T}\mathbf{v}) - \operatorname{tr}(\mathbf{T}\mathbf{L}) + \mathbf{b}\cdot\mathbf{v} = \rho\mathbf{a}\cdot\mathbf{v}, \tag{1.12}$$

where $\mathbf{L} \equiv \operatorname{grad}\dot{\mathbf{x}} = \dot{\mathbf{F}}\mathbf{F}^{-1}$. I remember that $\operatorname{grad}\mathbf{v} = \mathbf{D} + \mathbf{W}$, where $\mathbf{D} = \frac{1}{2}\left(\mathbf{L} + \mathbf{L}^T\right)$ and $\mathbf{W} = \frac{1}{2}\left(\mathbf{L} - \mathbf{L}^T\right)$.

The integral version of this scalar relation is

$$\frac{d}{dt}\int_{\mathcal{P}}\frac{1}{2}\rho\mathbf{v}\cdot\mathbf{v}dv + \int_{\mathcal{P}}\operatorname{tr}(\mathbf{T}\mathbf{D})\,dv = \int_{\partial\mathcal{P}}\mathbf{t_n}\cdot\mathbf{v}da + \int_{\mathcal{P}}\mathbf{b}\cdot\mathbf{v}dv. \tag{1.13}$$

where we have taken into account that the symmetry of the Cauchy stress tensor implies $\operatorname{tr}(\mathbf{T}\mathbf{L}) \equiv \operatorname{tr}(\mathbf{T}\mathbf{D})$.

In the reference configuration considering $dv = JdV$, and the Nanson's formula

$$\mathbf{F}^T da = JdA,$$

we have

$$\frac{d}{dt}\int_{\mathcal{P}_0}\frac{1}{2}\rho_0\mathbf{v}\cdot\mathbf{v}dV + \int_{\mathcal{P}_0}\operatorname{tr}\left(\mathbf{T}_R\dot{\mathbf{F}}\right)dV = \int_{\partial\mathcal{P}_0}(\mathbf{T}_R^T\mathbf{N})\cdot\mathbf{v}dA + \int_{\mathcal{P}_0}\mathbf{b}_0\cdot\mathbf{v}dV. \tag{1.14}$$

Usually the stress power per unit of volume of the nominal stress tensor is not an exact differential. However, it may be that there exist a scalar function, $W(\mathbf{X}, t)$, such that:

$$\frac{dW}{dt}(\mathbf{X}, t) = \operatorname{tr}(\mathbf{T}_R\dot{\mathbf{F}}^T) \equiv \mathbf{T}_R\cdot\dot{\mathbf{F}}. \tag{1.15}$$

In this case we say that the solid is hyperelastic and the elastic potential energy is given by the strain energy function

$$W(\mathbf{X}, t) = W(\mathbf{F}(\mathbf{X}, t), \mathbf{X}). \tag{1.16}$$

Use of (1.16) in (1.15) yields the identity $(\mathbf{T}_R - \partial W/\partial\mathbf{F})\cdot\dot{\mathbf{F}} = 0$, which must hold for all $\dot{\mathbf{F}}$. Hence, the principle of mechanical energy balance and the assumption (1.15) yield the following general constitutive equation relating the engineering stress and the deformation of a hyperelastic solid[2]:

$$\mathbf{T}_R = \frac{\partial W(\mathbf{F})}{\partial \mathbf{F}}. \tag{1.17}$$

Use of (1.9) in (1.17) provides an alternative form of the constitutive equation relating the Cauchy stress to the deformation of a hyperelastic solid:

$$\mathbf{T} = J^{-1}\frac{\partial W(\mathbf{F})}{\partial \mathbf{F}}\mathbf{F}^T. \tag{1.18}$$

Therefore *the time rate of change of the total mechanical energy for an arbitrary part of a given body is balanced by the total mechanical power.*

To have a complete picture of rubber elasticity we point out that the kinetic energy of a body is only a part of its energy; the remainder is called the internal energy density, e. The statement that we have derived straightforwardly from the balance of linear momentum, must be modified to take into account this fact. This is done as follows

$$\begin{aligned} \tfrac{d}{dt}\int_{\mathcal{P}}\left(\tfrac{1}{2}\rho\mathbf{v}\cdot\mathbf{v}+e\right)dv \quad &+ \quad \int_{\mathcal{P}}\operatorname{tr}(\mathbf{TD})\,dv \\ &= \quad \int_{\partial\mathcal{P}}(\mathbf{t_n}\cdot\mathbf{v}-\mathbf{q}\cdot\mathbf{n})\,da + \int_{\mathcal{P}}\mathbf{b}\cdot\mathbf{v}dv. \end{aligned} \tag{1.19}$$

In the RHS of (1.19) we have introduced the heat-flux vector $\mathbf{q}$. This is an example of energy different of the mechanical one which can enters in the balance (1.19) but many other energies may be considered (radiation, electromagnetic, ...).

The standard argument used to have a local form for (1.19) and the use of (1.5) allows us to write the balance of energy as

$$\rho\frac{de}{dt} = \operatorname{tr}(\mathbf{TD}) - \operatorname{div}\mathbf{q}. \tag{1.20}$$

Usually the heat-flux vector $\mathbf{q}$ is related by means to a constitutive equation to the temperature θ. In thermomechanics the rationale for constitutive equations cannot be understood without considering the role of the Clausius-Duhem inequality. Considering the sources of entropy η this inequality rewritten using the Helmotz free energy

$$\phi = e - \eta\theta. \tag{1.21}$$

The Clausius-Duhem inequality, which expresses the assumption that the local entropy production is non-negative, is given by

$$-\rho\left(\dot{e} - \eta\dot{\theta}\right) + \operatorname{tr}(\mathbf{TD}) + \left(\frac{1}{\theta}\right)\mathbf{q}\cdot\nabla\theta \geq 0. \tag{1.22}$$

[2]Here the explicit dependence on $\mathbf{X}$ has been removed, because we are considering bodies with uniform material properties.

Let us considering the constitutive assumptions

$$\mathbf{T} = \mathbf{T}(\mathbf{F}, \theta, \nabla\theta), \quad \mathbf{q} = \mathbf{q}(\mathbf{F}, \theta, \nabla\theta),$$
$$\eta = \eta(\mathbf{F}, \theta, \nabla\theta), \qquad \phi = \phi(\mathbf{F}, \theta, \nabla\theta). \tag{1.23}$$

The introduction (1.23) in (1.22) and requiring that the inequality hold for all processes, we obtain the following:

- only the response function, $\mathbf{q}$, may depend on $\nabla\theta$;
- the thermomechanical analogue of the (1.17)

$$\mathbf{T}_R = \frac{\partial\phi}{\partial\mathbf{F}};$$

- the entropy relation

$$\eta = -\frac{\partial\phi}{\partial\theta};$$

- the response function $\mathbf{q}$ obeys the heat conduction inequality

$$\mathbf{q}(F, \theta, \nabla\theta) \cdot \nabla\theta \geq 0. \tag{1.24}$$

In the molecular theory, at least in a first approximation, rubber-like materials are usually described by the theory of entropic elasticity. This means that the change of internal energy with deformation is zero and the internal energy is considered function only of the temperature. In the modified entropic elasticity the internal energy may also depend on the volume ratio through the term $\det\mathbf{F}$.

1.2 Material frame indifference

A change of frame from $\mathcal{R} = \{O; \mathbf{e}_k\}$ into $\widehat{\mathcal{R}} = \{\hat{O}; \hat{\mathbf{e}}_k\}$ is defined by

$$\hat{\mathbf{x}} = \mathbf{c}(t) + \mathbf{Q}(t)\mathbf{x}, \tag{1.25}$$

where $\mathbf{c}(t)$ is the position vector of O from $\hat{O}$ and $\mathbf{Q}(t)$ is a (proper orthogonal) rigid body rotation of $\mathcal{R}$ relative to $\widehat{\mathcal{R}}$. We recall that $\hat{\mathbf{x}} = \widehat{\mathbf{x}}(\mathbf{X}, t)$ and $\mathbf{x} = \chi(\mathbf{X}, t)$ describe the same motion of the material point $\mathbf{X}$ but referred to $\widehat{\mathcal{R}}$ and $\mathcal{R}$, respectively. We shall suppose that the reference configuration is the same for both observers whose frames may be chosen to coincide at the reference instant t_R, say, and that both observers use the same clock so that $\hat{t} = t$.

The deformation gradient $\mathbf{F} = \partial\mathbf{x}/\partial\mathbf{X}$ under the change of frame (1.25) transforms as $\hat{\mathbf{F}} = \mathbf{QF}$. Hence, we find that the right and left Cauchy-Green deformation tensors $\mathbf{B} = \mathbf{FF}^T$ and $\mathbf{C} = \mathbf{F}^T\mathbf{F}$, transform as

$$\hat{\mathbf{C}} = \mathbf{C}, \quad \hat{\mathbf{B}} = \mathbf{QBQ}^T,$$

the first of which reflects use of the same reference configuration to which both $\mathbf{C}$ and $\hat{\mathbf{C}}$ are referred.

We postulate that the strain energy of $\mathcal{B}$ is indifferent to the observer. This principle of material frame indifference applied to the strain energy function (1.16) implies that

$$W(\hat{\mathbf{F}}) = W(\mathbf{QF}) = W(\mathbf{F}) \tag{1.26}$$

must hold for all proper orthogonal $\mathbf{Q}$ and for all $\mathbf{F}$. Therefore, we may choose $\mathbf{Q} = \mathbf{R}^T$ in the polar decomposition $\mathbf{F} = \mathbf{RU}$ to deduce the necessary condition

$$W(\mathbf{F}) = W(\mathbf{U}) = \tilde{W}(\mathbf{C}), \tag{1.27}$$

wherein we recall that $\mathbf{C} = \mathbf{U}^2$. Conversely, if the strain energy has the form (1.27), upon replacing $\mathbf{F}$ by $\hat{\mathbf{F}}$ and recalling $\hat{\mathbf{C}} = \mathbf{C}$, we may recover (1.26) for all changes of frame. In consequence, the strain energy function for a hyperelastic material is frame indifferent if and only if it has the reduced form (1.27). Use of (1.27) in (1.17) and (1.18) yields the following equivalent reduced forms of the constitutive equation for a hyperelastic solid:

$$\mathbf{T}_R = 2\mathbf{F}\frac{\partial \tilde{W}(\mathbf{C})}{\partial \mathbf{C}}, \quad \mathbf{T} = 2J^{-1}\mathbf{F}\frac{\partial \tilde{W}(\mathbf{C})}{\partial \mathbf{C}}\mathbf{F}^T. \tag{1.28}$$

These rules provide the most general form of the constitutive equation that respects the principle of frame indifference, but it does not yet reflect any inherent structural symmetries of the material.

1.3 Material symmetry transformations

Let us consider all the special rigid rotations $\mathbf{Q}$, such that

$$\tilde{W}(\mathbf{C}) = \tilde{W}(\mathbf{Q}^T\mathbf{CQ}). \tag{1.29}$$

Such rotations are called material symmetry transformations. It is easy to show that the set of all material symmetry transformations at $\mathbf{X}$ form a subgroup $g_\mathbf{X}$ of the group $\mathcal{G}$ of proper orthogonal tensors. The group $g_\mathbf{X}$ is named the symmetry group at $\mathbf{X}$.

It is important to notice that the symmetry group depends on the reference configuration and therefore it is important to fix the reference configuration relative to which the material symmetry holds. Therefore, the undistorted state of vanishing deformation for which $\mathbf{F} = \mathbf{1}$ is chosen as the reference state that identifies the inherent symmetries of the material. If a material has no symmetries in its undistorted state at $\mathbf{X}$, then $g_\mathbf{X} = \{\mathbf{1}\}$, the smallest subgroup of $\mathcal{G}$; and the material is called triclinic at $\mathbf{X}$. On the other hand, if $g_\mathbf{X} = \mathcal{G}$, then every direction at $\mathbf{X}$ in the undistorted state is an axis of material symmetry. A material having this property is called isotropic at $\mathbf{X}$. A hyperelastic material that is isotropic at every material point in a global undistorted state is called an isotropic hyperelastic material.

The important case of a body capable of existing in different reference or natural configurations, or of a body for which an evolution in time of the natural configuration is possible is discussed by Rajagopal and Srinivasa (2000). Because this evolution is interesting in the framework of inelastic effects we shall not discuss this opportunity here, but we have to notice the basic importance of this idea.

1.4 Isotropic materials

For an isotropic hyperelastic material, the rule (1.29) must hold for all proper orthogonal tensors $\mathbf{Q}$. Therefore, we may take $\mathbf{Q} = \mathbf{R}^T$ in the polar decomposition of $\mathbf{F}$ and recall the relation $\mathbf{B} = \mathbf{RCR}^T$ to deduce for every isotropic material the necessary condition $\tilde{W}(\mathbf{C}) = \tilde{W}(\mathbf{B})$. This rule shows that for a given deformation the strain energy function has the same values whether $\mathbf{C}$ or $\mathbf{B}$ is used as the independent variable. For an arbitrary deformation, $\mathbf{C}$ and $\mathbf{B}$ generally are distinct symmetric tensors whose principal directions differ by a rigid rotation. On the other hand, their principal values always are the same; and hence the principal invariants $I_k(\mathbf{C})$ of $\mathbf{C}$ and $I_k(\mathbf{B})$ of $\mathbf{B}$ are the same for every deformation $\mathbf{F}$. Therefore, our previous relation suggests that the strain energy must be an isotropic scalar-valued function of these principal invariants alone:

$$\tilde{W}(\mathbf{C}) = \tilde{W}(\mathbf{B}) = W(I_1, I_2, I_3), \tag{1.30}$$

wherein

$$I_1 = \operatorname{tr}\mathbf{B}, \quad I_2 = \frac{1}{2}[I_1^2 - \operatorname{tr}(\mathbf{B}^2)], \quad I_3 = \det\mathbf{B}. \tag{1.31}$$

From $(1.30)_1$ and (1.28) the constitutive equation for an isotropic hyperelastic material may be written as

$$\mathbf{T}_R = 2\mathbf{F}\frac{\partial\tilde{W}}{\partial\mathbf{C}} = 2\frac{\partial\tilde{W}}{\partial\mathbf{B}}\mathbf{F}, \text{ or } \quad \mathbf{T} = 2J^{-1}\frac{\partial\tilde{W}}{\partial\mathbf{B}}\mathbf{B}. \tag{1.32}$$

The last of (1.32) is expressed entirely in terms of the symmetric tensor $\mathbf{B}$, whereas the first involves the local rigid body rotation $\mathbf{R}$. Use of (1.30) in the last of (1.32) allows to obtain the following representation of the general constitutive equation for an isotropic hyperelastic solid:

$$\mathbf{T} = \beta_0\mathbf{1} + \beta_1\mathbf{B} + \beta_{-1}\mathbf{B}^{-1}, \tag{1.33}$$

wherein the scalar coefficients $\beta_\Gamma = \beta_\Gamma(I_1, I_2, I_3)$, $\Gamma = 0, 1, -1$, called material response functions, are given in terms of the strain energy function by

$$\beta_0 = \frac{2}{\sqrt{I_3}}\left(I_2\frac{\partial W}{\partial I_2} + I_3\frac{\partial W}{\partial I_3}\right), \tag{1.34}$$

$$\beta_1 = \frac{2}{\sqrt{I_3}}\frac{\partial W}{\partial I_1}, \quad \beta_{-1} = -2\sqrt{I_3}\frac{\partial W}{\partial I_2}. \tag{1.35}$$

This is not the only representation of (1.33). Other representation exist and will be introduced if necessary.

It follows from (1.33) that in an undistorted state $\mathcal{C}_0$ in which $\mathbf{B} = \mathbf{1}$, the stress in an isotropic material is at most a hydrostatic stress $\mathbf{T}_0 = (\hat{\beta}_0 + \hat{\beta}_1 + \hat{\beta}_{-1})\mathbf{1}$, where $\hat{\beta}_\Gamma = \beta_\Gamma(3, 3, 1)$ are the values of the material response functions in $\mathcal{C}_0$. An undistorted reference configuration on which the stress $\mathbf{T}_0 = \mathbf{0}$ is called a stress-free or natural state of the material. Thus, in the natural state, the response functions must satisfy

$$\hat{\beta}_0 + \hat{\beta}_1 + \hat{\beta}_{-1} = 0. \tag{1.36}$$

A fundamental property of isotropic materials is the coaxiality between stress and strain i.e. the following relation holds

$$\mathbf{TB} = \mathbf{BT} \tag{1.37}$$

The (1.37) is composed by three scalar links between the stress components and the strain components which will be fundamental to be derive the so called **universal relations**. These are relations among stress components and shear strain which are valid for all elastic isotropic material.

1.5 Incompressibility

While compressible (or *unconstrained*) materials can experience various kinds of isochoric deformations, an incompressible material can sustain **only** isochoric deformations. Hence, the incompressibility constraint is that $J = dv/dV = \det \mathbf{F} = 1$ for all deformations of an incompressible material. This means also that $I_3 = 1$ must hold for all deformations of an incompressible material.

It is clear that no amount of all-around stress can deform an incompressible body. Consequently, the Cauchy stress in an incompressible material is determined by $\mathbf{F}$ only to within an arbitrary hydrostatic stress $-\tilde{p}(\mathbf{x}, t)\mathbf{1}$, say; and we thus write $\mathbf{T} = -\tilde{p}\mathbf{1} + \mathbf{T}_E(\mathbf{F})$. The extra stress $\mathbf{T}_E(\mathbf{F})$ reflects the elastic response in distortion of the material. For an isotropic material, the extra stress is described by (1.33), in which the dependence on I_3 is removed. Since the pressure is arbitrary, we may absorb the material function β_0 in a new pressure term $p \equiv \tilde{p} + \beta_0$, to write the constitutive equation for an incompressible, isotropic hyperelastic material as

$$\mathbf{T} = -p\mathbf{1} + \beta_1 \mathbf{B} + \beta_{-1} \mathbf{B}^{-1}, \tag{1.38}$$

in which the material response functions $\beta_\Gamma = \beta_\Gamma(I_1, I_2)$ are defined by

$$\beta_1 = 2\frac{\partial W}{\partial I_1}, \quad \beta_{-1} = -2\frac{\partial W}{\partial I_2}. \tag{1.39}$$

Another class of internally constrained materials is obtained by considering internal materials having inextensible fibers in a direction $\mathbf{E}(\mathbf{X})$ at a material point $\mathbf{X}$ in the reference configuration. These are called fiber-reinforced materials. In a deformation of the material this unit vector field is mapped into

$$\mathbf{e} = \mathbf{FE}, \tag{1.40}$$

the deformed fiber direction $\mathbf{e}(\mathbf{x})$ at the place $\mathbf{x}$ in the current configuration. The inextensibility constraint is thus provided by

$$\mathbf{e} \cdot \mathbf{e} = \mathbf{CE} \cdot \mathbf{E} = 1. \tag{1.41}$$

In this case

$$\mathbf{T}^R = q\mathbf{e} \otimes \mathbf{e}$$

and if the materials is also incompressible

$$\mathbf{T}^R = q\mathbf{e} \otimes \mathbf{e} - p\mathbf{I}.$$

Of course, it is possible to have more than one inextensible fiber in the material. For two such fibers, (1.40) and (1.41) hold for each fiber $\mathbf{E}_k(\mathbf{X})$ and

$$\mathbf{T}^R = q_1\mathbf{e}_1 \otimes \mathbf{e}_1 + q_2\mathbf{e}_2 \otimes \mathbf{e}_2.$$

A detailed survey of the mechanics of inextensible materials is provided by Spencer (1972).

1.6 The empirical inequalities

So far no restrictions have been imposed on the nature of the strain energy function besides these arising from frame indifference and material symmetry. It may be questioned that the material response functions cannot be fully arbitrary. At the least, to model real material behavior, the response functions β_Γ should be compatible with fairly general empirical results of mechanical response obtained from carefully controlled large deformation tests of isotropic materials of special kinds. This raises the related question on what reasonable restrictions on the form of the strain-energy must be imposed. A delicate and central matter sometimes refereed to as Truesdell's *Hauptproblem* and is related to the interplay between, stability, thermodynamics and the mathematical conditions which guarantee the existence and uniqueness of solutions in nonlinear elasticity.

It would be recognized, however, that the classical Hadamard notion, commonly used in linear elasticity, of well-posedness is too restrictive for the non-linear setting. A simple example of non-uniqueness is provided by a spherical cap cut from a tennis ball. It may be shown experimentally (in a home made laboratory) and theoretically that every uniform, compressible and isotropic spherical cap has an everted image in equilibrium under zero traction on the spherical boundary. This means that for a spherical shell under zero surface tractions we have at least two equilibrium solutions that differ by more than a rigid motion.

For these reasons I simply note that experimental data appear to support the following empirical inequalities:

$$\begin{array}{ll} \text{(a) Compressible:} & \beta_0 \leq 0, \;\; \beta_1 > 0, \;\; \beta_{-1} \leq 0, \\ \text{(b) Incompressible:} & \beta_1 > 0, \;\; \beta_{-1} \leq 0, \end{array} \qquad (1.42)$$

for compressible and incompressible materials. In fact, a variety of tests by Rivlin and Saunders, Treloar, and others on incompressible rubber-like materials support the relations in (1.42b); and this provides reasonable ground for accepting the same pair of inequalities in (1.42a) for compressible materials. The inequalities (1.42) are not a definitive solution to the Truesdell's Hauptproblem of the determination on the strain-energy function that guarantee reasonable behavior. It is only known that empirical inequalities are useful in some physical situations, because they reflect the intuitive behavior of materials. Other set of analytical restrictions have been postulate by many authors in nonlinear elasticity. Among the various possibilities the most celebrated are the Baker-Ericksen inequality and the Coleman and Noll inequality (C-N condition). For example the C-N condition states that the greatest stretch occurs in the direction of the greatest stress. At first sight, this is a reasonable expectation, but we shall realize in the next lectures that this *evidence* is in clear contradiction to some experimental findings.

Other requirements on the strain energy may come from the question of existence of a solution. In hyperelasticity this question is reduced to the existence of a minimum for a suitable functional: the potential energy (strain-energy minus work of exterior loads). We point out that the in general the strain-energy over the set of admissible functions to be defined is not convex (usually we require polyconvexity). Moreover, to ensure existence we must require a coerciveness inequality that means that we require not only an infinite amount of strain energy in order to expand a continuum body to the infinite range, but also that the growth condition must be sufficiently strong. Obviously this is a very hard condition to test experimentally because the elastic range of any material is finite. On the other hand the strain energy must also tend to infinity as the volume ratio, J, vanishes. When all these conditions are satisfied it is possible to show that for the functional potential there exist a minimizing sequence that weakly converges to a minimum that belongs to the set of admissible functions. From some very important mathematical results by L.C. Young (Pedregal, 2000) we know that there are the example of reasonable functionals for which a minimizing sequence does attain a minimum and that this situation may have also a physical interpretation in the appearance of microstructure. Therefore even non-existence may be an admissible and physically acceptable behavior.

The end of the story is that there are probably no restrictions at all on the strain-energy besides those arising from frame indifference and material symmetry and for this reason any additional condition must be consider critically.

2 General solutions of balance equations

Homogeneous deformations are described as

$$\mathbf{x} = \mathbf{F}\mathbf{X} + \mathbf{c}, \tag{2.1}$$

wherein $\mathbf{F}$ is a constant tensor and $\mathbf{c}$ is a constant vector. They play a fundamental role in the theory of finite elastic deformations. Not only are they the simplest deformations to treat in term of the mathematics involved, but also the only deformations which can be maintained in all homogeneous compressible bodies under the action of surface forces alone. Indeed, it easy to check that the stress corresponding to (2.1) is a constant tensor, and the balance equations are then trivially satisfied in absence of body forces. For this reason the deformations (2.1) are controllable for all uniform and isotropic elastic materials i.e. homogeneous deformations are *universal* solutions. Ericksen in 1955 proved that homogeneous deformations are the *only* controllable deformations possible in every unconstrained, uniform and isotropic hyperelastic material (Saccomandi, 2001).

Homogeneous deformations can be treated without specifying the constitutive equations. This is an important fact in the rational design of experimental tests. Before introducing the details on constitutive equations we consider some basic facts about these deformations.

2.1 Simple uniaxial extension

Let us consider the uniaxial extension of a rod whose lateral surfaces are traction-free. In this case, the Cartesian matrix of **B** is described by

$$B_{ij}=\text{diag}\{\lambda_1^2,\lambda_2^2,\lambda^2\}$$

in which $\lambda_3=\lambda$ is the uniaxial stretch, and λ_1, λ_2 are lateral stretches.

First consider an isotropic, compressible elastic material characterized by (1.33). We find by (1.33) that the lateral traction-free boundary conditions $T_{11}=T_{22}=0$ are satisfied provided that

$$\beta_0+\lambda_k^2\beta_1+\lambda_k^{-2}\beta_{-1}=0,\ k=1,2. \tag{2.2}$$

Forming the difference of (2.2), we obtain

$$(\lambda_1^2-\lambda_2^2)\left(\beta_1-\frac{1}{\lambda_1^2\lambda_2^2}\beta_{-1}\right)=0. \tag{2.3}$$

Accepting the empirical inequalities (1.42a), it is seen that (2.3) is satisfied if and only if $\lambda_1=\lambda_2$, in which case the principal invariants (1.31) reduce to functions of λ alone, so the material response functions β_Γ are functions of λ alone. In fact, (2.2) is now an equation that relates the lateral stretch λ_1 to the uniaxial stretch λ. When we can solve this equation uniquely for $\lambda_1=\lambda_1(\lambda)$, the extension is called simple; and the required uniaxial stress $T_{33}=T(\lambda)$ is thus related to the uniaxial stretch λ in accordance with

$$T(\lambda)=(\lambda^2-\lambda_1^2)\left(\beta_1-\frac{1}{\lambda^2\lambda_1^2}\beta_{-1}\right). \tag{2.4}$$

It is now evident that the stress is in the natural direction of the stretch if and only if the empirical inequalities (1.42a) hold. In particular, $\lambda>\lambda_1$ is a simple extension under a tensile stress $T(\lambda)>0$, whereas $\lambda<\lambda_1$ describes a simple compression due to a compressive stress $T(\lambda)<0$.

For an incompressible material, we have the lateral stretch relation $\lambda\lambda_1^2=1$ given by the incompressibility constraint $\det\mathbf{F}=1$. Now the equilibrium equations are satisfied provided the arbitrary pressure p in (1.38) is a constant. This constant can be chosen to satisfy the null lateral traction condition, and hence from(2.4) with $\lambda_1^2=\lambda^{-1}$ we have:

$$T(\lambda)=\left(\lambda^2-\frac{1}{\lambda}\right)\left(\beta_1-\frac{1}{\lambda}\beta_{-1}\right). \tag{2.5}$$

The Poisson function $\nu(\lambda)$ is a purely kinematical material function defined in a simple tension problem as the ratio of the lateral contraction to the longitudinal extension, namely,

$$\nu(\lambda)=-\frac{\lambda_1(\lambda)-1}{\lambda-1}. \tag{2.6}$$

For an incompressible material, we have $\lambda_1(\lambda)=\lambda^{-1/2}$ and hence the corresponding Poisson function is given by

$$\nu(\lambda)=\frac{1}{\sqrt{\lambda}\left(\sqrt{\lambda}+1\right)}. \tag{2.7}$$

In the natural state of an incompressible material, the Poisson ratio has the value $\nu_0 = \nu(1) = 1/2$; otherwise, (2.7) is a monotone decreasing function of the amount of uniaxial stretch.

To the Poisson function in finite elasticity Beatty and Stalnaker (1986) have devoted a fundamental paper. In a simple tension experiment we have that the (2.7) can be used to evaluate if the material is incompressible. Indeed the (2.7) is universal for any isotropic, uniform elastic materials which is incompressible. In the case of compressibility equation (2.6) is no more *universal* since the $\lambda_1(\lambda)$ depend on the special material we are considering. This fact introduces additional complexities in the study of unconstrained materials. It is necessary not only to model the stress-axial stretch data, but also the relationship between axial and lateral stretches. For example considering some foams Blatz and Ko proposed a power law relationship

$$\lambda_i = \lambda^{-n}, \quad i = 1, 2,$$

and Murphy and Rogerson the generalized relationship

$$\lambda_i^m - 1 = -n(\lambda^m - 1), \quad i = 1, 2,$$

where m and n are constants (Murphy and Rogerson, 2002).

2.2 Simple shear of a block

A simple shear deformation of amount $K = \tan\gamma$, where γ is the angle of shear, is described by

$$x = X + KY, \; y = Y, \; z = Z.$$

Here (x, y, z) are the current rectangular Cartesian coordinates of a material point initially at (X, Y, Z) in the same Cartesian frame $\mathcal{R} = \{O; \mathbf{i}_k\}$ at O. Hence, the deformation gradient tensor $\mathbf{F} = \partial\mathbf{x}/\partial\mathbf{X} = \mathbf{1} + K\mathbf{i}_1 \otimes \mathbf{i}_2$. Since $\det\mathbf{F} = 1$, a simple shear is isochoric (volume preserving). In this deformation we have that $\beta_\Gamma(I_1, I_2, I_3) = \beta_\Gamma(3+K^2, 3+K^2, 1) \equiv \tilde{\beta}_\Gamma(K^2)$. We thus find

$$\mathbf{T} = (\tilde{\beta}_0 + \tilde{\beta}_1 + \tilde{\beta}_{-1})\mathbf{1} + K\mu(K^2)\mathbf{i}_{12} + K^2\tilde{\beta}_1\mathbf{i}_{11} + K^2\tilde{\beta}_{-1}\mathbf{i}_{22}, \tag{2.8}$$

in which $\mathbf{i}_{jk} \equiv \mathbf{i}_j \otimes \mathbf{i}_k$ and, by definition,

$$\mu(K^2) \equiv \tilde{\beta}_1(K^2) - \tilde{\beta}_{-1}(K^2). \tag{2.9}$$

The shear stress in a finite simple shear, by (2.8), is given by

$$T_{12} = K\mu(K^2). \tag{2.10}$$

The constitutive functions combination $\mu(K^2)$ in (2.9) is denoted as the *generalized shear response function*; its value $\mu_0 \equiv \mu(0)$ in the natural state is the classical shear modulus. The empirical inequalities require that $\mu(K^2) > 0$ for all $K(\geq 0)$. Clearly, the shear stress is an *odd* function of the amount of shear, and therefore the shear stress is in the direction of the amount of shear. Shear stress does not suffice to determine a simple shear. It follows from (2.8) that additional normal stresses are required on all pairs of plane faces. In fact, the response functions $\tilde{\beta}_1$ and $\tilde{\beta}_{-1}$ are determined by the normal stress differences:

$$T_{11} - T_{33} = K^2\tilde{\beta}_1, \quad T_{22} - T_{33} = K^2\tilde{\beta}_{-1}, \tag{2.11}$$

where

$$T_{33} = \tilde{\beta}_0 + \tilde{\beta}_1 + \tilde{\beta}_{-1} \equiv K^2\tau(K^2). \tag{2.12}$$

The last of (2.11) determines $\tau(K^2)$, and hence $\tilde{\beta}_0(K^2)$. Because the response functions are even functions of K, the normal stresses are unchanged when the shear is reversed; and if these normal stresses are not applied, the block will tend to contract or to expand. Such normal stress effects are typical of problems in nonlinear elasticity.

An important aspect of the simple shear problem is that the shear stress is determined by the normal stress difference (Poynting effect):

$$KT_{12} = T_{11} - T_{22}. \tag{2.13}$$

In fact, T_{12} is determined in the same way for every homogeneous, isotropic hyperelastic material regardless of the form of the response functions. The rule (2.13) is an example of a universal relation valid for both compressible and incompressible materials. For an incompressible material, we replace $\tilde{\beta}_0$ with a constant pressure $-p$, which we may choose so that $T_{33} = 0$. In this case, we see that $T_{11} > 0$ and $T_{22} < 0$, if and only if the empirical inequalities hold.

The relation (2.13) is exactly the kind of universal result we were talking about when we have pointed out the coaxiality between stress and strain in isotropic elastic materials (1.37).

2.3 Biaxial stretch

Let us consider the general homogeneous deformation and the reference frame $\mathcal{R} = \{O; \mathbf{e}_i\}$ where the orthonormal vector basis $\mathbf{e}_i = \{\mathbf{e}_1, \mathbf{e}_2, \mathbf{e}_3\}$ is composed by the eigenvectors of $\mathbf{U}$ in such a way that

$$x_1 = \lambda_1 X_1, \quad x_2 = \lambda_2 X_2, \quad x_3 = \lambda_3 X_3, \tag{2.14}$$

where λ_1, λ_2, λ_3 are the corresponding eigenvalues. The Cartesian matrix of $\mathbf{B}$ is described by

$$B_{ij}=\text{diag}\{\lambda_1^2, \lambda_2^2, \lambda_3^2\}, \tag{2.15}$$

and its inverse

$$B_{ij}^{-1}=\text{diag}\{\lambda_1^{-2}, \lambda_2^{-2}, \lambda_3^{-2}\}. \tag{2.16}$$

Considering the stress tensor for an isotropic material we have that

$$T_{ij} = \text{diag}\left\{\beta_0 + \beta_1\lambda_i^2 + \beta_{-1}\lambda_i^{-2}\right\}. \tag{2.17}$$

As an example, of particular relevance to experiments, we can consider the biaxial stretching of a plane rectangular sheet by forces in its plane. The axes are chosen with the major faces perpendicular to the 3-direction and free of traction, so that

$$\mathbf{t}_3 = \mathbf{T}\mathbf{e}_3 = 0. \tag{2.18}$$

For compressible materials this conditions means that, for example, λ_3 must be determined from the equation

$$\beta_0 + \beta_1\lambda_3^2 + \beta_{-1}\lambda_3^{-2} = 0. \tag{2.19}$$

For incompressible materials

$$\lambda_1\lambda_2\lambda_3 = 1, \tag{2.20}$$

and in the stress the constitutive parameter β_0 is replaced by the reaction scalar $-p$ and from (2.18) we obtain

$$p = \beta_1\lambda_1^{-2}\lambda_2^{-2} + \beta_{-1}\lambda_1^2\lambda_2^2. \tag{2.21}$$

On the other hand, because the deformation is homogeneous, if we denote by f_1 and f_2 the forces per unit length on edges measured in the undeformed state imposed to stretch the sheet we have that the components T_{11} and T_{22} of the stress tensor are given by

$$T_{11} = \frac{\lambda_1 f_1}{H}, \quad T_{22} = \frac{\lambda_2 f_2}{H}, \tag{2.22}$$

where H is the thickness of the sheet in its undeformed state.

For incompressible isotropic hyperelastic materials this deformation is fundamental because in this case

$$\beta_1 = 2W_1, \quad \beta_{-1} = -2W_2.$$

From (2.22), (2.21) and (2.17) we deduce

$$W_1 = \frac{1}{2H\left(\lambda_1^2 - \lambda_2^2\right)}\left\{\frac{\lambda_1^3 f_1}{\lambda_1^2 - \lambda_1^{-2}\lambda_2^{-2}} - \frac{\lambda_2^3 f_2}{\lambda_2^2 - \lambda_1^{-2}\lambda_2^{-2}}\right\}, \tag{2.23}$$

and

$$W_2 = \frac{1}{2H\left(\lambda_2^2 - \lambda_1^2\right)}\left\{\frac{\lambda_1 f_1}{\lambda_1^2 - \lambda_1^{-2}\lambda_2^{-2}} - \frac{\lambda_2 f_2}{\lambda_2^2 - \lambda_1^{-2}\lambda_2^{-2}}\right\}. \tag{2.24}$$

These relations allow us to determine easily from an experiment the derivatives of the strain-energy.

When $f_1 = f_2$ the interesting phenomena, known as the of Treloar instability, arises. Indeed in this case because $T_{11} = T_{22}$ intuitively one expects $\lambda_1 = \lambda_2$. Treloar data suggest that for situations where $T_{11} = T_{22}$ it is possible to have deformations with $\lambda_1 \neq \lambda_2$ and these differences are too large to be attributed to experimental errors. These asymmetric solutions may be described theoretically by some classes of strain-energy densities (Ericksen, 1997).

We point out that in this case the Coleman-Noll inequality stating that the greatest stretch occurs in the direction of the greatest stress is clearly violated, because the greatest stretch does not occur in the direction of the greatest stress.

2.4 Incompressible materials

For incompressible materials, besides homogeneous universal deformation, we know five families of inhomogeneous deformations. The complete list of these deformation is given, for example, in (Saccomandi, 2001):

- **Family 1**: Bending, stretching and shearing of a rectangular block

$$r = \sqrt{2AX}, \qquad \theta = BY, \qquad z = \frac{Z}{AB} - BCY.$$

- **Family 2**: Straightening, stretching and shearing of a sector of a hollow cylinder

$$x = \frac{1}{2}AB^2R^2, \qquad y = \frac{\Theta}{AB}, \qquad z = \frac{Z}{B} + \frac{C\Theta}{AB}.$$

- **Family 3**: Inflation, bending, torsion, extension and shearing of an annular wedge

$$r = \sqrt{AR^2 + B}, \qquad \theta = C\Theta + DZ, \qquad z = E\Theta + FZ,$$

with $A(CF - DE) = 1$.

- **Family 4**: Inflation or eversion of a sector of a spherical shell

$$r = (\pm R^3 + A)^{1/3}, \qquad \theta = \pm\Theta, \qquad \phi = \Phi.$$

- **Family 5**:

$$r = \sqrt{AR}, \quad \theta = D\ln(BR) + C\Theta, \quad z = FZ,$$

Here A, B, C, D, E, F are constants.

These families of deformations are well known and they have been studied in many details. We point out that universal solutions are not the only kind of possible solutions for the balance equations. The search for exact and analytical solutions which are not universal (then with a geometry that depends on the particular choice of the constitutive equations) is a current topic of research and any advance in this direction is very precious. An example of non-universal solutions is discussed in Section 4.

Let us consider a circular cylindrical bar of radius R_0 and length L in its undeformed reference configuration $\mathcal{C}_t$. The bar is deformed by a uniform twist in which planes normal to the cylinder axis $\mathbf{E}_Z$ remain plane and undergo a rotation about $\mathbf{E}_Z$ proportional to their initial distance Z from one end of the bar. We suppose the bar remains cylindrical and retains its initial radius and length. Hence, the deformation is clearly isochoric. The pure torsion of a cylinder is a special member of Family 3 in which $A = 1,\ B = 0,\ C = 1,\ E = 0,\ F = 1:$

$$r = R, \quad \theta = \Theta + DZ, \quad z = Z, \tag{2.25}$$

where D is the angle of twist per unit length. Hence, the total angle of twist is $\psi = DL$, and the cylindrical surfaces are traction-free.

In physical components, we find $\mathbf{F} = \mathbf{1} + K(r)\mathbf{e}_\theta \otimes \mathbf{E}_Z$ in terms of the orthonormal bases $\mathbf{e}_k$ in κ and $\mathbf{E}_\alpha$ in κ_R, both referred to a common Cartesian frame $\mathcal{R} = \{O; \mathbf{i}_k\}$. Also, here we define the local amount of shear $K(r) \equiv rD$. Clearly, $I_3 \equiv 1$. We recall the constitutive equation for incompressible, isotropic elastic bodies, namely, $\mathbf{T} = -p(\mathbf{x})\mathbf{1} + \beta_1\mathbf{B} + \beta_{-1}\mathbf{B}^{-1}$, in which $\beta_\Gamma(I_1, I_2) = \beta_\Gamma(3 + K^2(r), 3 + K^2(r)) \equiv \tilde{\beta}_\Gamma(K^2(r))$ depend on r alone. The physical components of the Cauchy stress are thus given by

$$\mathbf{T} = (-p(r,\theta,z) + \tilde{\beta}_1 + \tilde{\beta}_{-1})\mathbf{1} + K\mu(K^2)\mathbf{e}_{\theta z} + K^2\tilde{\beta}_1\mathbf{e}_{\theta\theta} + K^2\tilde{\beta}_{-1}\mathbf{e}_{zz}. \tag{2.26}$$

Here $\mathbf{e}_{jk} \equiv \mathbf{e}_j \otimes \mathbf{e}_k$, and the shear response function is defined by

$$\mu(K^2) \equiv \tilde{\beta}_1(K^2) - \tilde{\beta}_{-1}(K^2). \tag{2.27}$$

If the empirical inequalities holds, the shear response function must be positive for all D. The analogy with a simple shear is evident and the shear stress in the pure torsion deformation, by (2.26), is given by

$$T_{\theta z} = K\mu(K^2) = rD\mu(r^2D^2), \tag{2.28}$$

which shows that the shear stress is in the direction of the twist. The normal stress differences obtained from (2.26) are

$$T_{\theta\theta} - T_{rr} = K^2\tilde{\beta}_1, \quad T_{zz} - T_{rr} = K^2\tilde{\beta}_{-1}, \quad T_{\theta\theta} - T_{zz} = KT_{\theta z}. \tag{2.29}$$

The last of these relations is the analogue of the universal relation (2.13).

Two of the equilibrium equations without body force are satisfied provided that $p = p(r)$, and the third reduces to $dT_{rr}/dr = (T_{\theta\theta} - T_{rr})/r (\equiv K^2\beta_1(K^2)/r)$, in which everything depends only on r. Because there are no lateral tractions on the surface $r = R_0$, $T_{rr}(R_0) = 0$ and integration of the last equilibrium equation yields

$$T_{rr}(r) = -p(r) + \tilde{\beta}_1 + \tilde{\beta}_{-1} = -D^2 \int_r^{R_0} r\tilde{\beta}_1(D^2r^2)dr. \tag{2.30}$$

The traction vector $\mathbf{t_n} = \mathbf{Tn}$ on the end faces of the cylinder for which $\mathbf{n} = \pm\mathbf{e}_z = \pm\mathbf{E}_Z$ is given by $\mathbf{t_n} = \pm(T_{\theta z}\mathbf{e}_\theta + T_{zz}\mathbf{e}_z)$, where $\mathbf{e}_\theta = -\sin\theta\mathbf{i} + \cos\theta\mathbf{j}$ in φ. Integration over the end face $\mathbf{n} = \mathbf{e}_z$ shows that a normal force $N = 2\pi\int_0^{R_0} T_{zz} r dr$ must be applied to constrain the length L of the bar in a pure torsion. With the aid of $(2.29)_2$ and upon integrating by parts the term in T_{rr}, we find that

$$N = -\pi D^2 \int_0^{R_0} r^3[\tilde{\beta}_1(r^2D^2) - 2\tilde{\beta}_{-1}(r^2D^2)]dr. \tag{2.31}$$

The empirical inequalities require that N be compressive. If this end force is not applied, the cylinder will elongate by torsion (Poynting effect in a pure torsion). The total torque about the cylinder axis applied to an end face of the cylinder is given by

$$M = 2\pi D \int_0^{R_0} r^3\mu(r^2D^2)dr. \tag{2.32}$$

The empirical inequalities imply that the torque must be in the direction of the twist. Notice that existence of a strain energy is not essential to any of the results. However, the integrations can be evaluated only for specific kinds of materials.

It is very important to point out that when $\beta_{-1} = 0$, i.e. we are considering $W = W(I_1)$, it is possible to obtain a new global universal relation. Indeed, from (2.31) and (2.32)

$$N = -\pi D^2 \int_0^{R_0} r^3\tilde{\beta}_1(r^2D^2)dr, \quad M = 2\pi D \int_0^{R_0} r^3\tilde{\beta}_1(r^2D^2)dr. \tag{2.33}$$

and we recover the *global universal relation*

$$N = -\frac{D}{2}M. \tag{2.34}$$

To the best of my knowledge all the published experimental data on simple torsion are in contradiction with (2.34) (see for example Horgan and Saccomandi, 1999) and this suggests that if any isotropic, incompressible elastic material *exists* it cannot be modelled by a strain-energy $W = W(I_1)$. Therefore it seems that to have a reliable and realistic model of material strain energies the dependence with respect I_2 is necessary.

3 The choice of the strain energy

One of the main problems encountered in the applications of mechanics of continua is the complete and accurate determination of the constitutive relations necessary for the mathematical description of the behavior of real materials. Some people believe that constitutive equations are the meeting point about theory and experiments and for this reason constitutive requirements are a sort of grey zone where commonly we believe that rational mechanics ends its goal and *applied* mechanics begins its action. This point of view contains some naivete as Truesdell and Noll have declared in their monumental work *The Non-Linear Field Theories of Mechanics* :

> While laymen and philosopher of science often believe, contend, or at least hope, that physical theories are directly inferred from experiments, anyone who has faced the problem of discovering a good constitutive equation or anyone who has sought and found the historical origin of the successful field theories knows how childish is such a prejudice. The task of the theorist to bring order into the chaos of the phenomena can be described efficiently and simply. Here is the place for "intuition", and here the old preconception, common among natural philosophers, that nature is simple and elegant, has led to many great successes. Of course, physical theory must be based on experience, but experiment comes after, not before, theory. Without theoretical concepts one would neither know what experiments to perform nor be able to interpret their outcome.

These words may be of some help to understand why the colossal work on the foundations of the mechanics of continua which has begun after the Second World War has allowed important insights on this problem. New achievements have been realized mainly by means of:

- a systematic and rigorous use of the restrictions imposed by thermodynamics;
- the derivation and use of the representation formulae, a consequence of the rigorous definitions of the different material behaviors and of the concept of material symmetry;

- the use of *a priori restrictions,* consequences of the requirement of mathematical well posedness for the models considered.

All these ideas have helped to focus the goal of experimental tests, and to a certain extent to derive a *rational and reasonable theory* of constitutive equations. Obviously we believe that the ultimate factor in determining the *right* constitutive equations is only the laboratory.

In applications there is the need of a *specific* constitutive equation, from which we are able to obtain *numbers* for the various quantities of physical interest and/or specific qualitative insights. This cannot be done only with theorems and general hypothesis, but this cannot be done without a general theory to frame experiments. First of all because we must be able to know the exact link between the various measured quantities in order to decide which values of the constitutive mappings or which quantities associated with them can be determined in the experiment, and then to provide the experimentalist with formulae expressing these in terms of quantities which he can measure. The experimentalist must be provided with key theoretical results that allows the *interpretation* of such a model. For example because it is not possible to have local (point) measurements both of the loads and of the deformations we must know a priori which resultants, if any, have the principal influence upon the deformation. Moreover because measurements can be done only on the boundary of the test specimen, we must know *a priori* the kind of deformation that we want to reproduce experimentally in order to know what quantities can be effectively measured (this is the reason for the importance of exact or approximate solutions v/s numerical solutions of a mathematical problem).

We must also point out that rubber-like elasticity has had one of the longest and distinguished histories in all of polymer science and however, the complex molecular nature of elastomeric materials and the necessity to use a nonlinear theory have complicated the development of realistic mathematical models. This is a serious problem because users and manufacturers of rubber now have access to sophisticated numerical techniques for design calculations, but these methods can only be as good as the constitutive equations used. For this reason recently there has been an increasing effort in proposing new models for rubbery materials. Here the idea of a mathematical model must be distinguished from the ideas of a simulation model. Sometimes this is an hard point to explain, but Piero Villaggio in the preface of his Mathematical Models of Elastic Structures (Cambridge University Press, 1997) clearly states this difference:

> The simulation of a phenomenon increases in usefulness with the quantity of specific details incorporated. The mathematical models should instead include as few details as possible, but preserve the essential outline of the problem. The simulation is concretely descriptive, but applies to only one case; the mathematical model is abstract and universal.

People working with rubber know very well that the mechanical behavior of this material is very complex and outside of the forecast possibilities of nonlinear elasticity. In any case the first step to a correct modelling of rubber mechanics is to have a precise idea of what is happening in the primary loading path and this is the main scope of nonlinear elasticity. This is because we are searching for a mathematical model of rubber-like mechanics.

In the following we present a theoretical framework for the analysis and derivation of constitutive equations. This is not a survey of all the explicit forms of strain-energy densities that have been previously proposed in the literature. Here we concentrate on some general methods to obtain a suitable approximation of the strain energy. These methods are mainly based on polynomial and rational approximations. Strictly speaking the following presentation of the material does not rely on experimental data, but I remember we are in the grey zone where theory and experiments mix together to build our experience.

3.1 From infinitesimal to finite deformations

In the framework of isotropic materials a first idea in deriving a constitutive equation for nonlinear elastic materials is to consider a linear (affine) constitutive equation between stress and strain. In such a way we hope to obtain elastic theories appropriate to small but finite deformations, because we do not consider the approximation of the strain measure as in linear elasticity but the exact geometrical measure. In so doing we discover clearly a fundamental difference from the linear theory. Linear behavior is clearly defined up to some constants: the constitutive parameters. Dealing with nonlinear theories we have to face a problem of multiple choice. In nonlinear elasticity the multiple choice is with respect different stress tensor measures and strain tensor measures.

For example, choosing $\mathbf{T}$ (the Cauchy stress tensor) and $\mathbf{B}$ (the left Cauchy-Green strain tensor) the *metaphor* of the linear isotropic law gives

$$\mathbf{T} = \frac{\lambda}{2}\left(\operatorname{tr}(\mathbf{B}-\mathbf{I})\right)\mathbf{I} + \mu\left(\mathbf{B}-\mathbf{I}\right), \tag{3.1}$$

for compressible materials and

$$\mathbf{T} = -p\mathbf{I} + \mu\left(\mathbf{B}-\mathbf{I}\right), \tag{3.2}$$

for incompressible materials. Here λ and μ are the *Lamé constants.* Equation (3.2) it is equivalent to the Gaussian basic model obtained from statistical mechanics treatments of rubber elasticity and usually known as the neo-Hookean model.

Signorini has proposed a linear model for the Cauchy stress but using the Almansi-Hamel strain tensor $\mathbf{A} = (\mathbf{I} - \mathbf{B}^{-1})/2$

$$\mathbf{T} = \lambda\left(\operatorname{tr}\mathbf{A}\right)\mathbf{I} + 2\mu\mathbf{A}. \tag{3.3}$$

To point out the difference between (3.1) and (3.2) it is sufficient to consider the principal stresses in terms of principal stretches of the deformation (which in both case are similar). In the first case we have

$$t_i = \frac{\lambda}{2}\left(\lambda_1^2 + \lambda_2^2 + \lambda_3^2 - 3\right) + \mu\left(\lambda_i^2 - 1\right), \tag{3.4}$$

in the second case

$$t_i = \frac{\lambda}{2}\left(3 - \left(\lambda_1^{-2} + \lambda_2^{-2} + \lambda_3^{-2}\right)\right) + \mu\left(1 - \lambda_i^{-2}\right). \tag{3.5}$$

Many other linear relations have been proposed in the literature. An interesting one is due to Hencky where the Kirchoff stress tensor $\mathbf{S}(= J\mathbf{T})$ is related to the Hencky strain $\mathbf{e} = \ln \mathbf{V}$ by

$$\mathbf{S} = \lambda\,(\mathrm{tr}\,\mathbf{e})\,\mathbf{I} + 2\mu\mathbf{e}. \tag{3.6}$$

Using (3.6) and coming back to Cauchy stress principal values we have

$$t_i = \left(\lambda\,(\ln\lambda_1 + \ln\lambda_2 + \ln\lambda_3) + 2\mu\ln\lambda_i\right)(\lambda_1\lambda_2\lambda_3)\,. \tag{3.7}$$

From these examples it is clear that although all these models are equivalent from a theoretical point of view, since there is a one-to-one correspondence between all the objects we have used, because a phenomenological theory selects the functional form to be fitted to stress-strain data, different approaches may yield very different results. Therefore, a clear and simple constitutive requirement such as a linear relationship between stress and strain leads to multiple choice in the domain of finite elasticity. To overcome this difficulty we can use the Rivlin expansion procedure.

3.2 Rivlin-Signorini method

The scheme used by Murnaghan (1937), Rivlin (1960) and Signorini (1955), was to approximate the material response functions by polynomials in the appropriate invariants. In this way, a particular material is then characterized by the constant coefficients of the polynomial rather than by functions. Because the field equations are not approximated the solutions based on this method are approximate solutions for the general hyperelastic material and are exact for special classes of materials. The Rivlin-Signorini method has also been used by Singh (1967) in the study of elastic dielectrics and by Martin and Carlson (1977) for elastic heat conductors.

The Rivlin-Signorini method is one of the few rational ways to obtain from the axiomatic corpus of continuum mechanics specific representations for explicit constitutive equations independently of the choice of the stress or strain tensor. The idea proposed by Murnaghan and formalized by Signorini and Rivlin is to expand in a power series of the invariants the strain energy. To have a method which does not depend on the choice of the invariants the requirement is to search for the Nth-order approximation in the principal stretches.

We remember that given two quantities f and g:

- we write $f = O(a)$ as $a \to 0$ if there exist constants $k > 0$ and $\varepsilon > 0$ such that $|f| < k\,|a|$ whenever $|a| < \varepsilon$;
- we say that f is of *order* N and we write $\mathrm{ord}f = N$ as $a \to 0$ if $N = max\,\{n \in \mathbb{N}\}$ such that $f = O(a^n)$ as $a \to 0$;
- we say g is an Nth-order approximation to f if $\mathrm{ord}(f - g) = N + 1$ as $a \to 0$.

Considering polynomials for small a the lowest degree term is the dominant one. Obviously if $f = O(a)$ then $\lim_{a\to 0} f/a$ =constant.

On introducing the notation

$$\lambda_1 = 1 + u_1, \quad \lambda_2 = 1 + u_2, \quad \lambda_3 = 1 + u_3, \tag{3.8}$$

for the principal stretches of the deformation, it is convenient (so that $u_i \to 0$ as $\lambda_i \to 1$) to introduce the new set of invariants

$$J_1 = I_1 - 3,$$
$$J_2 = (I_2 - 3) - 2(I_1 - 3), \tag{3.9}$$
$$J_3 = (I_3 - 1) - (I_2 - 3) + (I_1 - 3).$$

From (3.9) and the expression for I_i in terms of λ_i, we find that

$$J_1 = O(u_i), \quad J_2 = O(u_i^2), \quad J_3 = O(u_i^3), \quad i = 1, 2, 3. \tag{3.10}$$

Therefore with respect to u_i we have

$$\text{ord} J_1 = 1, \quad \text{ord} J_2 = 2, \quad \text{ord} J_3 = 3. \tag{3.11}$$

Let us expand the strain energy as a power series in the new invariants obtaining

$$W = \sum_{\alpha,\beta,\gamma} A_{\alpha,\beta,\gamma} J_1^\alpha J_2^\beta J_3^\gamma. \tag{3.12}$$

If only terms of order two or less in the u_i are retained considering (3.10) we obtain

$$W = A_{020} J_2 + A_{200} J_1^2, \tag{3.13}$$

whereas if we retain all terms of order three or less we get

$$W = A_{020} J_2 + A_{200} J_1^2 + A_{001} J_3 + A_{300} J_1^3 + A_{110} J_1 J_2. \tag{3.14}$$

For incompressible materials we have that $I_3 = 1$ therefore

$$(1 + u_1)^2 (1 + u_2)^2 (1 + u_3)^2 = 1. \tag{3.15}$$

Because we have that (3.15) is in force by a direct computation it is possible to check that

$$\text{ord} J_1 = 2. \tag{3.16}$$

Introducing

$$\widetilde{J}_2 = I_2 - I_1, \tag{3.17}$$

it is possible to check that

$$\text{ord} \widetilde{J}_2 = 3. \tag{3.18}$$

If we expand W in a power law as we have done in (3.12), require that the free energy, stress and the entropy vanish in the reference state, and retain only terms of order three or less (using the principal invariants) we obtain

$$W = a_1 (I_1 - 3) + a_2 (I_2 - 3). \tag{3.19}$$

Here we use a_j a compact notation for the coefficients $A_{\alpha\beta 0}$. The classical Mooney-Rivlin material and the neo-Hookean material of incompressible elasticity are embedded in (3.19). This shows that this methodology not only provides a general framework for the approximation of general elastic materials, but also a method to obtain particular strain energies.

It is interesting to point out that for incompressible materials many heuristic strain energies may be recovered by this procedure. For example if we consider generalized neo-Hookean materials in the isothermal and incompressible case we have that

$$W = \sum_{\alpha} A_{\alpha 00} J_1^{\alpha}. \tag{3.20}$$

For example let us consider the Knowles (1977) power-law strain-energy

$$W = \frac{\mu}{2b}\left[\left(1 + \frac{b}{n}(I_1 - 3)\right)^n - 1\right], \tag{3.21}$$

where μ is the shear modulus, and b and n are dimensionless positive material constants. It is easy to show that (3.21) can be obtained from (3.20). Obviously here we are restricting our attention to $n \in \mathbb{N}$. In the same way it is possible to obtain the Yeoh polynomial strain-energy (Arruda and Boyce, 2000)

$$W = C_{10}(I_1 - 3) + C_{20}(I_1 - 3)^2 + C_{30}(I_1 - 3)^3. \tag{3.22}$$

We know that the Mooney-Rivlin material is a good model in the range of moderate deformations. The neo-Hookean material is not able to describe experimental data either quantitatively or qualitatively in any range of deformation. This point must be clearly stated because in the literature it is usually claimed that this model captures experimental data for some limited range of deformations, but this is not true. This is clearly deduced by considering the reduced tensile force in a simple extension experiment. This point has been already discussed into details by Han, Horkay and McKenna (1999). Rivlin and Saunders (1951) have concluded from their celebrated experiments that

$$W = C_1 I_1 + f(I_2 - 3), \tag{3.23}$$

where the form of the function f depend on the material at hand. We shall come back on these points in the next Sections.

3.3 Valanis-Landel approach

Although from a theoretical point of view any complete set of invariants is equivalent, it has been observed by several authors that the approach used by Rivlin considering the principal invariants it is not very practical in fitting experimental data, because of the possible propagation of experimental errors[3]. Therefore it may be interesting to consider

[3] I think that the first paper where this observation appears is by Valanis and Landel (1967). The authors report that it was very difficult to design a careful experiment in which one can vary I_1 keeping I_2 constant. They attributes this difficulty to the fact that the invariants depend on both the principal stretches simultaneously and they find the equations determining the first derivative with respect the principal strain invariants very sensitive to experimental error especially for small values of the invariants.

the possibility of expressing the strain energy directly in terms of the principal stretches λ_i as $W(\lambda_1, \lambda_2, \lambda_3)$. In this case isotropy requires that

$$W(\lambda_1, \lambda_2, \lambda_3) = W(\lambda_1, \lambda_3, \lambda_2) = W(\lambda_3, \lambda_1, \lambda_2). \tag{3.24}$$

Obviously this symmetry requirement is very hard to manage searching for a phenomenological explicit expression. To overcome this difficulty Valanis and Landel (1967) have postulated that

$$W(\lambda_1, \lambda_2, \lambda_3) = w(\lambda_1) + w(\lambda_2) + w(\lambda_3), \tag{3.25}$$

in such a way all symmetry relations (3.24) imposed by isotropy are directly satisfied. At first sight there is no experimental reason to postulate (3.25), only convenience and experience may suggest (3.25). Indeed Valanis and Landel (1967) says:

> This postulated form is not fortuitous but is a natural generalization of more particular forms that already exist in the literature.

On the basis of the experimental data Valanis and Landel have proposed for incompressible rubber-like materials the explicit functional form

$$W = 2\mu \sum_{i=1}^{3} \lambda_i \left(\log \lambda_i - 1\right) \tag{3.26}$$

a form which was improved on the basis of data in simple extension some years later by Peng and Landel (1972)

$$W = 2\mu \sum_{i=1}^{3} \left[\lambda_i - 1 - \log \lambda_i - \frac{1}{6}\left(\log \lambda_i\right)^2 + \frac{1}{18}\left(\log \lambda_i\right)^3 - \frac{1}{216}\left(\log \lambda_i\right)^4\right]. \tag{3.27}$$

The Valanis-Landel hypothesis has been studied in detail by Ogden. The work of Ogden starts from an examination of biaxial data considering

$$W(\lambda_1, \lambda_2) = \widehat{W}(\lambda_1, \lambda_2, \lambda_1^{-1}\lambda_2^{-1}) \tag{3.28}$$

and representing data on the basis of the equation

$$t_1 - t_2 = \lambda_1 \frac{\partial W}{\partial \lambda_1} - \lambda_2 \frac{\partial W}{\partial \lambda_2}. \tag{3.29}$$

The key issue of Ogden's discussion is the postulate of a shape independence of (3.29) by a change in the value of λ_2. In the book of Ogden (page 493) it is clear that this shape independence means that the curves $t_1 - t_2$ plotted as functions of λ_1 are merely subjected to a vertical translation by a change in λ_2. Considering isotropy requirements this means that

$$\widehat{W}(\lambda_1, \lambda_2, \lambda_1^{-1}\lambda_2^{-1}) = w(\lambda_1) + w(\lambda_2) + \widehat{w}(\lambda_1\lambda_2), \tag{3.30}$$

where at this stage the functions w and $\widehat{w}$ are different. Several arguments are then proposed by Ogden to show that it must be

$$\widehat{w}(\lambda_1\lambda_2) \equiv w(\lambda_1^{-1}\lambda_2^{-1}) \equiv w(\lambda_3). \tag{3.31}$$

The simplest argument is that if in the experiment we reverse the role of λ_1 and λ_3 (or λ_2 and λ_3) we find an equivalent form of (3.30) from which (3.31) follows. Therefore for Ogden the Valanis-Landel hypothesis follows from the postulate of shape invariance, but there is no physical or theoretical reason for this postulate, this only a metaphoric interpretation of a special class of experimental data.

In any case the polynomial strain energy proposed by Ogden, given by

$$W = \sum_{m=1}^{N} \frac{\mu_m}{\alpha_m} \left(\lambda_1^{\alpha_m} + \lambda_2^{\alpha_m} + \lambda_3^{\alpha_m} - 3\right), \tag{3.32}$$

where $\sum_{m=1}^{N} \mu_m \alpha_m = 2\mu$, is one of the more widely used explicit forms of strain energies. We point out that also in this case we have a polynomial expansion in the principal stretches of the strain-energy.

The neo-Hookean and Mooney-Rivlin materials are particular cases of the Ogden strain energy and this is another important consideration. From (3.32) in the very simple case of $N = 1$, we recover also the Varga strain-energy

$$W = 2\mu\left(\lambda_1 + \lambda_2 + \lambda_3 - 3\right). \tag{3.33}$$

This very simple strain-energy has been deeply studied recently because of its mathematical feasibility (Hill, 2001). Other models that can be written in the form (3.32) are the liquidlike model by DiMarzio (Han, Horkay and McKenna, 1999)

$$W = C_1\left(\lambda_1^2 + \lambda_2^2 + \lambda_3^2 - 3\right) + C_2\left(\lambda_1^{-1} + \lambda_2^{-1} + \lambda_3^{-1} - 3\right),$$

and the localization model by Gaylord and Douglas (Han, Horkay and McKenna, 1999)

$$W = C_1\left(\lambda_1^2 + \lambda_2^2 + \lambda_3^2 - 3\right) + C_2\left(\lambda_1 + \lambda_2 + \lambda_3 - 3\right).$$

3.4 A generalization of the Rivlin scheme

An interesting generalization of Rivlin-Signorini scheme may be obtained by considering the more general set of approximating functions of rational functions (Horgan and Saccomandi, 2003). We point out that the Rivlin and Signorini methodology considers an approximation of the strain energy, but in experiments one does not measure energies, but rather stresses or strains. For this reason the more natural approach is to approximate directly the response coefficients. These two approaches may give rise to very different journeys. If we start from a polynomial strain-energy density we obtain polynomials response coefficients, but the converse it is not true when we consider compressible materials:

$$\mathbf{T} = \beta_0\mathbf{I} + \beta_1\mathbf{B} + \beta_{-1}\mathbf{B}^{-1}. \tag{3.34}$$

A first remarkable examples of this situation is obtained considering the follo simple, set of response coefficients

$$\beta_0 = \beta_1 = \rho_0 \mu, \quad \beta_{-1} = 0,$$

where μ is constant. This is a sort of one parameter trivial and natural *genera* finite strain of the law of *linear* elasticity. From the relationships

$$\beta_0 = 2\rho_0 \left(I_2 \frac{\partial W}{\partial I_2} + I_3 \frac{\partial W}{\partial I_3} \right), \quad \beta_1 = 2\rho_0 \frac{\partial W}{\partial I_1}, \quad \beta_{-1} = -2\rho_0 I_3 \frac{\partial W}{\partial I_2},$$

it is easily computed that the *non-polynomial* strain energy-associated with th tion (3.35) is

$$W = \frac{1}{2}\mu(I_1 - 3) + \mu \ln \sqrt{I_3}.$$

The strain energy (3.37) has been already introduced by Simo and Pister Arruda, Grosh, 2001).

To give another remarkable example we consider the simple and comm obtain compressible formulations of hyperelasticity by appending a bulk modu energy term

$$W_H(I_3),$$

to W. The (3.38) term adds to (3.37) a spherical term

$$\mathbf{T}_H = \beta_{H0} \mathbf{I}.$$

If we consider a simple linear bulk-term

$$\beta_{H0} = k I_3$$

we obtain

$$\frac{\partial W_H}{\partial I_3} = \frac{k}{2\rho_0} \frac{I_3 - 1}{I_3},$$

which yields

$$W_H = \frac{k}{\rho_0} \left(\frac{I_3 - 1}{2} - \ln \sqrt{I_3} \right).$$

which is a non-polynomial bulk term used by several investigators.

In the framework of incompressible materials because

$$\mathbf{T} = -p\mathbf{I} + \beta_1 \mathbf{B} + \beta_{-1} \mathbf{B}^{-1},$$

and

$$\beta_1 = 2\partial W / \partial I_1, \quad \beta_{-1} = -2\partial W / \partial I_2$$

polynomial strain-energies give rise to polynomial response coefficients and als verse is true. On the other hand if we consider a more general class of appr functions we see that also in this case approximations of the strain energy response functions may give different results.

In the following we develop a Rivlin-Signorini scheme for incompressible[4] hyperelastic materials by considering rational approximations of the response functions, i.e. approximations of the form

$$\frac{1}{2}\beta_1 = \frac{\sum_{\alpha,\beta} A_{\alpha\beta} J_1^\alpha \widetilde{J}_2^\beta}{\sum_{\delta,\nu} B_{\delta\nu\sigma} J_1^\delta \widetilde{J}_2^\nu}, \quad \frac{1}{2}\beta_{-1} = \frac{\sum_{\alpha,\beta} C_{\alpha\beta\gamma} J_1^\alpha \widetilde{J}_2^\beta}{\sum_{\delta,\nu} D_{\delta\nu\sigma} J_1^\delta \widetilde{J}_2^\nu}. \tag{3.44}$$

We have the trivial relations

$$\frac{1}{2}\beta_1 = \frac{\partial W}{\partial J_1}\frac{dJ_1}{dI_1} + \frac{\partial W}{\partial \widetilde{J}_2}\frac{d\widetilde{J}_2}{dI_1} = \frac{\partial W}{\partial J_1} - \frac{\partial W}{\partial \widetilde{J}_2}, \tag{3.45}$$

and

$$\frac{1}{2}\beta_{-1} = -\left(\frac{\partial W}{\partial J_1}\frac{dJ_1}{dI_2} + \frac{\partial W}{\partial \widetilde{J}_2}\frac{d\widetilde{J}_2}{dI_2}\right) = -\frac{\partial W}{\partial \widetilde{J}_2}. \tag{3.46}$$

When $\beta_{-1} \neq 0$, requiring the material to be hyperelastic means that

$$W = \int \frac{\sum_{\alpha,\beta} C_{\alpha\beta} J_1^\alpha \widetilde{J}_2^\beta}{\sum_{\delta,\nu} D_{\delta\nu} J_1^\delta \widetilde{J}_2^\nu} d\widetilde{J}_2 + \Phi(J_1), \tag{3.47}$$

and introducing

$$\Phi(J_1) = \frac{\sum_\alpha \widehat{A}_\alpha J_1^\alpha}{\sum_\delta \widehat{B}_\delta J_1^\delta}, \tag{3.48}$$

we obtain β_1 by simple differentiation.

On the other hand if $\beta_{-1} = 0$, i.e. we are considering generalized neo-Hookean materials, we have that

$$\frac{1}{2}\beta_1 = \frac{\sum_\alpha A_\alpha J_1^\alpha}{\sum_\delta B_\delta J_1^\delta}, \tag{3.49}$$

and the computation of the corresponding free energy is straightforward.

For example, if we consider

$$\beta_1 = \frac{a_0}{b_0 + b_1 J_1}, \tag{3.50}$$

we obtain an interesting result. Because (3.50) is the same as

$$\beta_1 = \frac{1}{b_0}\frac{a_0}{1 + \frac{b_1 J_1}{b_0}}, \tag{3.51}$$

by a simple integration we get

$$W = \frac{a_0}{b_1}\ln\left(1 + \frac{b_1 J_1}{b_0}\right). \tag{3.52}$$

[4]It is clear that the same kind of approximation may be extended to compressible materials with no difficulties. Here we consider incompressible materials for the sake of the simplicity.

The model (3.52) is equivalent to the Gent model (1996)

$$W = -\frac{\mu}{2} J_m \ln\left(1 - \frac{I_1 - 3}{J_m}\right). \tag{3.53}$$

Another example can be obtained following Pucci and Saccomandi (2002) where the model

$$W(I_1, I_2) = -\frac{\mu}{2} J_m \ln\left(1 - \frac{I_1 - 3}{J_m}\right) + C_2 \ln\left(\frac{I_2}{3}\right), \tag{3.54}$$

has been proposed as a simple modification of the Gent strain energy. The predictions of (3.54) in simple extension are in very good agreement with experimental data. To recover (3.54) we have to consider that

$$W(I_1, I_2) \equiv W(J_1, \widetilde{J}_2) = -\frac{\mu}{2} J_m \ln\left(1 - \frac{J_1}{J_m}\right) + C_2 \ln\left(\frac{\widetilde{J}_2 + J_1 + 3}{3}\right), \tag{3.55}$$

therefore

$$\beta_1 = \frac{\partial W}{\partial J_1} - \frac{\partial W}{\partial \widetilde{J}_2} = \frac{\mu}{1 - \frac{J_1}{J_m}}, \tag{3.56}$$

and

$$\beta_{-1} = -2\frac{\partial W}{\partial \widetilde{J}_2} = \frac{6C_2}{\widetilde{J}_2 + J_1 + 3}. \tag{3.57}$$

Therefore also this strain energy may be easily recovered by the procedure here proposed.

It is important to point out the relationship between (3.53) and molecular constitutive models for rubber based on non-Gaussian statistics that involve the inverse Langevin function. Such molecular models are widely used since they successfully capture the typical strain-hardening at large strains. The strain-energy density in the Gent model depends only on the first invariant I_1 of the Cauchy-Green strain tensor, is a simple logarithmic function of I_1 and involves just two material parameters, the shear modulus μ and a parameter J_m which measures a limiting value for $I_1 - 3$ reflecting limiting chain extensibility. Horgan and Saccomandi (2002b) have shown that the Gent phenomenological model is a very accurate approximation to a molecular based stretch averaged full-network model involving the inverse Langevin function and that the constants μ and J_m in the Gent model are related to microscopic properties of the material. Since the Gent model is remarkably simple, and since analytic closed-form solutions to several benchmark boundary-value problems have been obtained recently on using this model, it is thus an attractive alternative to the comparatively complicated molecular models for incompressible rubber involving the inverse Langevin function. (For a discussion about limiting chain extensibility we refer the reader to Horgan and Saccomandi 2002a).

3.5 Orthogonal stress response terms

Recently a novel constitutive formulation has been developed for finitely hyperelastic materials that exhibit isotropic or laminar response by Criscione and others. Here we restrict our attention only to the isotropic case. In Criscione et al. (2000) and (2002) a special choice of invariants is preferred. These invariants are directly related to the natural or Hencky strain

$$\mathbf{e} = \ln \mathbf{V}, \tag{3.58}$$

and for this reason they introduce naturally some log terms. The strain measure (3.58) separates in an additively manner the dilatation from distortion. This because

$$\mathbf{V} = \left(J^{1/3}\mathbf{I}\right)\left(J^{-1/3}\mathbf{V}\right) \tag{3.59}$$

and therefore

$$\ln \mathbf{V} = \ln\left(J^{1/3}\mathbf{I}\right) + \ln\left(J^{-1/3}\mathbf{V}\right). \tag{3.60}$$

The invariants considered in Criscione et al, labelled by K_1, K_2, and K_3, are defined as

$$\begin{aligned} K_1 &= \operatorname{tr}\mathbf{e} = \ln(J), \\ K_2 &= \sqrt{\ln\left(J^{-1/3}\mathbf{V}\right) : \ln\left(J^{-1/3}\mathbf{V}\right)}, \\ K_3 &= 3\sqrt{6}\det\left(K_2^{-1}\ln\left(J^{-1/3}\mathbf{V}\right)\right). \end{aligned} \tag{3.61}$$

Using the principal stretches these invariants may be represented as

$$\begin{aligned} K_1 &= \ln\left(\lambda_1\lambda_2\lambda_3\right), \\ K_2 &= \sqrt{\left(\ln\lambda_1 - \tfrac{1}{3}K_1\right)^2 + \left(\ln\lambda_2 - \tfrac{1}{3}K_1\right)^2 + \left(\ln\lambda_3 - \tfrac{1}{3}K_1\right)^2}, \\ K_3 &= 3\sqrt{6}\left(\tfrac{\ln\lambda_1 - \frac{1}{3}K_1}{K_2}\right)\left(\tfrac{\ln\lambda_2 - \frac{1}{3}K_1}{K_2}\right)\left(\tfrac{\ln\lambda_3 - \frac{1}{3}K_1}{K_2}\right). \end{aligned} \tag{3.62}$$

In the case of incompressible materials $K_1 = 0$. The interesting geometrical properties of these invariants are described into details in the paper by Criscione et al. (2000, 2002). In any case we note that

$$\operatorname{tr}\left(\frac{\ln\left(J^{-1/3}\mathbf{V}\right)}{K_2}\right) = 0, \quad \operatorname{tr}\left(\frac{\ln\left(J^{-1/3}\mathbf{V}\right)\ln\left(J^{-1/3}\mathbf{V}\right)}{K_2^2}\right) = 1. \tag{3.63}$$

Using these invariants we have that in the case of incompressible materials[5]

$$\mathbf{T} = -p\mathbf{I} + \frac{1}{K_2}\frac{\partial W}{\partial K_2}\ln\mathbf{V} + \frac{1}{K_2}\frac{\partial W}{\partial K_3}\mathbf{Y}, \tag{3.64}$$

where

$$\mathbf{Y} = 3\sqrt{6}\frac{\ln\left(\mathbf{V}\right)^2}{K_2^2} - \sqrt{6}\mathbf{I} - 3K_3\frac{\ln\left(\mathbf{V}\right)}{K_2}. \tag{3.65}$$

[5]It is interesting to note that $\partial W/\partial K_2 \to 0$ and $\partial W/\partial K_3 \to 0$ as $K_2 \to 0$.

The important feature of this special choice of invariants is that in this case

$$\frac{\ln\left(J^{-1/3}\mathbf{V}\right)}{K_2}:\mathbf{I}=\mathbf{0},\quad \mathbf{Y}:\mathbf{I}=\mathbf{0},\quad \frac{\ln\left(J^{-1/3}\mathbf{V}\right)}{K_2}:\mathbf{Y}=\mathbf{0}. \tag{3.66}$$

This orthogonality condition is the main point of the choice by Criscione et al. Equations (3.66) are valid also in the case of compressible materials where

$$\mathbf{Y}=3\sqrt{6}\ln\left(J^{-1/3}\mathbf{V}\right)^2-\sqrt{6}\mathbf{I}-3K_3\ln\left(J^{-1/3}\mathbf{V}\right).$$

If we consider the ratio of correlation among the various tensor that appears in the representation formula it is well known from elementary statistics that the higher the correlation the higher is the propagation of the error. Because of the orthogonality condition (3.66) the use of the invariants K_1, K_2 and K_3 minimizes this correlation, whereas if we use the principal invariants this correlation is maximum. Therefore, to avoid unnecessary errors (especially in the low strain zone) the invariants K_1, K_2 and K_3 are a smart choice. This choice is smart and it explains clearly the experimental difficulties encountered by many authors as in Valanis and Landel (1967). We point out, that because we are in a finite dimensional space all choices of a set of invariants are theoretically equivalent. This means that modulo a suitable transformation is always possible to avoid the difficulties encountered in managing the experimental and this using any set of invariants. In any case it is true that the minimum property of the ratio of correlation using K_1, K_2 and K_3 is obtained without any artificial transformation of the experimental data

3.6 Modelling uniaxial extension for incompressible materials

We come back to the deformation of uniaxial (or simple) extension (and compression). In this case

$$\lambda_1=\lambda,\qquad \lambda_2=\lambda_3=\frac{1}{\sqrt{\lambda}}. \tag{3.67}$$

Requiring $t_2=t_3=0$ because the lateral surfaces are traction-free, we find

$$p=2\left(\frac{1}{\lambda}\frac{\partial W}{\partial I_1}-\lambda\frac{\partial W}{\partial I_2}\right)$$

and the tensile force per unit of unstrained cross-sectional area, f, necessary to obtain the extension ratio, λ, is

$$f=\frac{t_1}{\lambda}=2\left(\lambda-\frac{1}{\lambda^2}\right)\left(\frac{\partial W}{\partial I_1}+\frac{1}{\lambda}\frac{\partial W}{\partial I_2}\right). \tag{3.68}$$

The standard requirements on (3.68) are that $f(1)=0$, $f(\lambda)>0$ when $\lambda>1$ and $f(\lambda)<0$ when $\lambda<1$.

Related to (3.68) we have the *reduced tensile force*

$$[f](z)=\frac{\partial W}{\partial I_1}+z\frac{\partial W}{\partial I_2}, \tag{3.69}$$

where $z = 1/\lambda$. Equation (3.69) is obtained from (3.68) using the universal geometrical factor $2\left(\lambda - 1/\lambda^2\right)$ as a scale factor i.e.

$$[f] \equiv \frac{f}{2\left(\lambda - 1/\lambda^2\right)}.$$

The plot of (3.69), usually named the *Mooney plot*, is widely used in the experimental literature to compare the different models, because it is very sensitive to relative errors (Han, Horkay and McKenna, 1999).

For Mooney-Rivlin materials $[f]\,(z)$ is a linear function, and for any polynomial strain-energy density function the reduced tensile force is not singular in the range of stretch of interest for the uniaxial extension (i.e. $z \in (0,1)$). For the Gent Material we have

$$[f]\,(z) = \frac{\mu}{2}\frac{J_m}{J_m - \left(z^{-2} + 2z - 3\right)}, \tag{3.70}$$

then

$$\frac{d\,[f]}{dz} = \frac{\mu J_m(z^3 - 1)}{[2z^3 - (J_m + 3)z^2 + 1]^2},$$

for $z \in (1/\lambda_{\max}, 1)$, the tensile force associated with the Gent model is a monotone decreasing function with a vertical asymptote for $z \to 1/\lambda_{\max}$.

On observing the classical data of Treloar for the simple extension experiment and roughly speaking, it is possible to say that these experimental data lie in a monotonic curve with two typical features. First of all around $\lambda \approx 4$ the presence of an inflexion point it is evident: the beginning of the peculiar strain hardening zone. The second important feature is that in the hardening zone the tensile force seems to grow very fast. The Mooney plot of the uniaxial data is must be calculated from the data for the tensile force by scaling with a factor which goes to zero for $\lambda \to 1$. This means that the use of the reduced tensile force will correspond to a magnification of any error around the undeformed state.

From the Mooney plot it is clear that the neo-Hookean model is not suitable for describing the experimental data, either from a *qualitative* or from a *quantitative* point of view. If we wish always to consider a strain-energy $W = W(I_1 - 3)$ the relation (3.68) shows that the tensile force is the product of two factors. The universal factor $\partial I_1/\partial\lambda$ and the response coefficient $\partial W/\partial I_1$. The universal factor is a monotone increasing function in $[1, +\infty)$. When $W(I_1 - 3)$ the second factor is exactly the reduced tensile force. Using a higher order polynomial strain-energy as the one proposed by Yeoh and Fleming (1997), i.e.

$$W = \sum_{i=1}^{3} C_{1,i}\left(I_1 - 3\right)^i, \tag{3.71}$$

a better agreement with experimental data is obtained, at least at first sight, indeed the reasons for this better agreement are not only the different functional form considered, but also the larger number of constitutive parameters introduced. For this reason we have to change the functional class used for the approximation. We have already explained that functions with singularities are a principal target of rational approximations. For this reason it is interesting to explore what happens if we consider the simplest rational approximation for the reduced tensile force. This simplest rational form is given exactly by the Gent model. The fitting of the Gent model with uniaxial data is very good for large values of the strain, but poor for small values of the extension ratio. The reason for this is that these models cannot represent the upturn point. The upturn point is a minimum point in the range $z \in (1/\lambda_{\max}, 1)$ for the reduced tensile force and we have shown that $[f](z)$ for the Gent material in this case is monotone. Thus, for the Gent model, it is impossible to obtain good predictions, both from a qualitative and a quantitative point of view, over all the range of the extension ratios.

It is well known that there are some strong objections to using a strain-energy based only on the first invariant (at least when we are considering general deformations). These objections date back to the classical experimental work of Rivlin (Pucci e Saccomandi, 2002), and a quantitative way to measure the departure of experimental data from the theoretical predictions of generalized neo-Hookean models has been given in the analysis of the torsion deformation

For this reason it is worthwhile considering also the dependence on I_2. In so doing the main problem is that it is no longer possible to recover the general form of the strain-energy from the tensile force associated with a simple tension experiment. This is because we cannot have a one-to-one correspondence between the tensile force, which is an expression containing only the principal extension λ, and the strain-energy function which depends on I_1 and I_2. To overcome this problem we consider the assumption of additive separability, a hypothesis which obviously is strictly related to our chosen representation for the strain energy. For example, the celebrated *Rivlin-Saunders* model

$$W(I_1-3, I_2-3) = \widehat{W}(I_1-3) + F(I_2-3), \tag{3.72}$$

where $\widehat{W}$ and F are an arbitrary functions, is a separable model.

A first trial is to generalize the Gent model by considering a term linear in the second invariant, i.e., $F = C_2(I_2-3)$, where C_2 is a constitutive parameter

$$W = -\frac{\mu}{2} J_m \ln\left(1 - \frac{I_1-3}{J_m}\right) + C_2\left(I_2-3\right). \tag{3.73}$$

In this case the reduced tensile force is given by

$$[f] = \frac{\mu}{2}\frac{J_m}{J_m - I_1 + 3} + \frac{C_2}{\lambda}. \tag{3.74}$$

Now from the empirical inequalities we have

$$\mu > 0, \quad C_2 > 0$$

and so $f(\lambda) \geq 0$ if $\lambda \geq 1$.

Rewriting (3.74) as

$$[f](z) = -\frac{\mu}{2}\frac{(\lambda_{\max}^3 + 2 - 3\lambda_{\max})z^2}{(\lambda_{\max}^3 + 2)z^2 - \lambda_{\max}(1+2z^3)} + C_2 z \tag{3.75}$$

it is possible to check that

$$\lim_{z\to 1/\lambda_{\max}} \frac{d[f]}{dz} = -\infty, \qquad \lim_{z\to 1} \frac{d[f]}{dz} = C_2. \tag{3.76}$$

Then, when $C_2 > 0$ at least a minimum point for (3.75) is in the good range and the upturn can be modelled by (3.74).

From the mathematical point of view first we have to require that the $F(I_2)$ term introduces the upturn, then, because the fit to the Treloar data obtained with the Gent model is very good in the hardening zone the contribution of this term for large values of λ must be not significant. This second requirement suggests that we consider the functional form

$$F(I_2) \equiv \widehat{C}_2 \ln\left(\frac{I_2}{3}\right),$$

first proposed by Gent and Thomas (1958). Indeed from the model

$$W = -\frac{\mu}{2} J_m \ln\left(1 - \frac{I_1 - 3}{J_m}\right) + \widehat{C}_2 \ln\left(\frac{I_2}{3}\right), \tag{3.77}$$

we obtain the following reduced tensile force

$$[f](z) = \frac{\mu}{2} \frac{(\lambda_{\max}^3 + 2 - 3\lambda_{\max})z^2}{(\lambda_{\max}^3 + 2)z^2 - \lambda_{\max}(1 + 2z^3)} + \widehat{C}_2 \frac{z^2}{z^3 + 2}, \tag{3.78}$$

which is a $[5, 6]$ rational function. In this form the term multiplying $\widehat{C}_2$ is the functional form with the slowest growth possible for $z \to 0$.

The empirical inequalities for the model (3.77), when $J_m > 0$, lead to the very simple restrictions

$$\mu > 0, \quad \widehat{C}_2 \geq 0,$$

and it is possible to show the existence of the upturn point because

$$\lim_{z\to 1/\lambda_{\max}} \frac{d[f]}{dz} = -\infty, \qquad \lim_{z\to 1} \frac{d[f]}{dz} = \frac{\widehat{C}_2}{3}.$$

Equation (3.78) leads to a very good qualitative and quantitative agreement with the experimental data for all the range of deformations in the uniaxial experiment. The relative errors in the extension range are always under 5%. Obviously, here, our claims are restricted to a deformation of uniaxial tension.

4 A boundary value problem

The aim of this Section is to give a non technical discussion about the issue of existence and uniqueness in nonlinear elasticity. We are trying to give this account in a simple way and to this end we use the antiplane shear class of deformations defined by

$$x_1 = X_1, \quad x_2 = X_2, \quad x_3 = X_3 + u(X_1, X_2). \tag{4.1}$$

Here $\mathbf{X}$ denote the reference coordinates, $\mathbf{x}$ the current coordinates of the body and the displacement $u(X_1, X_2)$ is a twice continuously differentiable function.

To have an idea of what happens in the general framework of three dimensional deformations we need mathematical tools too much sophisticated for our lectures notes, therefore in this Section we have the restricted goal to illustrate the main ingredient of the classical theory of existence (i.e *convexity* and *growth* conditions) for a special and simple boundary value problem. A more general and complete, but also mathematical more oriented, survey of these topics may be found, for example, in the booklet by Pedregal (2000) and a detailed survey of the role of antiplane shear in linear and nonlinear elasticity may be found in (Horgan, 1995).

Coming back to (4.1) we have that the balance equations for an isotropic incompressible elastic material, in this case, reduce the following system of three overdetermined system of three differential equations in the two unknowns $p(X_1, X_2, X_3)$ and $u(X_1, X_2)$

$$p_{,3} - \left[2\left(W_1 + W_2\right) u_{,\beta}\right]_{,\beta} = 0, \tag{4.2}$$

$$p_{,\alpha} - \left[2W_1 + (2+k^2)W_2\right]_{,\alpha} - \left[2u_{,\alpha}u_{,\beta}W_2\right]_{,\beta} = 0, \tag{4.3}$$

where the Greek subscripts runs from 1 to 2,

$$W_i = \left.\frac{\partial W}{\partial I_i}\right|_{I_1=I_2=3+k^2},$$

and the amount of shear, k, is defined as

$$k^2 = \left|\nabla u\right|^2. \tag{4.4}$$

Usually the overdetermined system (4.2) and (4.3) does not allow a solution for every choice of the strain-energy density function. Only for special deformations the *formal* solution of this system is possible for any regular strain energy and only for special classes of materials this system is compatible for any antiplane shear deformation. For example, in the case of a rectilinear shear deformation

$$x_1 = X_1, \quad x_2 = X_2, \quad z_1 = Z_1 + f(X_1). \tag{4.5}$$

the system (4.2) and (4.3) reduces to

$$\left[2\left(W_1 + W_2\right) f'\right]' - p_{,3} = 0,$$

$$p_{,1} - \left[2W_1 + (2+k^2)W_2\right]' - 2\left[f'^2 W_2\right]' = 0,$$

$$p_{,2} = 0,$$

where the amount of shear is given by

$$k^2 = f'^2.$$

In this case we have that

$$p(X_1, X_3) = \left[2W_1 + 2(1+f'^2)W_2\right]' + \pi X_3, \tag{4.6}$$

and

$$[2(W_1+W_2)f']' = \pi. \tag{4.7}$$

Here π it is the constant gradient of pressure in the X_3–direction. In this case a formal solution of the balance equation is possible for any smooth strain-energy density function. A similar situation occurs if we are considering an axial shear deformation, i.e. if $u(X_1, X_2) = w(X_1^2 + X_2^2)$.

On the other hand if, for example, we consider generalized neo-Hookean materials such that $W = W(I_1)$ the overdetermined system (4.2) and (4.3) reduces to

$$p_{,3} - [2W_1 u_{,\beta}]_{,\beta} = 0, \tag{4.8}$$

and

$$p_{,\alpha} - [2W_1]_{,\alpha} = 0. \tag{4.9}$$

Therefore, by defining

$$p(X_1, X_2, X_3) = 2W_1 + \pi X_3, \tag{4.10}$$

(4.8) and (4.9) reduce to a single quasilinear second-order partial differential equation

$$[M(k)u_{,\beta}]_{,\beta} = \pi, \tag{4.11}$$

where $M(k) = [2W_1]_{I_1=3+k^2}$ is the secant modulus of shear, and then a formal solution of the balance equations is possible.

A typical cartoon of a general antiplane shear deformation may be obtained by considering an infinite cylindrical tube[6] loaded axially on the external mantle (i.e. the external traction has the same direction of the cylinder generatrix). There is no mechanical reasons because such deformation cannot be possible in the real world, but the mathematics *seems* to tell us that only for special materials this deformation is possible. From a mathematical point of view the fact that a *pure* antiplane shear deformation cannot be sustained in an elastic material means that the overdetermined differential system (4.2) and (4.3), corresponding to the strain energy we are using to model the *real* material do not have common solutions. Therefore mathematics says that the geometry and load condition of the problem do not allow a **pure** antiplane shear deformation but, obviously, it may be possible a pure antiplane shear deformation coupled to a *secondary* (for example in the plane of the section) deformation. It is possible that this secondary deformation is more difficult to be appreciated (for example in an experiment) than the antiplane one but it must be taken into account to satisfy the balance equations. Because, it is clear that in real world situations it is impossible that a mathematical condition is exactly fulfilled, the special models that allow for a pure antiplane state are only strong idealizations and there are good reasons to believe that secondary deformations are always present and the true point is to know if they are or are not negligible. (Here we have another striking example of the fundamental importance of universal results in nonlinear elasticity).

To the best of my knowledge only the case of the axial shear in the annular region between two eccentrically placed cylinders (Mollica and Rajagopal, 1997) or the case

[6]The case of circular tube is trivial.

of antiplane shear deformation of cylinders with non circular cross section (Fosidck and Kao, 1978) has been studied into details (for a discussion on the coupling of an antiplane and a plane deformation see Horgan and Saccomandi 2004). In a cartesian co-ordinate system (X, Y, Z) the equations of the two eccentrically placed cylinders whose radii are R_1 and R_2 $(R_1 < R_2)$, are

$$X^2 + Y^2 = R_2^2, \quad (X - |\mathbf{O}_1\mathbf{O}_2|)^2 + Y^2 = R_1^2,$$

or introducing the eccentricity $\varepsilon = |\mathbf{O}_1\mathbf{O}_2| / R_1$ and using cylindrical coordinates (R, Θ, Z)

$$R^2 = R_2^2, \quad R^2 - 2RR_1\varepsilon\cos\Theta - R_1^2(1 - \varepsilon^2) = 0.$$

When $\varepsilon = 0$, the problem have axial symmetry, therefore the antiplane shear problem may be solved considering $u(X_1, X_2) = w(X_1^2 + X_2^2)$. In this case the equation of balance reduces to a single ordinary differential equation and this for any incompressible isotropic elastic material. If $\varepsilon \neq 0$, the equation of balance for antiplane shear reduce to an overdetermined system of partial differential equations, which is compatible only in special cases. In this case when a general material is considered we have to couple a *in-plane* deformation to the antiplane one, i.e. using polar coordinates (r, θ, z) also in the actual configuration,

$$r = R + v(R, \Theta), \quad \theta = \Theta + w(R, \Theta), \quad z = Z + g(R, \Theta).$$

Now for every incompressible elastic material the balance equations reduce to a determined system of partial differential equations. For $\varepsilon \ll 1$ some preliminary information may be obtained by considering a perturbation scheme

$$v(R, \Theta) = \varepsilon v_1(R, \Theta) + O(\varepsilon^2), \quad w(R, \Theta) = \varepsilon w_1(R, \Theta) + O(\varepsilon^2),$$

and

$$g(R, \Theta) = f(R) + \varepsilon g_1(R, \Theta) + O(\varepsilon^2).$$

In a such a way when $\varepsilon = 0$, we have that the solution is exactly the classical axial shear deformation. The evaluation of the in-plane deformation and stresses is a question of technical relevance in many technical applications as for example in the design of bush mountings.

4.1 Rectilinear shear deformation

The rectilinear shear deformation is a simple good one to illustrate by using the semi-inverse approach, the solution to simple boundary value problems. A standard boundary problem in this case is given by a slab bonded at both $X_1 = 0$ and $X_1 = H$ to rigid surfaces. The rigid surfaces may be fixed; in this case the shearing is due to the existence of a pressure gradient generated by applied traction at $X_3 = \pm\infty$ and π must be different from zero.

Another opportunity is that a rigid surface, say the one at $X_1 = H$, may be translated by a certain fixed amount in the X_1-direction (now π may also be null). Modulo a rigid displacement the surface $X_1 = 0$ may be, without lack of generality, fixed.

In the case that both surfaces at $X_1 = 0$ and $X_1 = H$ are fixed the appropriate boundary conditions are

$$f(0) = 0, \quad f(H) = 0. \tag{4.12}$$

Let $X = X_1/H$, $\overline{f} = f/H$, $\overline{W} = W/\mu$ and integrating (4.7) the BVP we are considering is given by

$$2\left(\overline{W}_1 + \overline{W}_2\right)\frac{d\overline{f}}{dX} = \frac{\pi H}{\mu}\left(X + C_1\right), \tag{4.13}$$

and

$$\overline{f}(0) = 0, \quad \overline{f}(1) = 0. \tag{4.14}$$

If the dimensionless secant modulus of shear $\overline{M}(\overline{k}) = M(k/H)/\mu$ is assumed to be positive for all $k > 0$, $df(X)/dX$ is zero only for $X = -C_1$. If we are searching for regular solutions of this BVP this means that $-C_1 \in (0,1)$. We remember that $\overline{M}(\overline{k})$ is an even function around $X = -C_1$ and by squaring (4.13) we see that $(df(X)/dX)^2$ also is even. This means that $d\overline{f}(X)/dX$ is odd and $\overline{f}(X)$ is even. Therefore, $C_1 = -1/2$ and the original BVP has been translated into a standard IVP. Therefore, our problem is reduce to the solution of the equation

$$2\left(\overline{W}_1 + \overline{W}_2\right)\frac{d\overline{f}}{dX} = \frac{\pi H}{\mu}\left(X - \frac{1}{2}\right), \tag{4.15}$$

subject to

$$\overline{f}(0) = 0, \tag{4.16}$$

in the range $(0, 1/2)$. The solution in the range $(1/2, 1)$ is then obtained by symmetry. Equation (4.15) is not in normal form and therefore standard existence theorems for ordinary differential equations cannot be applied. The conditions to recast (4.15) in normal form can be obtained by considering

$$X\left(\overline{k}\right) \equiv \frac{1}{\widetilde{\pi}}\left[2\left(\overline{W}_1 + \overline{W}_2\right)\overline{k}\right] + \frac{1}{2}, \tag{4.17}$$

where $\widetilde{\pi} = (\pi H)/\mu$. We have to show that the codomain of the $X\left(\overline{k}\right)$ contains the $(0,1)$ interval. This is ensured if in $(0, 1/2)$ we have that $X(\overline{k})$ is a monotone function and

$$\frac{d}{d\overline{k}}\left(\left(\overline{W}_1 + \overline{W}_2\right)\overline{k}\right) > 0, \tag{4.18}$$

and

$$\lim_{\overline{k}\to\infty}\left(\left(\overline{W}_1 + \overline{W}_2\right)\overline{k}\right) = \infty. \tag{4.19}$$

This very simple framework reveals that the essential ingredients for the existence of solutions in nonlinear elasticity are the monotonicity properties, i.e. (4.18), and coerciveness properties, i.e. (4.19). The uniqueness of solutions is related to convexity. Indeed if we consider the equation

$$\Xi \equiv 2\left(\overline{W}_1 + \overline{W}_2\right)\overline{k}\Big|_{X=0} = -\frac{\widetilde{\pi}}{2}, \tag{4.20}$$

from which we determine the initial value for the derivative of the shear strain, uniqueness fails when this equation admits more than one solution and this may be the case when Ξ is non-convex, an opportunity that may have a physical interpretation in the framework of phase-transitions (Aberayatne and Knowles, 2001).

We point out that if the IVP $\{(4.15)\cup(4.16)\}$ is suitable for theoretical investigation, but for numerical applications the best strategy is to consider the second order normal form equation

$$\left(\overline{W}_1+\overline{W}_2\right)\frac{d^2\overline{f}}{dX^2}+2\left(\overline{W}_{11}+2\overline{W}_{12}+\overline{W}_{22}\right)\left(\frac{d\overline{f}}{dX}\right)^2=\frac{\pi H}{2\mu},$$

subject to the initial conditions

$$\overline{f}\big|_{X=0}=0,\quad \left.\frac{d\overline{f}}{dX}\right|_{X=0}=f^*,$$

where f^* is the solution of the equation (4.20).

4.2 The power law material

Let us consider the case of generalized power-law material (3.21) therefore the (4.15) is rewritten as

$$\left(1+\frac{b}{n}\left(\frac{d\overline{f}}{dX}\right)^2\right)^{n-1}\frac{d\overline{f}}{dX}=\widetilde{\pi}\left(X-\frac{1}{2}\right). \tag{4.21}$$

Exact solutions for this equation are possible for some values of n. In the table I we are considering some of the values of n for which is possible to compute explicitly the positive branch of (4.21). The BVP problem here considered has been already studied numerically by Zhang and Rajagopal (1992), but our qualitative analysis is original.

TABLE I

$n=1/2$	$\left(\frac{d\overline{f}}{dX}\right)^2=-\frac{\widetilde{\pi}^2(2X-1)^2}{2\left(\widetilde{\pi}^2(2X-1)^2-2\right)},$
$n=3/4$	$\left(\frac{d\overline{f}}{dX}\right)^2=\frac{\widetilde{\pi}^2(1-2X)}{24}\left(\begin{array}{c}(\widetilde{\pi}^4(16X^4-32X^3+24X^2-8X+1)^2\\ +36)^{1/2}+\widetilde{\pi}^2(2X-1)^2\end{array}\right),$
$n=1$	$\frac{d\overline{f}}{dX}=\widetilde{\pi}\left(X-\frac{1}{2}\right),$
$n=3/2$	$\left(\frac{d\overline{f}}{dX}\right)^2=\frac{\sqrt{3}}{4}\left(\sqrt{2a^2(2x-1)^2+3}-3\right).$

We point out that

$$\overline{W}_1\overline{k} = \left(1 + \frac{\overline{k}^2}{n}\right)^{n-1} \overline{k}$$

(from now on we set $b = 1$) and

$$\frac{d\left(\overline{kM}\right)}{d\overline{k}} = \left[\frac{2n-1}{n}\overline{k}^2 + 1\right]\left(1 + \frac{b}{n}\overline{k}^2\right)^{n-2}$$

Therefore for $n < 1/2$ we have not the required monotony property required and in the case $n = 1/2$ we have not the required coercivity property. Therefore in this case ($n = 1/2$) the solution does not exists for any value of $\widetilde{\pi}$. This is clear from the first example of table *I*. In this case existence implies

$$\widetilde{\pi}^2 (2X - 1)^2 - 2 < 0$$

for all $X \in (0, 1/2)$ therefore a necessary condition is

$$\widetilde{\pi}^2 < 2.$$

It is very interesting to point out the value of the derivatives at zero for different values of n (see Table *II*)

TABLE II

$n = 1/2$	$\left(\left.\frac{d\overline{f}}{dX}\right\|_{X=0}\right)^2 = \frac{\widetilde{\pi}^2}{2(2-\widetilde{\pi}^2)}$,
$n = 3/4$	$\left(\left.\frac{d\overline{f}}{dX}\right\|_{X=0}\right)^2 = \frac{\widetilde{\pi}^2}{24}\left(\sqrt{\widetilde{\pi}^4 + 36} + \widetilde{\pi}^2\right)$,
$n = 1$	$\left(\left.\frac{d\overline{f}}{dX}\right\|_{X=0}\right)^2 = \frac{\widetilde{\pi}^2}{4}$,
$n = 3/2$	$\left(\left.\frac{d\overline{f}}{dX}\right\|_{X=0}\right)^2 = \frac{\sqrt{3}}{4}\left(\sqrt{2\widetilde{\pi}^2 + 3} - 3\right)$.

We point out once again that in the case $n = 1/2$ the derivative in $X = 0$ blows up as $\widetilde{\pi}^2 \to 2$. In the other cases we have a very different behavior if n is greater or smaller than 1. To point out this difference, from (4.21), we consider the function

$$\widetilde{\pi}(y, n) = -2\left(1 + \frac{y^2}{n}\right)^{n-1} y, \tag{4.22}$$

in the domain $(y, n) \in (-\infty, \infty) \times \left(\frac{1}{2}, \infty\right)$.

The basic properties of (4.22) are

$$\widetilde{\pi}(0,n)=0, \quad \widetilde{\pi}(y,n)=-\widetilde{\pi}(-y,n), \quad \frac{d\widetilde{\pi}}{dy}<0, \quad \lim_{y\to\pm\infty}\widetilde{\pi}(y,n)=\mp\infty$$

and

$$\frac{d^2\widetilde{\pi}}{dy^2}=\begin{cases}\text{positive} & \text{if} \quad \frac{1}{2}<n<1\\ 0 & \text{if} \quad n=1\\ \text{negative} & \text{if} \quad 1<n<\infty.\end{cases}$$

Therefore when $y>0$ the (4.22) is a convex function if $1/2<n<1$ and a concave function otherwise. Moreover, the order of $\widetilde{\pi}(y,n)$ as $y\to\infty$ is $2n-1$ i.e.

$$\widetilde{\pi}\sim y^{2n-1}.$$

for large y. Collecting these information we derive for the inverse function $y=y(\widetilde{\pi},n)$ the following properties

$$y(0,n)=0, \quad y(\widetilde{\pi},n)=y(-\widetilde{\pi},n), \quad \frac{dy}{d\widetilde{\pi}}>0, \quad \lim_{y\to\pm\infty}\widetilde{\pi}(y,n)=\infty,$$

and for $\widetilde{\pi}\to\pm\infty$ we have that y approach ∞ faster and faster as $n\to 1/2$.

Fixed a value $\widetilde{\pi}$, it is possible to solve (4.22) for y and compute the value of the first derivative $d\overline{f}/dX\big|_{X=0}$. It is clear that the solution of (4.22) blows up as $n\to 1/2$.

To have a more detailed picture of this situation let us consider, using elementary asymptotic analysis (Murray, 1984), what happens as $\widetilde{\pi}\to\infty$ or $\varepsilon=\widetilde{\pi}^{-1}\to 0$ near $X=1/2$. For $b=1$, introducing the new variables

$$X=\xi+1/2, \ n=m+1 \text{ and } \overline{f}'^2=d\overline{f}/d\xi,$$

we rewrite (4.21) as

$$\varepsilon\left(1+\frac{1}{n}\overline{f}'^2\right)^m\overline{f}'=\xi.$$

If $m>0$ $(n>1)$ rescaling by τ the spatial variable to zoom what happens when $\xi\to 0$, we have

$$\varepsilon\left(\tau^2+\frac{1}{n}\overline{f}'^2\right)^m\overline{f}'=\tau^{2+2m}\xi,$$

and considering $\varepsilon\sim\tau^{2+2m}$ and $\overline{f}=f_0+\tau f_1+...$ we find

$$\frac{1}{n^m}f_0'^{2m+1}=\xi.$$

If, for example, $2m\in\mathbb{N}^+$ the solution of the last equation at first order is given by

$$f_0=\frac{2n-1}{2}\frac{n^{n^{\frac{1}{2n-1}}}}{n^{\frac{2n}{2n-1}}}\xi^{\frac{2n}{2n-1}}.$$

On the other hand if $m < 1$ $(n < 1)$

$$\varepsilon \overline{f}' = \tau^{2-2|m|}\xi\left(\tau^2 + \frac{1}{n}\overline{f}'^2\right)^{|m|},$$

balancing the LHS and the RHS of this equation and considering $\overline{f} = f_0 + \tau f_1 + ...$ we find the solution at first order to be constant: f_0 =constant.

This simple analysis reveals in a direct way the possibility of localization of the solution near $X = 1/2$ if n is large enough, and of the localization of the solution at the external boundary if n is near to $1/2$ (please note that for $n = 1/2$ small ε is meaningless because we have not existence of the solution for large $\widetilde{\pi}$).

To have a quantitative idea of the thickness of boundary layer structure we consider the limiting case $n = 1/2$ and the numerical value $\widetilde{\pi}^2 = 1.8$. In this case we have

$$\left(\frac{d\overline{f}}{dX}\right)^2 = -\frac{(1.8)(2X-1)^2}{2\left((1.8)(2X-1)^2 - 2\right)}$$

Let us introduce the following global measure of the variation of strain in the interval $(0, \alpha)$

$$\Gamma(\alpha) = \int_0^\alpha \left(\frac{d\overline{f}}{dX}\right)^2 dX,$$

$(0 < \alpha \leq 1/2)$. This is a natural measure equivalent to the one introduced in (Chen and Rajagopal, 2001).

By a simple computation it is easy to check that

$$\Gamma(1/2) = 0.2292, \quad \Gamma(0.05) = 0.11951 \quad \Gamma(0.1) = 0.16733.$$

Therefore in the subinterval $(0, 0.05)$ (the 10% of the interval $(0, 1/2)$) we record the 52% of the total variation of the strain and in the subinterval $(0, 0.1)$ (the 20% of the whole interval) we record the 73% of the total variation.

The simple boundary value problem here analyzed not only allows a clear introduction of some typical features of nonlinear elasticity but may be used as a benchmark for numerical methods. The presence of boundary layers (or localization) is a very good test for boundary value methods because we generate multiscale problems (for more details we refer to Horgan, Saccomandi, Sgura, 2002).

5 Concluding remarks

In this self-contained survey of hyperelasticity I have searched to give an overview of useful theoretical tools for the mathematical study of rubber-like mechanics. From the pure mathematical point of view there are still many outstanding open problems especially with respect the questions of existence and uniqueness of the solutions (see for example Ball, 2002). On the other hand, from the point of view of theoretical mechanics we still do not know how to choose in some *rational* way the specific form of the strain-energy function. This is an essential step in any industrial application of phenomenological models. In these lecture notes we have tried to point out some general idea about this problem, but we are far away to be sure that the implementation of the methods here presented will generate a family of strain energies that allows to have accurate models for any deformation field.

These open problems may have heavy drawbacks in applications and therefore they are not only of academic interest. Indeed, numerical codes to investigate complex deformation fields may not be more accurate than the strain-energy forms they implement. If we are not aware of the mathematical problems we may encounter in the solution procedure, the numbers we compute may be only a non-sense list.

For these reasons, my opinion is to concentrate our attention to specific simple problems that allow to obtain exact solutions or at least a detailed qualitative theory. These problems may be used not only to have more insights into the physics, but also to have a control of the accuracy of complex numerical solutions. In some sense with this program we are trying to follow the advice of Adhémar-Jean Claude Barré de Saint-Venant (Benvenuto, 1998) i.e. *connaître la solution exacte d'un problème fort rapproché à celui qui est proposé* indeed only in this way (always quoting Barré de Saint-Venant) *l'usage des mathématiques cessera de s'attirer des reproches* and mathematics will be useful to understand rubber-like mechanics in real world applications.

ACKNOWLEDGEMENTS
This research was partially supported by GNFM of Italian INDAM and by ex-60% research grant of the University of Lecce.

Aberayatne, R. and Knowles, J. (2001) in: Y. Fu, R. Ogden (Eds.), *Non-linear Elasticity*, Cambridge University Press, Cambridge.

Arruda, E. M. and Boyce, M. C. (1993) A three dimensional constitutive model for the large deformation stretch behavior of rubber elastic materials, *J. Mech. Phys. Solids* **41**, 389-412.

Ball, J. M. (2003) Some open problems in elasticity, in: P. Newton, P. Holmes, A. Weinstein (Eds.) *Geometry, Mechanics and Dynamics*, Springer, New-York.

Benvenuto, E. (1998) A. J. C. Barrè de Saint-Venant: the man, the scientist, the engineer, in: *Il problema di de Saint-Venant: aspetti teorici ad applicativi*, Atti dei Convegni Lincei 140, Accademia dei Lincei, Roma.

Bischoff, E. J., Arruda, E. M. and Grosh, K. (2001) A new constitutive model for the compressibility of elastomers at finite deformations, *Rubber Chemistry and Technology* **74**, 541-559.

Boyce, M. C. and Arruda, E. M. (2000) Constitutive models of rubber elasticity: a review, *Rubber Chemistry and Technology* **73**, 504-523.

Chen, Y. C. and Rajagopal, K. R.(2001) Boundary layer solutions in elastic solids, *J. of Elasticity* **62**, 203-216.

Coleman, B. D. and Noll, W. (1963) The thermodynamics of elastic materials with heat conduction and viscosity, *Arch. Rational Mech. Anal.* **13**, 245-261.

Criscione, J. C., Humphrey, J. D., Douglas, A. S. and Hunter, W. C. (2000) An invariant basis for natural strain which yields orthogonal stress response terms in isotropic hyperelasticity, *J. Mech. and Phys. of Solids*, **48**, 2445-2465.

Criscione, J. C., McCulloch, A. D., and Hunter, W. C. (2002) Constitutive framework optimized for myocardium and other high-strain laminar materials with one fiber family, *J. Mech. and Phys. of Solids*, **50**, 1681-1702.

Dorfmann, A. and Muhr, A., (Eds.)(1999) *Constitutive Models for Rubber*, Balkema, Rotterdam.

Ericksen, J. L. (1997) *Introduction to the Thermodynamics of Solids*, Springer, New-York.

Fosdick, R. L. and Kao, B. G. (1978) Transverse deformations associated with rectilinear shear in elastic solids, *J. of Elasticity*, **8** 117-142.

Gent, A. N. (1996) A new constitutive relation for rubber, *Rubber Chemistry and Technology*, **69**, 59-61.

Holzapfel, G. A. (2001) *Nonlinear Solid Mechanics*, Wiley, Chichester.

Horgan, C. O. (1995) Antiplane shear deformation in linear and nonlinear solid mechanics, *SIAM Rev.* **37**, 53-81

Horgan, C. O. and Saccomandi, G. (1999) Simple torsion of isotropic, hyperelastic, incompressible materials with limiting chain extensibility, *J. of Elasticity,* **56** 159-170.

Horgan, C. O. and Saccomandi, G. (2002a) Constitutive modelling of rubber-like and biological materials with limiting chain extensibility, *Mathematics and Mechanics of Solids*, **7** 353-371.

Horgan, C. O. and Saccomandi, G. (2002b) A molecular-statistical basis for the Gent constitutive model of rubber elasticity, *Journal of Elasticity*, **68** 167-176.

Horgan, C. O. and Saccomandi, G. (2003) Finite thermoelasticity with limiting chain extensibility, *Journal of Mechanics and Physics of Solids*, **75**, 839-851.

Horgan, C. O. and Saccomandi, G. (2004) Coupling of anti-plane shear deformations with plane deformations in generalized neo-Hookean isotropic, incompressible, hyperelastic materials, submitted.

Horgan, C. O., Saccomandi, G. and Sgura, I. (2003) Finite thermoelasticity with limiting chain extensibility, *SIAM Journal on Applied Mathematics*, **65**, 1712-1727.

Knowles, J. K. (1977) The finite anti-plane shear field near the tip of a crack for a class of incompressible elastic solids, *Int. J. of Fracture* **13**, 611-639.

Martin, S.E. and Carlson, D. E. (1977) The behavior of elastic heat conductors with second order response functions, *ZAMP* **28**, 311-329.

Müller, I. (1985) *Thermodynamics*, Pitman, Boston-London-Melbourne.

Murnaghan, F. D. (1937) Finite deformation of an elastic solid, *Amer. J. Math.* **59**, 235-260.

Murphy J. and Rogerson G. (2002) A method to model simple tension experiments using finite elasticity theory with an application to some polyurethane foams, *Int. J.of Engngr. Sci* **40**, 499-510.

Mollica, F. and Rajagopal, K. R. (1997) Secondary deformations due to axial shear of the annular region between two eccentricaly place cylinder, *J. of Elasticity* **48** 103-123.

Murray, J. D. (1984) *Asymptotic Analysis,* Springer, New-York.

Ogden, R. W. (1972) Large deformation isotropic elasticity: on the correlation of theory and experiment for compressible rubberlike solids, *Proc. Roy. Soc. London A* **328,** 567-584.

Ogden, R. W. (1984) *Non-Linear Elastic Deformations,* Ellis Horwood, Chichester. Reprinted by Dover (1997).

Ogden, R.W. (1986) Recent advances in the theory of phenomenological theory of rubber elasticity, *Rubber Chemistry and Technology,* 59, 361-383

Pedregal, P. (2000) *Variational methods in nonlinear elasticity,* SIAM, Philadelphia.

Peng, T. J. and Landel, R. F. (1972) Stored energy function of rubberlike materials derive from tensile data, *J. of Appl. Phys.* **43**, 3064-3067.

Perrin, G. (2000) Analytic stress-strain relationship for isotropic network model of rubber elasticity, *C.R. Acad. Sci. Paris* **328, Série II**, 5-10.

Pucci, E. and Saccomandi, G. (1999) Some remarks on the Gent model of rubber elasticity, *CanCNSM proceedings* (edited by Elena M. Croitoro), University of Victoria Press, Victoria, Canada, 163-172.

Pucci, E. and Saccomandi, G. (2002) A note on the Gent model for rubber-like materials, to appear in *Rubber Chemistry and Technology,* **75** 839-851.

Rajagopal, K. R. and Srinivasa, A. R. (2000) A thermodynamic frame work for rate type fluid models, *J. Non_Newtonian Fluid. Mech.* **88**, 207-227.

Rivlin, R. S. (1960) Some topics in finite elasticity in *Proc. of the First Naval Symp. on Structural Mech.*, pp. 169-198, Pergamon Press, New York.

Saccomandi, G. (2001) Universal results in finite elasticiy, in: Y. Fu, R. Ogden (Eds.), *Non-linear Elasticity,* Cambridge University Press, Cambridge.

Signorini, A. (1955) Trasformazioni termoelastiche finite, *Ann. di Mat. Pura Appl.* **39**, 147-201.

Singh, S. (1967) Small finite deformations of elastic dielectrics, *Quart. Appl. Math.* **25**, 275-284.

Spencer. A. J. M. (1972) *Deformations of fibre-reinforced materials,* Oxford University Press.

Valanis, K. C. and Landel, R. F. (1967) The strain-energy function of a hyperelastic materials in terms of extension ratios, *J. of Appl. Phys.*, **38**, 2997-3002.

Zhang, J. P. and Rajagopal, K. R. (1992) Some inhomogeneous motions and deformations within the context of a nonlinear elastic solid, *Int. J. Engngr. Sci.*, **30** 919-938.

Elasticity and Inelasticity of Rubber

R.W. Ogden
Department of Mathematics, University of Glasgow
Glasgow G12 8QW, UK
Email: rwo@maths.gla.ac.uk

Abstract. In this chapter we focus on the isothermal phenomenological theory of the elasticity and inelasticity of rubber. First, we describe experimental results that characterize the elastic behaviour of rubber, in particular of vulcanized natural rubber. This is followed by illustrations of how the behaviour departs from the purely elastic; we examine stress softening associated with the Mullins effect, and the different degrees of stress softening for different rubbers are highlighted. Other inelastic effects such as hysteretic stress-strain cycling following pre-conditioning of the material (to remove the Mullins effect) are also described.

With this background established we then begin the process of mathematical modelling of these behaviours. We describe the theory of elasticity necessary for the modelling of the elastic behaviour, and for simple homogeneous deformations we illustrate the good agreement between theory and experiment. We then move on from elasticity and discuss the modelling of stress softening and the Mullins effect. For this purpose the (quasi-static) theory of pseudo-elasticity is used since this represents a relatively simple extension of the well-established theory of elasticity and is able to capture the Mullins effect both qualitatively and quantitatively. The theory is described and then used to fit some basic data on the Mullins effect.

Finally, we examine briefly the effects of time and rate dependence and associated with viscoelastic behaviour, and some outstanding problems in the modelling of the inelastic behaviour of rubber are discussed with particular reference to viscoelasticity.

1 Background: stress-strain data for rubber

1.1 Stress-strain data for natural rubber

In Figure 1 we show typical stress-stretch results for loading of unfilled vulcanized natural rubber in simple tension based on data of Treloar (1944), depicted by the solid circles. This illustrates the characteristic sigmoidal shape of curve associated with rubberlike materials. Figures 2(a) and 2(b) show corresponding results for other commonly used experimental tests, namely pure shear and equibiaxial tension respectively. In these two tests the largest stretch achievable is less than for simple tension. The general characteristics of the results shown in Figures 1 and 2 are found for many synthetic rubbers, but in some cases the largest stretch obtained can be much larger than for natural rubber (as much as 10–15 in simple tension for example).

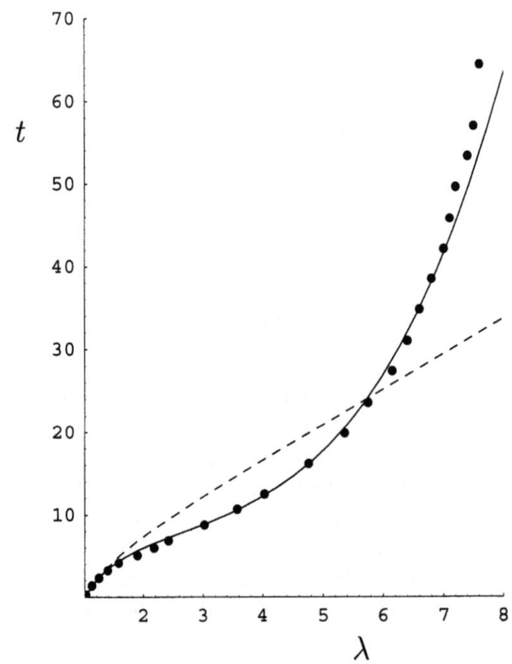

Figure 1. Treloar's data (1944) for simple tension (circles), with the nominal stress t (units kg cm^{-2}) plotted against the stretch λ. The continuous and dashed curves correspond to the predictions of theoretical models to be discussed in Sections 2 and 3.

Figures 1 and 2 show only the loading curves. In general, the loading curve is not re-traced exactly on unloading but for natural rubber (and for many synthetic rubbers) there is only a small stress softening or hysteresis effect if the strains are not too large, and the materials are therefore modelled as perfectly elastic (i.e. the unloading curve re-traces the loading curve exactly). We remark that the tests are conducted at a relatively low strain rate so that the curves can be regarded as corresponding essentially to quasi-static behaviour.

1.2 The Mullins effect and stress softening

A different picture emerges, however, when unloading is properly accounted for. Figure 3 shows the typical simple tension behaviour of a specimen of unfilled natural rubber loaded from the virgin state to a stretch $\lambda = 2$, unloaded, then loaded to a stretch of 3, unloaded, and so on up to a stretch of 6 and finally unloaded. As can be seen, there is significant stress softening beyond a stretch of about $\lambda = 4$, i.e. the stress on unloading is less than that on loading at the same value of the stretch. For sufficiently large stretches a small residual strain is also noticeable. This stress softening, associated with deformation from the virgin state, is known as the Mullins effect, which was examined in detail by Mullins (1947, 1969) and Mullins and Tobin (1957) and in other papers.

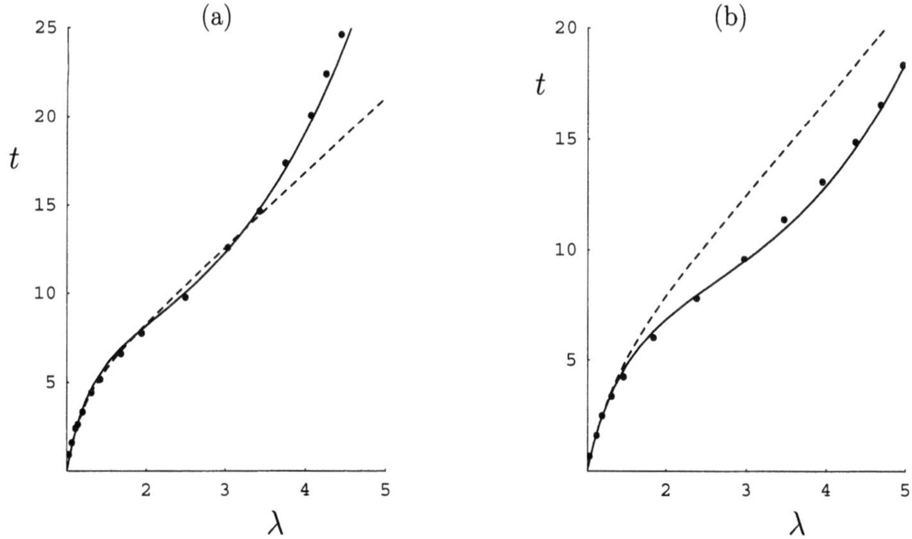

Figure 2. Treloar's data (1944) for (a) equibiaxial tension and (b) pure shear (circles), with the nominal stress t (units kg cm^{-2}) plotted against the stretch λ. The continuous and dashed curves correspond to predictions of theoretical models to be discussed in Sections 2 and 3.

However, according to a recent paper by Marckmann *et al.* the effect had been recorded in the literature much earlier by Bouasse and Carrière (1903).

Figure 4 shows the corresponding picture for a filled natural rubber with a significant proportion (50–60% by volume) of carbon black filler, the maximum stretch in this case being $\lambda = 5$. Here, the stress-softening effect is much more pronounced, even at relatively small stretches. The residual strain is also much more noticeable.

A somewhat different situation is evident for a (filled) liquid silicone rubber, results for which are shown in Figure 5. In this case there is again a very significant stress softening, but the residual strain is very small. Moreover, the reloading curves are very close to the unloading curves. This picture is very close to what may be described as the *idealized Mullins effect*, which is depicted in Figure 6.

Consider, in Figure 6, the loading path abb' from the virgin state, where b' an arbitrary point at which unloading is initiated. On unloading the path $b'Ba$ is followed. The latter path is retraced when the material is reloaded, and on further loading beyond b' the path $b'c$ is traced in continuation of the initial loading path $abb'cc'd$ (which is the path that would be followed in the absence of unloading). If loading is next terminated at c' then the path $c'Ca$ is followed on unloading and retraced back to c' on reloading. If no further loading beyond c' is applied then the curve aCc' represents the subsequent material response, which is then purely elastic. For loading beyond c' the primary path

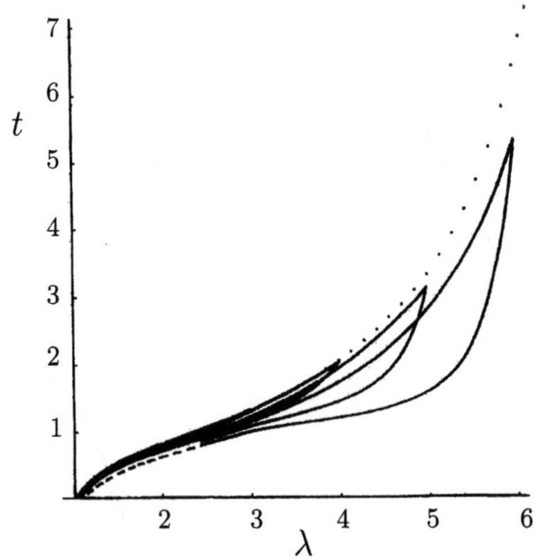

Figure 3. Nominal stress t (in MPa) versus stretch λ in simple tension for an unfilled natural rubber (EDS 19). The dotted curve corresponds to loading of a virgin test piece without unloading; the continuous curves are for loading to different stretch levels followed by unloading and reloading to successively higher stretches (based on Figure 2 in Muhr *et al.*, 1999).

is again followed and the pattern of unloading/reloading is repeated. The key features of this idealized version of the Mullins effect are: (i) the *stress softening* on unloading relative to the primary loading path; (ii) reloading follows the unloading path until the previous maximum stretch is reached and then follows the virgin loading curve; (iii) there is no residual strain.

It is this idealized picture which has been the basis for most modelling in the literature.

1.3 Hysteretic stress-strain cycling

As indicated above, the Mullins effect is associated with deformation from the virgin state of the material. Once the material has been deformed there is then (at least in the idealized situation) no further stress softening generated provided the material is not strained beyond the maximum value of the stretch already achieved. This observation is the basis for pre-conditioning the material: in pre-conditioning the material is strained to a level beyond that expected to be achieved in service conditions so that the Mullins effect is no longer operative. In fact, the Mullins effect is largely a temporary effect and, after straining, the material recovers (or nearly recovers) its virgin condition in a period of time at normal room temperatures (hours, days, weeks or even months depending on the material in question). This 'healing' process is accelerated significantly at higher

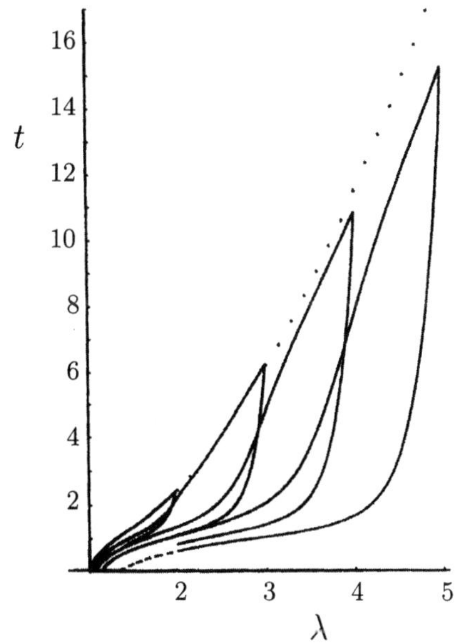

Figure 4. Nominal stress t (in MPa) versus stretch λ in simple tension for a filled natural rubber (EDS 16). The dotted curve corresponds to loading of a virgin test piece without unloading; the continuous curves are for loading to different strain levels followed by unloading and reloading to successively higher stretches (based on Figure 3 in Muhr *et al.*, 1999).

temperatures.

Figure 7 shows experimental stress-strain cycling results in simple tension for a filled rubber that has been pre-conditioned up to a stretch $\lambda = 3$. The material is first loaded from the virgin state to $\lambda = 3$ and then unloaded. A residual stretch is generated, i.e. when the stress is removed the stretch has not returned to its initial value of unity. This residually-strained configuration is taken as the reference configuration for measurements of the stretch in subsequent loading. The material is then reloaded up to the original maximum stretch value and unloaded again. This process is repeated several times and essentially no further residual strain is generated beyond that in the initial loading. After 5 or 6 such cycles the results settle into a repeatable pattern, and there is a loading-unloading hysteresis pattern in which the unloading curve is always below the loading/reloading curve (whereas for the Mullins effect the unloading and reloading curves are the same). Similar behaviour can be found in compression tests and in other deformations, although data for other deformations are very limited.

A further illustration of the response of rubber is given in Figure 8. Following the pre-conditioning described above and once the repeatable pattern is established the re-

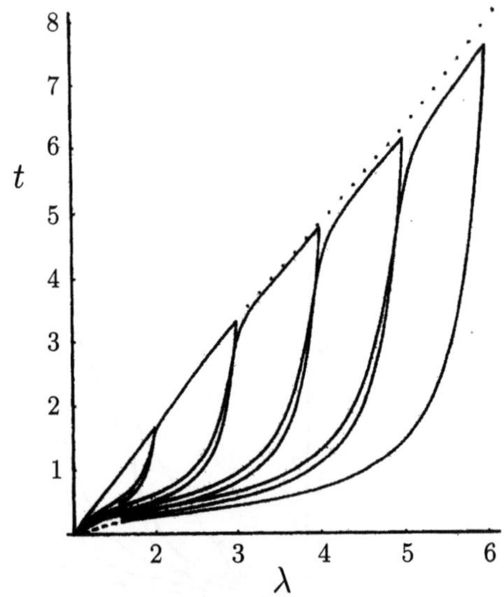

Figure 5. Nominal stress t versus stretch λ in simple tension for a liquid silicone rubber (LSR). The dotted curve corresponds to loading of a virgin test piece without unloading; the continuous curves are for loading to different strain levels followed by unloading and reloading to successively higher stretches (based on Figure 4 in Muhr *et al.*, 1999).

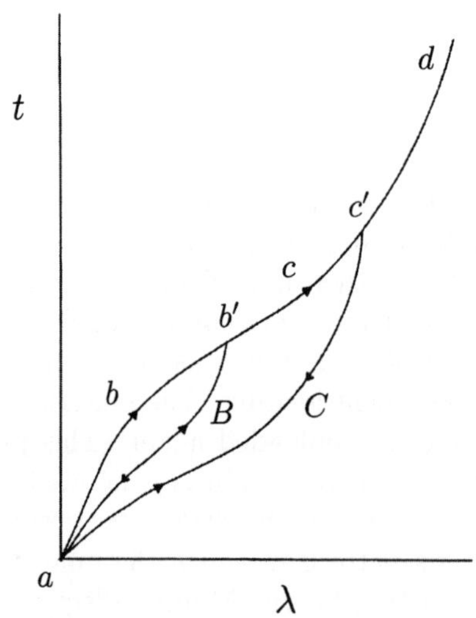

Figure 6. Schematic of the idealized Mullins effect showing loading/unloading curves in simple tension with nominal stress t plotted against the stretch λ.

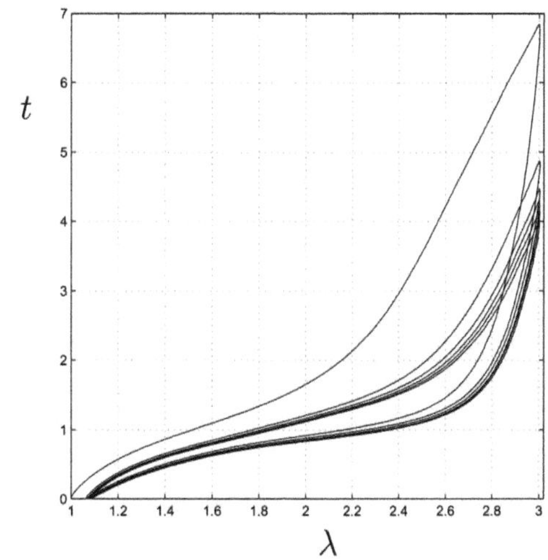

Figure 7. Pre-conditioning of a carbon-black reinforced rubber in simple tension with nominal stress t plotted against stretch λ (based on Figure 1 in Dorfmann and Ogden, 2003).

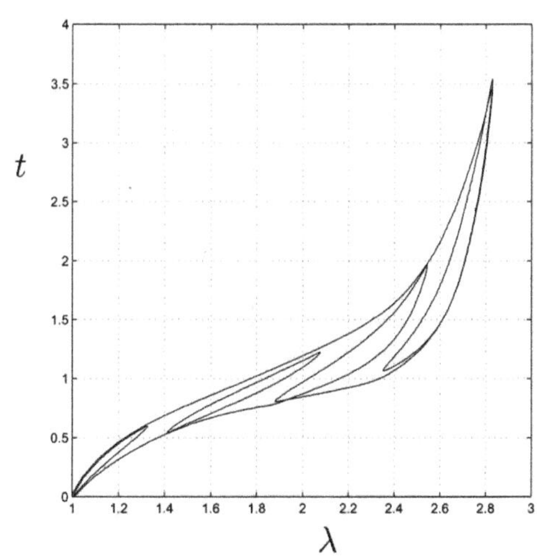

Figure 8. Experimental data for partial reloading-unloading from the (pre-conditioned) unloading path in simple tension for a carbon-black filled rubber with nominal stress t plotted against stretch λ (based on Figure 2 in Dorfmann and Ogden, 2003).

sulting loading/unloading curve is plotted with the reference configuration for the stretch shifted so that the residual stretch is discounted. Reloading from different points on the unloading curve is then carried out. Each reloading is continued until the loading curve is reached (or almost reached) and then unloading in initiated and followed until the unloading curve is reached again. In each case there is a local reloading-unloading hysteresis cycle.

Thus far we have not discussed time dependence (creep, relaxation and recovery, for example) or rate dependence. These topics are examined briefly in Section 6 in the context of viscoelasticity.

1.4 The modelling problem

The question now to be addressed is how we model these different and complex phenomena. We shall start with the simplest situation — that of pure elasticity — and then try to develop and extend the theory of elasticity so as to account for these phenomena by gradually incorporating more complicated effects into the model.

As far as the *elasticity* of rubber is concerned there are a number of fundamental assumptions that are adopted in the phenomenological theory. These are that the material is

(a) HYPERELASTIC,

(b) ISOTROPIC,

(c) INCOMPRESSIBLE.

Hyperelasticity means that the properties of the material are described in terms of a strain-energy function.

Isotropy (relative to a stress-free configuration) is a very good approximation in most circumstances and is almost invariably used by practitioners. It is, of course, essential to account for anisotropy in fibre-reinforced rubber materials, but there may also be, for example, significant anisotropy induced by damage (this is different from the strain-induced anisotropy normally associated with finite elasticity), and such anisotropy will need to be accounted for in material models (this aspect of the theory has hardly been touched on thus far in the literature.

Incompressibility is an idealization. In practice the shear modulus of the material is very much smaller than the bulk modulus (typically the ratio is of order 10^{-4}) and volume changes can be neglected except in extreme situations where the hydrostatic stress is very large — an example of where compressibility might be important is where rubber in a car tyre is highly constrained near the rim of the wheel. For the most part incompressibility is assumed in practical applications.

2 The equations of elasticity

In this section we summarize briefly the necessary equations of *incompressible* nonlinear elasticity, with particular emphasis on isotropy. The corresponding equations for compressible elasticity are described in the chapter by Saccomandi in this volume, although it should be pointed out that there are some minor differences in notation. More details of both the incompressible and compressible theories can be found in, for example, Ogden (1997), Fu and Ogden (2001) and Hayes and Saccomandi (2001).

2.1 Kinematics

Let an arbitrary point in the stress-free reference configuration, denoted $\mathcal{B}_r$, be labelled by its position vector $\mathbf{X}$ relative to some chosen origin. The boundary of $\mathcal{B}_r$ is denoted by $\partial\mathcal{B}_r$. Now suppose that the body is deformed quasi-statically from $\mathcal{B}_r$ so that it occupies a new configuration, $\mathcal{B}$ say, with boundary $\partial\mathcal{B}$. We refer to $\mathcal{B}$ as the *current* or *deformed configuration* of the body. The deformation is represented by the mapping $\chi : \mathcal{B}_r \to \mathcal{B}$ which takes points $\mathbf{X}$ in $\mathcal{B}_r$ to points $\mathbf{x}$ in $\mathcal{B}$. Thus,

$$\mathbf{x} = \chi(\mathbf{X}), \qquad \mathbf{X} \in \mathcal{B}_r, \tag{2.1}$$

where $\mathbf{x}$ is the position vector of the point $\mathbf{X}$ in $\mathcal{B}$. The mapping χ is called the *deformation* from $\mathcal{B}_r$ to $\mathcal{B}$.

For simplicity we consider only Cartesian coordinate systems and let $\mathbf{X}$ and $\mathbf{x}$ respectively have coordinates X_α and x_i, where α, $i \in \{1,2,3\}$, so that $x_i = \chi_i(X_\alpha)$. Greek and Roman indices refer, respectively, to $\mathcal{B}_r$ and $\mathcal{B}$ and the usual summation convention for repeated indices will be adopted in the following sections where appropriate.

The *deformation gradient tensor*, denoted $\mathbf{F}$, is given by

$$\mathbf{F} = \operatorname{Grad} \mathbf{x}, \tag{2.2}$$

where Grad is the gradient operator in $\mathcal{B}_r$. The Cartesian components of $\mathbf{F}$ are $F_{i\alpha} = \partial x_i/\partial X_\alpha$. Here we consider only *incompressible* materials, so that the deformation is volume preserving (*isochoric*) and all deformation gradients $\mathbf{F}$ must therefore satisfy

$$\det \mathbf{F} = 1. \tag{2.3}$$

The deformation gradient can be decomposed according to the *polar decompositions*

$$\mathbf{F} = \mathbf{R}\mathbf{U} = \mathbf{V}\mathbf{R}, \tag{2.4}$$

where $\mathbf{R}$ is a proper orthogonal tensor and $\mathbf{U}$, $\mathbf{V}$ are positive definite and symmetric tensors. Each of the decompositions in (2.4) is unique. Respectively, $\mathbf{U}$ and $\mathbf{V}$ are called the *right* and *left stretch tensors*. For $\mathbf{U}$ we have the *spectral decomposition*

$$\mathbf{U} = \sum_{i=1}^{3} \lambda_i \mathbf{u}^{(i)} \otimes \mathbf{u}^{(i)}, \tag{2.5}$$

where $\lambda_i > 0$, $i \in \{1,2,3\}$, are the *principal stretches*, $\mathbf{u}^{(i)}$, the (unit) eigenvectors of $\mathbf{U}$, are called the *Lagrangian principal axes* and $\otimes$ denotes the tensor product. The stretches are of fundamental importance in rubber elasticity. Similarly, $\mathbf{V}$ has the spectral decomposition

$$\mathbf{V} = \sum_{i=1}^{3} \lambda_i \mathbf{v}^{(i)} \otimes \mathbf{v}^{(i)}, \tag{2.6}$$

and we note that

$$\mathbf{v}^{(i)} = \mathbf{R}\mathbf{u}^{(i)}, \quad i \in \{1, 2, 3\}. \tag{2.7}$$

It follows from (2.3), (2.4) and (2.5) that

$$\det \mathbf{U} = \det \mathbf{V} = \lambda_1 \lambda_2 \lambda_3 = 1. \tag{2.8}$$

On use of the polar decompositions (2.4), we may form the tensor measures of deformation defined by

$$\mathbf{C} = \mathbf{F}^T \mathbf{F} = \mathbf{U}^2, \qquad \mathbf{B} = \mathbf{F}\mathbf{F}^T = \mathbf{V}^2, \tag{2.9}$$

where ^{T} denotes the transpose of a tensor. Respectively, $\mathbf{C}$ and $\mathbf{B}$ are called the *right* and *left Cauchy-Green deformation tensors.*

The principal invariants of $\mathbf{C}$ and $\mathbf{B}$ are the same and defined by

$$I_1 = \operatorname{tr} \mathbf{C}, \quad I_2 = I_3 \operatorname{tr}(\mathbf{C}^{-1}) = \frac{1}{2}\left((\operatorname{tr} \mathbf{C})^2 - \operatorname{tr}(\mathbf{C}^2)\right), \quad I_3 = \det \mathbf{C}, \tag{2.10}$$

but, since $\det \mathbf{C} = (\det \mathbf{F})^2$, we have $I_3 = 1$ and there remain only two independent invariants.

In terms of the stretches, we have

$$I_1 = \lambda_1^2 + \lambda_2^2 + \lambda_3^2, \quad I_2 = \lambda_1^{-2} + \lambda_2^{-2} + \lambda_3^{-2}. \tag{2.11}$$

These two invariants played a very important role in the development of the theory of finite elasticity and its application to rubber elasticity.

2.2 Stress tensors and equilibrium equations

The surface force per unit area (or *stress vector*) on a vector area element $\mathrm{d}\mathbf{a} = \mathbf{n}\mathrm{d}a$ in the deformed body, where $\mathbf{n}$ is the unit outward normal to the body surface, is denoted by $\mathbf{t}$. It depends on $\mathbf{n}$ according to the formula

$$\mathbf{t} = \boldsymbol{\sigma}^T \mathbf{n}, \tag{2.12}$$

where $\boldsymbol{\sigma}$, a second-order tensor independent of $\mathbf{n}$, is the *Cauchy stress tensor.* According to Nanson's formula $\mathbf{n}\mathrm{d}a$ is related to the reference area element $\mathbf{N}\mathrm{d}A$ for an incompressible material by

$$\mathbf{n}\mathrm{d}a = \mathbf{F}^{-T}\mathbf{N}\, \mathrm{d}A. \tag{2.13}$$

This enables us to write

$$\mathbf{t}\mathrm{d}a = \mathbf{S}^T \mathbf{N}\mathrm{d}A, \tag{2.14}$$

where $\mathbf{S}$ is the *nominal stress tensor*, from which the traction in the deformed configuration per unit *reference* area is calculated. For an incompressible material $\mathbf{S}$ is related to $\boldsymbol{\sigma}$ by

$$\mathbf{S} = \mathbf{F}^{-1}\boldsymbol{\sigma}, \tag{2.15}$$

and we note that $\mathbf{S}^T$ is identified as the *first Piola-Kirchhoff stress tensor.*

In the absence of body forces the equilibrium equations may be written in either of the equivalent forms

$$\operatorname{div}\boldsymbol{\sigma} = \mathbf{0}, \qquad \operatorname{Div}\mathbf{S} = \mathbf{0}, \tag{2.16}$$

where div and Div denote the divergence operators in $\mathcal{B}$ and $\mathcal{B}_r$ respectively.

Balance of the moments of the forces acting on the body yields simply $\boldsymbol{\sigma}^T = \boldsymbol{\sigma}$, which may also be written as

$$\mathbf{S}^T\mathbf{F}^T = \mathbf{F}\mathbf{S}. \tag{2.17}$$

A detailed discussion of stress and strain tensors can be found in Ogden (1997), for example.

2.3 Hyperelasticity

In the theory of *hyperelasticity* there exists a *strain-energy function* (or *stored-energy function*), per unit volume, here denoted $W = W(\mathbf{F})$ and defined on the space of deformation gradients such that

$$\mathbf{S} = \frac{\partial W}{\partial \mathbf{F}} - p\mathbf{F}^{-1}, \quad \boldsymbol{\sigma} = \mathbf{F}\frac{\partial W}{\partial \mathbf{F}} - p\mathbf{I}, \qquad \det\mathbf{F} = 1, \tag{2.18}$$

where p is a Lagrange multiplier associated with the incompressibility constraint (2.3) and is referred to as an *arbitrary hydrostatic pressure.* Equations (2.18) are *stress-deformation relations for an incompressible elastic material* for which there exists a strain-energy function W.

Objectivity and Material Symmetry The elastic stored energy is required to be independent of superimposed rigid deformations and therefore

$$W(\mathbf{QF}) = W(\mathbf{F}) \tag{2.19}$$

for *all* rotations $\mathbf{Q}$. A strain-energy function satisfying this requirement is said to be *objective* (or frame-indifferent).

Use of the polar decomposition (2.4) and the choice $\mathbf{Q} = \mathbf{R}^T$ in (2.19) shows that

$$W(\mathbf{F}) = W(\mathbf{U}). \tag{2.20}$$

Thus, W depends on $\mathbf{F}$ only through the stretch tensor $\mathbf{U}$ and may therefore be defined on the class of *positive definite symmetric tensors.* This leads to the introduction of the *Biot stress tensor*, denoted here by $\mathbf{T}$ and defined by

$$\mathbf{T} = \frac{\partial W}{\partial \mathbf{U}} - p\mathbf{U}^{-1}, \qquad \det\mathbf{U} = 1 \tag{2.21}$$

(for an incompressible material). Note that $\mathbf{T}$ is symmetric. Note also that when expressed as a function of $\mathbf{U}$ the strain energy automatically satisfies the objectivity requirement since it is a Lagrangian tensor and is unaffected by a rotation in the deformed configuration.

From the mathematical point of view, there is as yet no restriction on the form that the function W may take other than (2.19) (although it is usually assumed that W and the stresses vanish in the reference configuration). This leaves considerable flexibility in the choice of the functional form of W. However, the predictions of W must be acceptable for the description of the elastic behaviour of real materials and must also make mathematical sense (i.e. lead to existence of solutions of well-posed boundary-value problems, for example).

Isotropic hyperelasticity Further restrictions on the form of W are generated if the material possesses symmetries in the configuration $\mathcal{B}_r$. For a hyperelastic material such a symmetry requirement has the form

$$W(\mathbf{FP}) = W(\mathbf{F}) \tag{2.22}$$

for *every* deformation gradient $\mathbf{F}$ (satisfying (2.3) for an incompressible material) and for all second-order tensors $\mathbf{P}$ in some set. Thus, the strain-energy function is unaffected by a change of reference configuration with deformation gradient $\mathbf{P}$, and the set of such $\mathbf{P}$ forms a multiplicative group, the so-called *symmetry group* of the material *relative to* $\mathcal{B}_r$.

For the description of the elastic behaviour of rubberlike solids it is appropriate to restrict attention to *isotropic elasticity*, in which case the symmetry group is the *proper orthogonal group*. Then, we have

$$W(\mathbf{FQ}) = W(\mathbf{F}) \tag{2.23}$$

for *all* rotations $\mathbf{Q}$. Bearing in mind that the $\mathbf{Q}$'s appearing in (2.19) and (2.23) are independent the combination of these two equations yields

$$W(\mathbf{QUQ}^T) = W(\mathbf{U}) \tag{2.24}$$

for all rotations $\mathbf{Q}$. Equation (2.24) states that W is an *isotropic function* of $\mathbf{U}$. It follows from the spectral decomposition (2.5) that W depends on $\mathbf{U}$ only through the principal stretches $\lambda_1, \lambda_2, \lambda_3$. To avoid introducing additional notation we express this dependence as $W(\lambda_1, \lambda_2, \lambda_3)$; by selecting appropriate values for $\mathbf{Q}$ in (2.24) we may deduce that W depends symmetrically on $\lambda_1, \lambda_2, \lambda_3$, i.e.

$$W(\lambda_1, \lambda_2, \lambda_3) = W(\lambda_1, \lambda_3, \lambda_2) = W(\lambda_2, \lambda_1, \lambda_3). \tag{2.25}$$

An important consequence of isotropy is that the Biot stress tensor $\mathbf{T}$ is *coaxial* with $\mathbf{U}$ and hence, in parallel with (2.5), we have

$$\mathbf{T} = \sum_{i=1}^{3} t_i \mathbf{u}^{(i)} \otimes \mathbf{u}^{(i)}, \tag{2.26}$$

where t_i, $i \in \{1,2,3\}$ are the *principal Biot stresses.* For an incompressible material the principal Biot stresses are given by

$$t_i = \frac{\partial W}{\partial \lambda_i} - p\lambda_i^{-1}, \qquad \lambda_1\lambda_2\lambda_3 = 1. \tag{2.27}$$

We note here that the corresponding principal Cauchy stresses σ_i, $i \in \{1,2,3\}$, are given by

$$\sigma_i = \lambda_i \frac{\partial W}{\partial \lambda_i} - p, \qquad \lambda_1\lambda_2\lambda_3 = 1. \tag{2.28}$$

In (2.28) there is no summation over the repeated index i.

The Cauchy stress tensor $\boldsymbol{\sigma}$ is coaxial with $\mathbf{V}$ and given by

$$\boldsymbol{\sigma} = \sum_{i=1}^{3} \sigma_i \mathbf{v}^{(i)} \otimes \mathbf{v}^{(i)}, \tag{2.29}$$

while the nominal stress $\mathbf{S}$ has the decomposition

$$\mathbf{S} = \sum_{i=1}^{3} t_i \mathbf{u}^{(i)} \otimes \mathbf{v}^{(i)}, \tag{2.30}$$

and we note the connection $\mathbf{S} = \mathbf{T}\mathbf{R}^T$.

If W is now regarded as a function of the two (independent) invariants I_1, I_2, we may write

$$W(\lambda_1, \lambda_2, \lambda_3) = \bar{W}(I_1, I_2), \tag{2.31}$$

recalling (2.11), the overbar on $\bar{W}$ signifying that I_1, I_2 are to be regarded as the independent deformation variables.

We then require the expressions

$$\frac{\partial I_1}{\partial \mathbf{F}} = 2\mathbf{F}^T, \quad \frac{\partial I_2}{\partial \mathbf{F}} = 2\mathbf{F}^T(I_1\mathbf{I} - \mathbf{B}), \tag{2.32}$$

and from (2.18) we calculate the Cauchy stress in the form

$$\boldsymbol{\sigma} = 2\frac{\partial \bar{W}}{\partial I_1}\mathbf{B} + 2\frac{\partial \bar{W}}{\partial I_2}(I_1\mathbf{I} - \mathbf{B})\mathbf{B} - p\mathbf{I}. \tag{2.33}$$

In the reference configuration we have $I_1 = I_2 = 3$ and we require the strain energy and the stress to vanish there. This leads to the restrictions

$$\bar{W}(3,3) = 0, \quad 2\frac{\partial \bar{W}}{\partial I_1}(3,3) + 4\frac{\partial \bar{W}}{\partial I_2}(3,3) - p_0 = 0, \tag{2.34}$$

where p_0 is the value of p in the considered reference configuration. Furthermore, for compatibility with the classical theory of incompressible isotropic elasticity we must have

$$\frac{\partial \bar{W}}{\partial I_1}(3,3) + \frac{\partial \bar{W}}{\partial I_2}(3,3) = \frac{1}{2}\mu, \tag{2.35}$$

where $\mu\,(> 0)$ is the shear modulus of the material in the reference configuration.

Examples of strain-energy functions Here we just mention, for purposes of illustration, a few examples of strain-energy functions. As already noted the strain energy of an incompressible isotropic elastic solid can be regarded either as a symmetric function of the principal stretches or as a function of two independent invariants, such as I_1, I_2. An important example is the *Mooney-Rivlin* form of strain energy, which is linear in I_1 and I_2 and defined by

$$W = C_1(I_1 - 3) + C_2(I_2 - 3) \equiv C_1(\lambda_1^2 + \lambda_2^2 + \lambda_3^2 - 3) + C_2(\lambda_1^{-2} + \lambda_2^{-2} + \lambda_3^{-2} - 3), \quad (2.36)$$

where C_1, C_2 are constants and $\lambda_1\lambda_2\lambda_3 = 1$. Note that it follows from (2.35) that $2C_1 + 2C_2 = \mu$. When $C_2 = 0$ this reduces to the *neo-Hookean* strain energy

$$W = \frac{1}{2}\mu(I_1 - 3) \equiv \frac{1}{2}\mu(\lambda_1^2 + \lambda_2^2 + \lambda_3^2 - 3), \quad (2.37)$$

where C_1 has been replaced by $\mu/2$. These two forms of energy function played key roles in the development of the subject of finite elasticity, particularly in respect of its connection with rubber elasticity. For reviews of this aspect we refer to Ogden (1982, 1986) in which more details are given of different forms of strain-energy functions appropriate for rubberlike solids.

Each of the strain-energy functions (2.36)–(2.37) is of the *separable form*

$$W = w(\lambda_1) + w(\lambda_2) + w(\lambda_3), \quad (2.38)$$

which was introduced by Valanis and Landel (1967). Equivalent to (2.38) is the expansion

$$W = \sum_{m=1}^{\infty} \mu_m(\lambda_1^{\alpha_m} + \lambda_2^{\alpha_m} + \lambda_3^{\alpha_m} - 3)/\alpha_m \quad (2.39)$$

in terms of powers of the stretches, where each μ_m and α_m is a material constant, the latter not necessarily being integers (Ogden, 1972, 1982). For practical purposes the sum in (2.39) is restricted to a finite number of terms, while, for consistency with the classical theory, the constants must satisfy the requirement

$$\sum_{m=1}^{N} \mu_m\alpha_m = 2\mu, \quad (2.40)$$

where N is a positive integer and μ is again the shear modulus of the material in the undeformed stress-free (natural) configuration. The counterpart of (2.40) for the Valanis-Landel material is

$$w''(1) + w'(1) = 2\mu. \quad (2.41)$$

In respect of (2.38) the principal Cauchy stresses are obtained from (2.28) in the form

$$\sigma_i = \lambda_i w'(\lambda_i) - p \quad (2.42)$$

(no summation over i) and the specializations appropriate for (2.36), (2.37) and (2.39) are then easily read off. For the neo-Hookean solid, for example, we have

$$\boldsymbol{\sigma} = \mu\mathbf{B} - p\mathbf{I}, \quad (2.43)$$

where **B** is the left Cauchy-Green deformation tensor defined in (2.9).

For detailed discussion of other forms of strain-energy function we refer to, for example, Ogden (1972, 1982, 1997) and Treloar (1975) and the recent review by Boyce and Arruda (2000). A wealth of information on the early developments in finite elasticity and its applications is contained in the two volumes of the collected papers of R. S. Rivlin edited by Barenblatt and Joseph (1996).

2.4 Homogeneous deformations

A *homogeneous deformation* is a deformation in which the deformation gradient $\mathbf{F}$ is independent of $\mathbf{X}$. A *pure homogeneous strain* is a homogeneous deformation of the form

$$x_1 = \lambda_1 X_1, \quad x_2 = \lambda_2 X_2, \quad x_3 = \lambda_3 X_3, \tag{2.44}$$

where the principal stretches $\lambda_1, \lambda_2, \lambda_3$ are independent of $\mathbf{X}$. For this deformation $\mathbf{F} = \mathbf{U} = \mathbf{V}, \mathbf{R} = \mathbf{I}$ and the principal axes of the deformation coincide with the Cartesian coordinate directions and are fixed as the values of the stretches change.

Since the material considered is incompressible, only two stretches can be varied independently and it is important to recognize that *biaxial* tests are sufficient to obtain a characterization of W. In this connection it is convenient to make use of the incompressibility constraint to express the strain energy as a function of two independent stretches, and we therefore introduce the notation defined by

$$\hat{W}(\lambda_1, \lambda_2) = W(\lambda_1, \lambda_2, \lambda_1^{-1}\lambda_2^{-1}). \tag{2.45}$$

This enables p to be eliminated from equation (2.28) and leads to

$$\sigma_1 - \sigma_3 = \lambda_1 \frac{\partial \hat{W}}{\partial \lambda_1}, \quad \sigma_2 - \sigma_3 = \lambda_2 \frac{\partial \hat{W}}{\partial \lambda_2}. \tag{2.46}$$

For the energy and stresses to vanish in the natural configuration and for consistency with the classical theory of incompressible isotropic elasticity we must have

$$\begin{aligned} &\hat{W}(1,1) = 0, \quad \hat{W}_1(1,1) = \hat{W}_2(1,1) = 0, \\ &\hat{W}_{12}(1,1) = 2\mu, \quad \hat{W}_{11}(1,1) = \hat{W}_{22}(1,1) = 4\mu, \end{aligned} \tag{2.47}$$

which are the counterparts of (2.34) and (2.35) for $\hat{W}$. In (2.47) the subscripts 1 and 2 signify differentiation with respect to λ_1 and λ_2 respectively.

It is important to note that, because of the incompressibility constraint, equations (2.46) are unaffected by the superposition of an arbitrary hydrostatic stress. Thus, without loss of generality, we may set $\sigma_3 = 0$ in (2.46). In terms of the principal Biot stresses we then have simply

$$t_1 = \frac{\partial \hat{W}}{\partial \lambda_1}, \quad t_2 = \frac{\partial \hat{W}}{\partial \lambda_2}, \tag{2.48}$$

which provides two equations relating λ_1, λ_2 and t_1, t_2 and therefore a simple basis for characterizing $\hat{W}$ from measured (homogeneous) biaxial data.

There are several special cases of the biaxial test which are of interest.

(i) *Simple tension*

For simple tension we set $t_2 = 0$ and $t_1 = t$. Then, by symmetry, the incompressibility constraint yields $\lambda_2 = \lambda_3 = \lambda_1^{-1/2}$ and we write $\lambda_1 = \lambda$. The strain energy may now be treated as a function of just λ, and we define

$$\tilde{W}(\lambda) = \hat{W}(\lambda, \lambda^{-1/2}), \tag{2.49}$$

and (2.48) reduces to

$$t = \tilde{W}'(\lambda), \tag{2.50}$$

where the prime indicates differentiation with respect to λ.

(ii) *Pure shear*

For pure shear we have $\lambda_2 = 1$ and $\lambda_3 = \lambda_1^{-1}$. Again we write $\lambda_1 = \lambda$, while the active stress is $t_1 = t$. We define

$$\tilde{W}(\lambda) = \hat{W}(\lambda, 1), \tag{2.51}$$

and (2.48) reduces to

$$t = \tilde{W}'(\lambda). \tag{2.52}$$

(iii) *Equibiaxial tension*

For equibiaxial tension we have $\lambda_1 = \lambda_2$ ($= \lambda$, say) and $\lambda_3 = \lambda_1^{-2}$ and $t_1 = t_2$ ($= t$, say). Then, we define

$$\tilde{W}(\lambda) = \hat{W}(\lambda, \lambda), \tag{2.53}$$

and (2.48) reduces to

$$t = \tilde{W}'(\lambda)/2. \tag{2.54}$$

Note that the definition of $\tilde{W}$ is different in each case and note also the factor 1/2 in (2.54).

Since there are only two independent stretches in the incompressible theory (for biaxial deformations) the range of possible deformations can be represented in the (λ_1, λ_2)-plane, and since the stretches are positive only the first quadrant of that plane is relevant. This is depicted in Figure 9, with curves corresponding to the specializations appropriate for simple tension (with either $t_1 = 0$ or $t_2 = 0$), pure shear and equibiaxial tension shown. To obtain a satisfactory characterization of $\hat{W}$, and hence of W, it is necessary to have data points covering as much of the (λ_1, λ_2)-space as possible. For this purpose the 'one-dimensional' tests discussed above are clearly inadequate. A suitable set of data can, however, be obtained from biaxial tests in which (for example) λ_2 is held at a series of fixed values while λ_1 is varied.

Figure 10 shows the corresponding (I_1, I_2)-space, in which the invariants I_1 and I_2 must satisfy the restrictions $I_1 \geq 3, I_2 \geq 3$ and

$$I_1^2 \geq 3I_2, \quad I_2^2 \geq 3I_1. \tag{2.55}$$

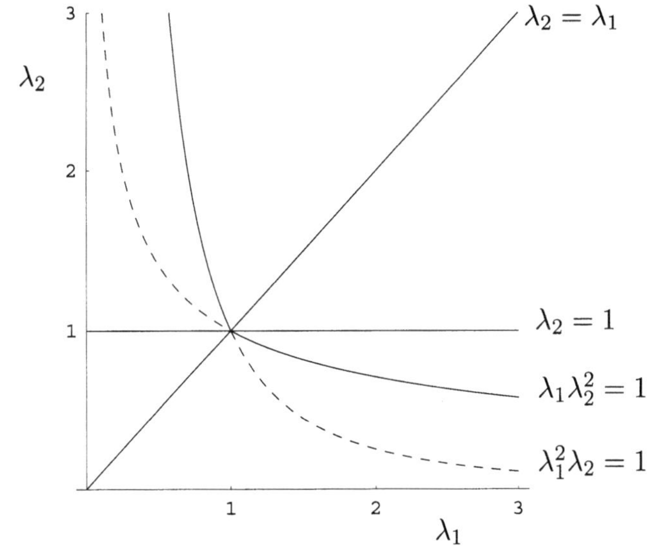

Figure 9. Curves in (λ_1, λ_2) space corresponding to pure shear ($\lambda_2 = 1$), equibiaxial deformation ($\lambda_2 = \lambda_1$), and simple tension with $t_1 = 0$ ($\lambda_1^2\lambda_2 = 1$) and $t_2 = 0$ ($\lambda_1\lambda_2^2 = 1$).

They must also satisfy the stronger inequality

$$I_1^2 I_2^2 - 4I_1^3 - 4I_2^3 + 18I_1I_2 - 27 \geq 0 \tag{2.56}$$

in order to ensure that given values of (I_1, I_2) yield three positive stretches. The curves $I_1^2 = 3I_2$ and $I_2^2 = 3I_1$ are shown as dashed curves in Figure 10, while the strict inequality (2.56) is satisfied in the region above and to the right of the two continuous curves shown in Figure 10 for $I_1 \geq 3$.

While in (λ_1, λ_2)-space the whole of the first quadrant is admissible, in (I_1, I_2)-space, because the material is incompressible, the only admissible deformations are those satisfying (2.56). Note that, on use of the incompressibility condition (2.8), the invariants (2.11) can be written in terms of two independent stretches as

$$I_1 = \lambda_1^2 + \lambda_2^2 + \lambda_1^{-2}\lambda_2^{-2}, \quad I_2 = \lambda_1^{-2} + \lambda_2^{-2} + \lambda_1^2\lambda_2^2. \tag{2.57}$$

Moreover, we record here that the line $I_2 = I_1$ necessarily corresponds to a plane strain deformation since it is easily established from (2.57) that one of the stretches is then unity.

In terms of the invariants we have $\bar{W}(I_1, I_2)$ and equations (2.46) can be rewritten as

$$\begin{aligned} \sigma_1 - \sigma_3 &= 2(\lambda_1^2 - \lambda_1^{-2}\lambda_2^{-2})\frac{\partial \bar{W}}{\partial I_1} + 2(\lambda_1^2\lambda_2^2 - \lambda_1^{-2})\frac{\partial \bar{W}}{\partial I_2}, \\ \sigma_2 - \sigma_3 &= 2(\lambda_2^2 - \lambda_1^{-2}\lambda_2^{-2})\frac{\partial \bar{W}}{\partial I_1} + 2(\lambda_1^2\lambda_2^2 - \lambda_2^{-2})\frac{\partial \bar{W}}{\partial I_2}. \end{aligned} \tag{2.58}$$

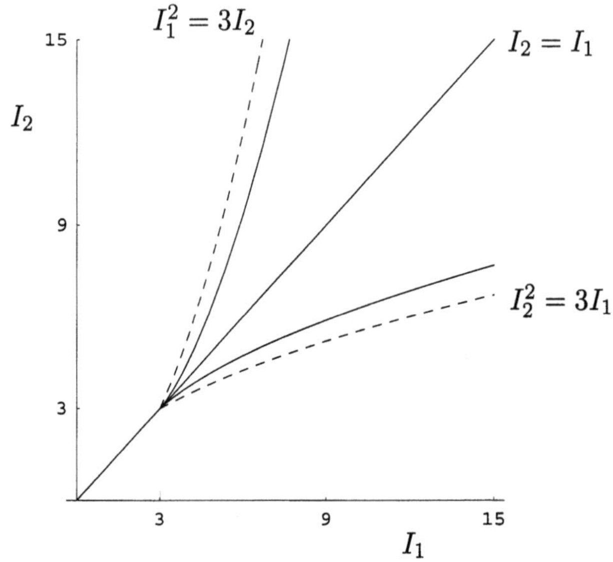

Figure 10. Region of (I_1, I_2)-space showing the (dashed) curves $I_1^2 = 3I_2, I_2^2 = 3I_1$, the line $I_2 = I_1$ corresponding to plane strain and the (continuous) curves defined by $I_1^2 I_2^2 - 4I_1^3 - 4I_2^3 + 18I_1I_2 - 27 = 0$.

We have seen that the strain energy may be written as a function of two independent deformation variables — either (λ_1, λ_2) or (I_1, I_2) — but these are not the only possibilities. Any pair of suitable invariants will serve the purpose and many other invariants are used in the literature in various contexts. For the most part, however, it is the deformation variables considered here that are used in comparing the theory with experimental data. In the following section we examine a representative set of biaxial data and compare this data with the theory based on the use of the stretches, in particular the equations given in (2.46).

3 Theory and experiment compared: elasticity

It turns out that it is instructive to represent biaxial data with the stress difference $\sigma_1 - \sigma_2$ considered as a function of λ_1 at fixed λ_2, and for this purpose we therefore use (2.46) to obtain

$$\sigma_1 - \sigma_2 = \lambda_1 \hat{W}_1 - \lambda_2 \hat{W}_2, \tag{3.1}$$

where the subscripts 1 and 2 on $\hat{W}$ indicate differentiation with respect to λ_1 and λ_2 respectively. Then, for the Valanis-Landel strain energy (2.38) equation (3.1) yields

$$\sigma_1 - \sigma_2 = \lambda_1 w'(\lambda_1) - \lambda_2 w'(\lambda_2). \tag{3.2}$$

For the Mooney-Rivlin specialization (2.36) this gives

$$\sigma_1 - \sigma_2 = \mu_1(\lambda_1^2 - \lambda_2^2) + \mu_2(\lambda_1^{-2} - \lambda_2^{-2}), \tag{3.3}$$

where C_1 and C_2 have been replaced by $\mu_1/2$ and $-\mu_2/2$ respectively in accordance with the notation in (2.39), while for the neo-Hookean strain energy this becomes

$$\sigma_1 - \sigma_2 = \mu(\lambda_1^2 - \lambda_2^2). \tag{3.4}$$

A representative set of biaxial data is that of Jones and Treloar (1975) for a vulcanized natural rubber, and this is plotted in Figures 11(a)–(e) for $\lambda_2 = 1, 1.502, 1.984, 2.295$ and 2.623 respectively. Also shown in these figures for comparison are the predictions of the neo-Hookean (dashed curves) and Mooney-Rivlin (continuous curves) energy functions based on (3.4) and (3.3) respectively. The fit to the data is reasonable only for small deformations when $\lambda_2 = 1$ and deteriorates rapidly as λ_2 increases.

Next, in Figure 12, we show the data from Figures 11(a)–(e) superimposed. The main thing to notice here is that the shapes of the curves for different values of λ_2 are essentially the same. Theoretically, this is reflected in the separability associated with the Valanis-Landel energy function. This means that the shapes of the curves can be determined using data for $\lambda_2 = 1$, for example, i.e. on the basis of the equation

$$\sigma_1 - \sigma_2 = \lambda_1 w'(\lambda_1) - w'(1). \tag{3.5}$$

The curves for $\lambda_2 \neq 1$ are then obtained by a vertical shift given by

$$w'(1) - \lambda_2 w'(\lambda_2). \tag{3.6}$$

Equation (3.2) is recovered by adding equations (3.5) and (3.6). For a more detailed discussion we refer to Ogden (1982, 1986).

For the strain-energy function (2.39) equation (3.5) gives

$$\sigma_1 - \sigma_2 = \sum_{n=1}^{N} \mu_n(\lambda_1^{\alpha_n} - 1). \tag{3.7}$$

This applies for $\lambda_1 \geq 1$ (and $\lambda_2 = 1$). The part of (λ_1, λ_2)-space where $\lambda_1 < 1$ (with $\lambda_2 = 1$) can be covered by noting that since $\sigma_3 = 0$

$$-\sigma_2 = \sum_{n=1}^{N} \mu_n(\lambda_3^{\alpha_n} - 1) \tag{3.8}$$

and $\lambda_3 \leq 1$. The data are then plotted with λ_3 replacing λ_1 on the horizontal axis.

The shift factor (3.6) is

$$\sum_{n=1}^{N} \mu_n(1 - \lambda_2^{\alpha_n}). \tag{3.9}$$

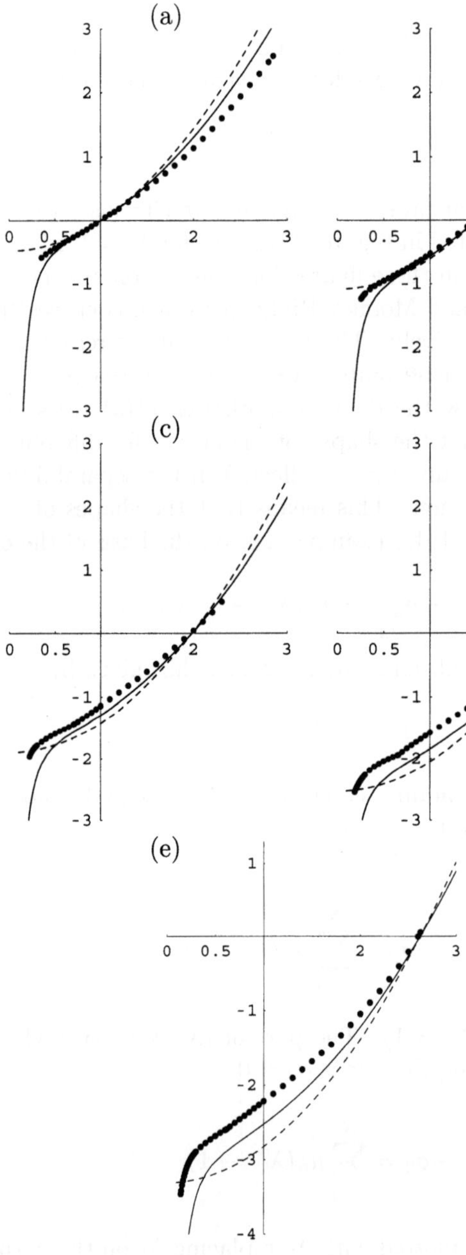

Figure 11. Plot of the biaxial data of Jones and Treloar (1975) (circles) with $\sigma_1 - \sigma_2$ (units N mm^{-2}) plotted against λ_1 compared with the neo-Hookean (dashed curve) and Mooney-Rivlin (continuous curve) theories: (a) $\lambda_2 = 1$; (b) $\lambda_2 = 1.502$; (c) $\lambda_2 = 1.984$; (d) $\lambda_2 = 2.295$; (e) $\lambda_2 = 2.623$.

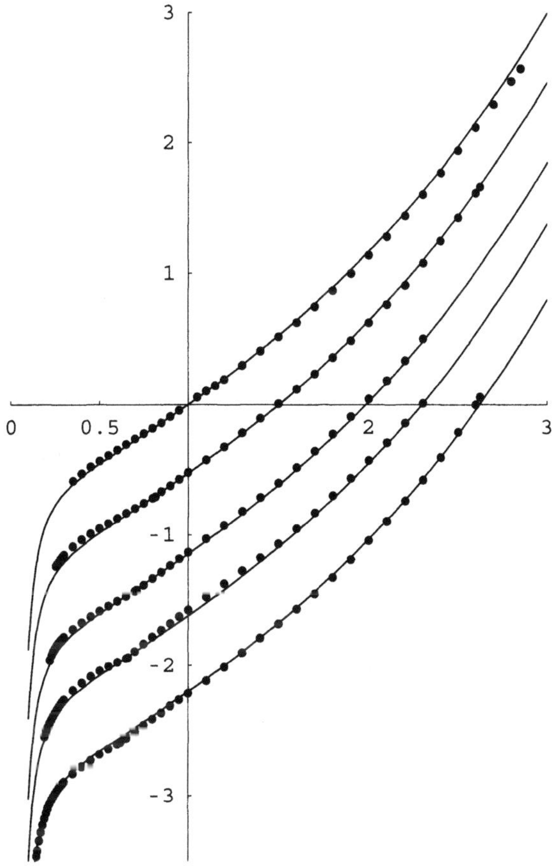

Figure 12. Plot of the biaxial data of Jones and Treloar (1975) (circles) with $\sigma_1 - \sigma_2$ (units N mm^{-2}) plotted against λ_1 with data from Figures 11(a)–(e) superimposed. The continuous curves are for a strain-energy function of the form (2.39) with material constants given by (3.10).

The data in Figures 11(a)–(e) have been fitted on the basis of equations (3.7)–(3.9) with $N = 3$ and the constants given by

$$\alpha_1 = 1.3, \quad \alpha_2 = 4.0, \quad \alpha_3 = -2.0,$$
$$\mu_1 = 0.69, \quad \mu_2 = 0.01, \quad \mu_3 = -0.0122, \tag{3.10}$$

the units of the μ's being N mm^{-2}. The continuous curves in Figure 12 correspond to this theoretical prediction.

The Jones-Treloar data were obtained from plane sheets of rubber. An alternative way to obtain biaxial data is from extension/inflation tests on thin-walled tubes. Data on this basis was obtained by Vangerko and Treloar (1978). Other references to biaxial studies are listed in Ogden (1982), for example.

Finally in this section we refer back to Figures 1–3. The theoretical curves in those figures are for a neo-Hookean strain energy (dashed curves) and a strain energy of type (2.39) (continuous curves), but with a different set of constants from (3.10) since the materials are different (see, for example, Ogden, 1997).

4 The equations of pseudo-elasticity

We now turn our attention to stress softening and the Mullins effect. Our aim here is to capture the essence of the Mullins effect with a relatively simple theory that requires only a slight modification of the theory of elasticity discussed so far. The theory is referred to as *pseudo-elasticity*, and the development presented here is based on the papers by Ogden and Roxburgh (1999a, b) and Ogden (2001).

As in Section 2 we consider the deformation of a continuous body which in its initial (virgin) stress-free configuration occupies the region $\mathcal{B}_r$, material points being identified by their position vectors $\mathbf{X}$ in $\mathcal{B}_r$. After deformation the body occupies the region $\mathcal{B}$, and the point $\mathbf{X}$ is relocated to the position $\mathbf{x}$. The deformation gradient relating $\mathcal{B}$ to $\mathcal{B}_r$ is again denoted by $\mathbf{F}$.

In *pseudo-elasticity* the strain-energy function $W(\mathbf{F})$ of elasticity theory is modified by incorporating an additional variable, here denoted η, into the function and we write

$$W = W(\mathbf{F}, \eta). \tag{4.1}$$

The variable η is referred to as a *damage* variable since this reflects the underlying mechanism of the Mullins effect. It may also be called a *softening variable.* It provides a mechanism for changing the form of the energy function during the deformation process and hence changing the description of the material properties. In a deformation process in which η changes, the overall response of the material is not elastic and we refer to $W(\mathbf{F}, \eta)$ as a *pseudo-energy function.*

The variable η may be active or inactive and may be switched from active to inactive (or conversely) according to some suitable criterion, thereby inducing a change in the material properties. This change may be either continuous or discontinuous. Here we consider only the case in which the change is continuous.

If η is inactive we set it to some constant value, which here we take as unity, and we write

$$W_0(\mathbf{F}) = W(\mathbf{F}, 1), \tag{4.2}$$

for the resulting *strain-energy function.* Again we restrict attention to incompressible materials so that we have the constraint

$$\det \mathbf{F} = 1. \tag{4.3}$$

The nominal stress associated with (4.2) is denoted by $\mathbf{S}_0$ and is given by

$$\mathbf{S}_0 = \frac{\partial W_0}{\partial \mathbf{F}}(\mathbf{F}) - p_0 \mathbf{F}^{-1}, \quad \det \mathbf{F} = 1, \tag{4.4}$$

where p_0 is the notation used for the Lagrange multiplier associated with the constraint (4.3) in this case. We attach a zero subscript to quantities associated with inactive η ($\eta = 1$).

When η is active it is taken to depend on the deformation gradient (the damage evolves with deformation) and we write this dependence in the form

$$C(\mathbf{F}, \eta) = 0. \tag{4.5}$$

This provides an implicit connection between η and $\mathbf{F}$ in which the function C can be regarded as a constraint on these variables. We associate with this constraint a Lagrange multiplier, q say, so that the nominal stress $\mathbf{S}$ may be written

$$\mathbf{S} = \frac{\partial W}{\partial \mathbf{F}}(\mathbf{F}, \eta) + q\frac{\partial C}{\partial \mathbf{F}}(\mathbf{F}, \eta) - p\mathbf{F}^{-1}, \quad \det \mathbf{F} = 1, \tag{4.6}$$

where p is the counterpart of p_0 for active η.

Here we consider functions C for which (4.5) can (in principle if not explicitly) be solved uniquely for η in terms of $\mathbf{F}$, and we write the solution as

$$\eta = \eta_e(\mathbf{F}), \tag{4.7}$$

where the subscript e indicates that (4.7) is an explicit solution of (4.5). Note that for a general C the solution may not be unique, in which case discontinuities may arise, but, as already mentioned, we do not admit this possibility here. Using (4.7) we define the *strain-energy function* w by

$$w(\mathbf{F}) = W(\mathbf{F}, \eta_e(\mathbf{F})), \tag{4.8}$$

and the nominal stress $\mathbf{S}$ is then given by

$$\mathbf{S} = \frac{\partial w}{\partial \mathbf{F}} - p\mathbf{F}^{-1} = \frac{\partial W}{\partial \mathbf{F}}(\mathbf{F}, \eta) + \frac{\partial W}{\partial \eta}(\mathbf{F}, \eta)\frac{\partial \eta}{\partial \mathbf{F}}(\mathbf{F}) - p\mathbf{F}^{-1}, \tag{4.9}$$

again subject to (4.3). Comparison of (4.6) and (4.9) then shows that

$$q = -W_\eta / C_\eta, \tag{4.10}$$

where use has been made of the connection

$$\frac{\partial \eta}{\partial \mathbf{F}} = -\frac{\partial C}{\partial \mathbf{F}}\frac{1}{C_\eta},$$

obtained from (4.5), and the subscript η signifies partial differentiation with respect to η.

The function C in (4.5) is essentially arbitrary, subject to the uniqueness restriction imposed above, and is required to satisfy only the appropriate objectivity condition (as discussed briefly in Section 4.1 below). Moreover, the dependence of W on η is also arbitrary and this flexibility ensures that there is no generality lost by setting $C = W_\eta$. This amounts to setting the Lagrange multiplier q to zero. The constraint (4.5) then takes the form

$$\frac{\partial W}{\partial \eta}(\mathbf{F}, \eta) = 0. \tag{4.11}$$

The expression (4.6), or equivalently (4.9), for the nominal stress then reduces to

$$\mathbf{S} = \frac{\partial w}{\partial \mathbf{F}}(\mathbf{F}) - p\mathbf{F}^{-1} = \frac{\partial W}{\partial \mathbf{F}}(\mathbf{F}, \eta) - p\mathbf{F}^{-1} \quad \det \mathbf{F} = 1, \tag{4.12}$$

where the right-hand side is evaluated for η given by (4.7), or implicitly by (4.11). The nominal stress is given by (4.12) whether η is active or inactive. In the latter case (4.12) reduces to (4.4).

Equation (4.11) defines a hypersurface in the 9-dimensional $(\mathbf{F}, \eta)$-space to which values of η must be restricted, bearing in mind that $\mathbf{F}$ satisfies (4.3). The hypersurface (4.11) identifies stationary points of $W(\mathbf{F}, \eta)$ with respect to η. When η is defined uniquely in terms of $\mathbf{F}$, as we have assumed, we may write the solution formally as (4.7), and we then use the notation w for the resulting (unique) strain-energy function, as defined by (4.8). The details of the restrictions on η depend on the particular model considered.

We may regard equation (4.11) as a field equation, which, in the absence of body forces, is coupled with the equilibrium equation in the form

$$\text{Div}\,\mathbf{S} = \mathbf{0} \quad \text{in } \mathcal{B}_r, \tag{4.13}$$

where, as in Section 2, Div denotes the divergence operator in $\mathcal{B}_r$.

Thus far we have not specified the form of the dependence of W on η, or, more particularly, the form of the function $\eta_e(\mathbf{F})$ in (4.7), i.e. we have not specified a particular model within the general framework of pseudo-elasticity. Thus, there remains considerable flexibility in the choice of specific models. Moreover, we have not identified a criterion for switching η on or off. Such considerations depend on the application to be considered. In the application to stress softening associated with the Mullins effect that we are considering here, η is taken to be inactive during loading and to switch on during unloading (with loading and unloading being defined relative to the energy expended on a loading path). In this example, the energy changes continuously from its value $W_0(\mathbf{F}) = W(\mathbf{F}, 1)$ to $W(\mathbf{F}, \eta)$ with η decreasing from the value 1 as unloading proceeds. Corresponding measures of stress also change continuously. This will be discussed in Section 5.

Note that for an incompressible material the Cauchy stress tensors, denoted by $\boldsymbol{\sigma}_0$ and $\boldsymbol{\sigma}$ for inactive and active η respectively, are related to $\mathbf{S}_0$ and $\mathbf{S}$, respectively, by

$$\boldsymbol{\sigma}_0 = \mathbf{F}\mathbf{S}_0, \qquad \boldsymbol{\sigma} = \mathbf{F}\mathbf{S}. \tag{4.14}$$

4.1 Objectivity

As in elasticity theory (Section 2.3.1) we require the constitutive law to be objective. This means that $W(\mathbf{F}, \eta)$ must be objective whether η is active or inactive. If η is inactive this just requires that

$$W_0(\mathbf{QF}) = W_0(\mathbf{F}) \tag{4.15}$$

for all rotations $\mathbf{Q}$ and every deformation gradient $\mathbf{F}$. When η is active it suffices to take it to be an objective (scalar) function of $\mathbf{F}$, so that

$$\eta_e(\mathbf{QF}) = \eta_e(\mathbf{F}) \tag{4.16}$$

for all rotations $\mathbf{Q}$ and every deformation gradient $\mathbf{F}$. It follows that for $W(\mathbf{F}, \eta)$ to be objective we must have

$$W(\mathbf{QF}, \eta) = W(\mathbf{F}, \eta) \tag{4.17}$$

for all rotations $\mathbf{Q}$ and every deformation gradient $\mathbf{F}$, with $\eta = \eta_e(\mathbf{F})$ satisfying (4.16).

Note that objectivity of $W_0(\mathbf{F})$ does not in general guarantee that of $W(\mathbf{F}, \eta)$, which has to be imposed separately. Equally, it is important to note that since we make no assumption about η in respect of material symmetry, any material symmetry ascribed to $W_0(\mathbf{F})$ is not in general inherited by $W(\mathbf{F}, \eta)$.

4.2 Material symmetry

Material symmetry for an elastic material with a strain-energy function $W(\mathbf{F})$ has been discussed in Section 2.3.1. In the present context we need to consider separately the material symmetries of a pseudo-elastic material when η is active and inactive and how these symmetries are related. In general, as indicated above, the symmetry of $W_0(\mathbf{F})$ is different from (and independent of) that of $w(\mathbf{F})$, which is defined in (4.8). To illustrate this point we consider the model pseudo-energy function given by

$$W(\mathbf{F}, \eta) = W_0(\mathbf{F}) + (\eta - 1)N(\mathbf{F}) + \phi(\eta). \tag{4.18}$$

In (4.18) the function ϕ, which depends only on η, is referred to as a *damage* or *softening function* and, for consistency with (4.2), it must satisfy

$$\phi(1) = 0, \tag{4.19}$$

while $N(\mathbf{F})$ is an objective function of $\mathbf{F}$, independent of W_0, whose (material) symmetry properties are in general different from those of $W_0(\mathbf{F})$. The specialization of (4.11) in this case is

$$\phi'(\eta) = -N(\mathbf{F}). \tag{4.20}$$

The nominal stress associated with (4.18) is

$$\mathbf{S} = \mathbf{S}_0 + (\eta - 1)\frac{\partial N}{\partial \mathbf{F}} - (p - p_0)\mathbf{F}^{-1}, \tag{4.21}$$

where $\mathbf{S}_0$ is given by (4.4) and if $\eta = 1$ then $p = p_0$.

For the special case in which $N = W_0$ equation (4.18) reduces to

$$W(\mathbf{F}, \eta) = \eta W_0(\mathbf{F}) + \phi(\eta), \tag{4.22}$$

and the (material) symmetry is the same whether η is active or inactive. The isotropic specialization of (4.22) was used by Ogden and Roxburgh (1999a, b) in their model of the Mullins effect and will be discussed in Section 5.

With $N = W_0$ equation (4.20) specializes to

$$\phi'(\eta) = -W_0(\mathbf{F}). \tag{4.23}$$

Equation (4.23), or, more generally, (4.20) enables η to be determined in terms of $\mathbf{F}$, at least in principle. Note, however, that in the case of (4.23) η depends on $W_0(\mathbf{F})$ and, as already intimated for (4.22), the symmetry is therefore unchanged when η is switched on or off.

Thus far in the literature applications of pseudo-elasticity have been based, for simplicity, on the assumption that the material response is isotropic relative to $\mathcal{B}_r$ and that it remains isotropic when the material properties change. In general, however, this may not be the case since, for example, the material response relative to a (stress-free) residually strained configuration may be anisotropic. In the formulation based on (4.7), $\eta_e(\mathbf{F})$ can be selected so that the material symmetry of $w(\mathbf{F})$ in (4.8) either inherits the material symmetry of $W_0(\mathbf{F})$ or has material symmetry independent of that of $W_0(\mathbf{F})$. Hence, switching on or off of η can change the material symmetry. Thus, for example, if the material is initially isotropic then the symmetry induced by the generation of residual strain will (in general) be locally orthotropic. Detailed discussion of the characterization of changes in material symmetry for a pseudo-elastic material will be given in the paper by Ogden (2003) and some initial discussion of this matter is contained in the forthcoming paper by Horgan *et al.* (2003). In the remainder of this section, however, we restrict attention to isotropic material response for simplicity.

4.3 Isotropic material response

When specialized to isotropic response (relative to $\mathcal{B}_r$) the pseudo-elastic energy function (4.1) takes the form

$$W(\lambda_1, \lambda_2, \lambda_3, \eta), \tag{4.24}$$

where $(\lambda_1, \lambda_2, \lambda_3)$ are the principal stretches associated with the deformation from $\mathcal{B}_r$. Since the material is incompressible the stretches satisfy the constraint

$$\lambda_1\lambda_2\lambda_3 = 1. \tag{4.25}$$

The Cauchy stress tensor $\boldsymbol{\sigma}$, for an incompressible material, is related to the nominal stress $\mathbf{S}$ by $(4.14)_2$ and for an isotropic material $\boldsymbol{\sigma}$ is coaxial with the Eulerian principal axes, which are the principal axes of $\mathbf{F}\mathbf{F}^T$ (recall Section 2.3.2). For the considered incompressible isotropic material it follows that the principal Cauchy stresses σ_i, $i = 1, 2, 3$, are given by

$$\sigma_i = \lambda_i \frac{\partial W}{\partial \lambda_i} - p, \quad i \in \{1, 2, 3\}, \tag{4.26}$$

whether or not η is active. Equation (4.11) specializes to

$$\frac{\partial W}{\partial \eta}(\lambda_1, \lambda_2, \lambda_3, \eta) = 0, \tag{4.27}$$

which gives η implicitly in terms of the stretches.

Following the pattern of Section 2.4, we define the modified pseudo-energy function $\hat{W}(\lambda_1, \lambda_2, \eta)$ by

$$\hat{W}(\lambda_1, \lambda_2, \eta) = W(\lambda_1, \lambda_2, \lambda_1^{-1}\lambda_2^{-1}, \eta) \tag{4.28}$$

so that, on elimination of p from (4.26),

$$\sigma_1 - \sigma_3 = \lambda_1 \hat{W}_1, \quad \sigma_2 - \sigma_3 = \lambda_2 \hat{W}_2, \tag{4.29}$$

where $\hat{W}_1$ and $\hat{W}_2$ denote the partial derivatives of $\hat{W}$ with respect to λ_1 and λ_2 respectively. Equation (4.27) is then modified to

$$\frac{\partial \hat{W}}{\partial \eta}(\lambda_1, \lambda_2, \eta) = 0 \tag{4.30}$$

so that η is now given implicitly in terms of λ_1 and λ_2 only.

When η is active we suppose that equation (4.30) can be solved uniquely for η and, using the notation from (4.7), we write

$$\eta = \eta_e(\lambda_1, \lambda_2) = \eta_e(\lambda_2, \lambda_1). \tag{4.31}$$

Then, an energy function for active η, symmetrical in (λ_1, λ_2) and denoted $\hat{w}(\lambda_1, \lambda_2)$, may be defined by

$$\hat{w}(\lambda_1, \lambda_2) \equiv \hat{W}\big(\lambda_1, \lambda_2, \eta_e(\lambda_1, \lambda_2)\big). \tag{4.32}$$

From equations (4.29), (4.30) and (4.32) it follows that

$$\sigma_\alpha - \sigma_3 = \lambda_\alpha \frac{\partial \hat{w}}{\partial \lambda_\alpha} = \lambda_\alpha \frac{\partial \hat{W}}{\partial \lambda_\alpha}, \qquad \alpha = 1, 2. \tag{4.33}$$

For inactive η we define the function $\hat{W}_0(\lambda_1, \lambda_2)$ via

$$\hat{W}_0(\lambda_1, \lambda_2) \equiv \hat{W}(\lambda_1, \lambda_2, 1), \tag{4.34}$$

which is the isotropic specialization of (4.2). From (4.29) the specialization (4.34) yields the stresses

$$\sigma_{0\alpha} - \sigma_{03} = \lambda_\alpha \hat{W}_{0\alpha}, \qquad \alpha = 1, 2, \tag{4.35}$$

where the subscript zero again refers to a deformation path on which η is not active, so that (4.30) is not operative. A subscript α following the subscript 0 on $\hat{W}$ indicates partial differentiation with respect to λ_α $(\alpha = 1, 2)$.

From equations (2.47) we deduce that $\hat{W}_0(\lambda_1, \lambda_2)$ satisfies

$$\hat{W}_0(1,1) = 0, \quad \hat{W}_{0\alpha}(1,1) = 0, \quad \hat{W}_{012}(1,1) = 2\mu, \quad \hat{W}_{0\alpha\alpha}(1,1) = 4\mu, \tag{4.36}$$

where $\mu\,(> 0)$ is the shear modulus of the material in $\mathcal{B}_r$ and the index α takes the value 1 or 2.

Simple tension For the simple tension specialization of the above equations we may take $\sigma_2 = \sigma_3 = 0$ and we write $\sigma_1 = \sigma$. We also write $\lambda_1 = \lambda$, so that $\lambda_2 = \lambda_3 = \lambda^{-1/2}$, and define $\tilde{W}$ by

$$\tilde{W}(\lambda, \eta) \equiv \hat{W}(\lambda, \lambda^{-1/2}, \eta). \tag{4.37}$$

Equations (4.33) and (4.30) then simplify to

$$\sigma = \lambda \tilde{W}_\lambda(\lambda, \eta) \equiv \lambda t, \quad \tilde{W}_\eta(\lambda, \eta) = 0, \tag{4.38}$$

wherein the principal Biot stress $t\,(= t_1)$ is defined and the subscripts signify partial derivatives.

From the second (implicit) equation in (4.38), η is (in principle) determined in terms of λ, and from equations (4.38) it follows that

$$\frac{dt}{d\lambda} = \tilde{W}_{\lambda\lambda} - (\tilde{W}_{\lambda\eta})^2/\tilde{W}_{\eta\eta}. \tag{4.39}$$

Equation (4.39) shows how the inclusion of the variable η modifies the stiffness $\tilde{W}_{\lambda\lambda}$ of the material appropriate for the case of inactive η.

By defining

$$\tilde{W}_0(\lambda) = \tilde{W}(\lambda, 1), \tag{4.40}$$

we may deduce from (4.36) the specializations

$$\tilde{W}_0(1) = \tilde{W}_0'(1) = 0, \quad \tilde{W}_0''(1) = 3\mu, \tag{4.41}$$

where the prime signifies differentiation with respect to λ.

This simple tension specialization will be examined in detail in connection with the modelling of the Mullins stress softening effect in Section 5.

Pure shear For a pure shear deformation we set $\lambda_2 = 1$ and $\sigma_3 = 0$ and use the notation

$$\lambda = \lambda_1, \qquad \lambda_3 = \lambda^{-1}. \tag{4.42}$$

The pseudo-energy function is then given in terms of λ by the definition

$$\tilde{W}(\lambda, \eta) \equiv \hat{W}(\lambda, 1, \eta), \tag{4.43}$$

and equations (4.33) and (4.30) respectively specialize to

$$\sigma_1 = \lambda\tilde{W}_\lambda, \qquad \tilde{W}_\eta = 0, \tag{4.44}$$

where, once more, the subscripts denote partial derivatives. Note that in this case there is a lateral stress, σ_2, which is given by $\sigma_2 = \hat{W}_2(\lambda, 1)$.

Equibiaxial deformations The appropriate specialization of equations (4.30) and (4.33) for equibiaxial deformations are now considered. We set $\lambda_2 = \lambda_3 = \lambda, \lambda_1 = \lambda^{-2}$ so that $\sigma_2 = \sigma_3$, and we define the function $\tilde{W}$ by

$$\tilde{W}(\lambda, \eta) = \hat{W}(\lambda^{-2}, \lambda, \eta). \tag{4.45}$$

It follows from (4.33) and (4.45) that

$$2(\sigma_2 - \sigma_1) = \lambda\tilde{W}_\lambda(\lambda, \eta) \tag{4.46}$$

and $\hat{W}_2 = 0$, while (4.30) simplifies to

$$\tilde{W}_\eta(\lambda, \eta) = 0, \tag{4.47}$$

where the subscripts λ and η again denote partial derivatives.

We now write

$$\tilde{W}_0(\lambda) = \tilde{W}(\lambda, 1) \tag{4.48}$$

for the case in which η is inactive. Then, the analogues of (4.41) are

$$\tilde{W}_0(1) = \tilde{W}_0'(1) = 0, \quad \tilde{W}_0''(1) = 12\mu. \tag{4.49}$$

It should be emphasized that the definition of $\tilde{W}(\lambda, \eta)$ is different for each of the above three deformations.

5 Theory and experiment compared: the Mullins effect

5.1 A model for stress softening

The idealized Mullins effect We recall that Figure 6 in Section 1 depicts the idealized Mullins effect in simple tension. This effect was described in Section 1, but some supplementary comments are appropriate here. We emphasize that the curve abcd in Figure 6 models the material response on a *loading* path from the unstressed configuration of the virgin material if there is no unloading (this is referred to as the *primary loading path*). The curve b′Ba is a typical unloading curve from a point on the primary loading path. The idealized Mullins effect requires that each unloading path passes through the origin, i.e. the stretch λ returns to 1 when the stress t is removed ($t = 0$). Moreover, if the material is reloaded after the initial unloading the unloading path is retraced until the primary loading curve is met, after which the primary loading path is re-joined on further loading. This is the second main aspect of the idealization. This description applies whichever point on the primary loading path the unloading begins from. For further discussion see Ogden and Roxburgh (1999a), which contains detailed references. The original work of Mullins is contained in a number of papers, including Mullins (1947, 1969) and Mullins and Tobin (1957).

In this section we focus on a specific form of the pseudo-elastic constitutive law, that used by Ogden and Roxburgh (1999a) to model the idealized Mullins effect. The material is taken to be incompressible and isotropic and we use the pseudo-energy function (4.22) in its isotropic form, subject to (4.19), together with the notation of Section 4.3. We consider homogeneous biaxial deformations with (λ_1, λ_2) taken as the independent deformation variables. The considered pseudo-energy function is

$$\hat{W}(\lambda_1, \lambda_2, \eta) = \eta \hat{W}_0(\lambda_1, \lambda_2) + \phi(\eta), \tag{5.1}$$

with

$$\phi(1) = 0. \tag{5.2}$$

From (4.35), (4.33) and (5.1), the Cauchy stress differences are calculated as

$$\sigma_\alpha - \sigma_3 = \eta \lambda_\alpha \frac{\partial \hat{W}_0}{\partial \lambda_\alpha} = \eta(\sigma_{0\alpha} - \sigma_{03}), \qquad \alpha = 1, 2, \tag{5.3}$$

and equation (4.23) becomes

$$\phi'(\eta) = -\hat{W}_0(\lambda_1, \lambda_2), \tag{5.4}$$

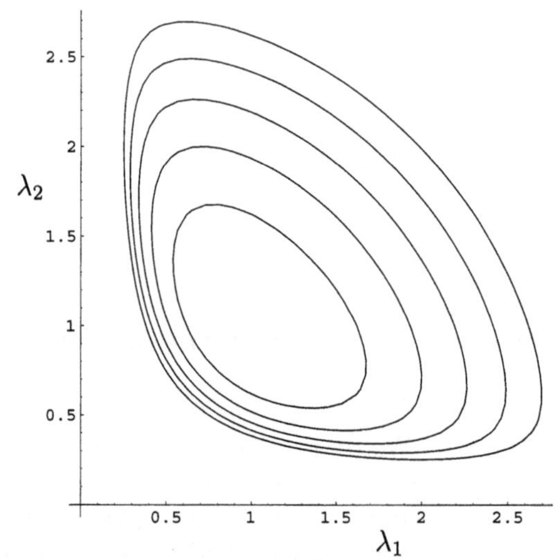

Figure 13. Contours of constant strain energy in (λ_1, λ_2)-space for the neo-Hookean strain-energy function (5.5) with (dimensionless) values $\hat{W}_0/\mu = 0.5, 1, 1.5, 2, 2.5$.

which defines the damage parameter η (implicitly) in terms of the stretches. We emphasize that (5.1) is a very simple specialization of the general pseudo-elastic energy model.

We recall that $\hat{W}_0$ satisfies the properties (4.36). We assume, additionally, that $\hat{W}_0$ has a *global minimum* $(= 0)$ at $(1, 1)$ and has no other stationary points in (λ_1, λ_2)-space. This is reasonable since, for example, in the deformation of a thin sheet with $\sigma_{03} = 0$, any other stationary point requires that both (in-plane) stresses σ_{01} and σ_{02} in (5.3) vanish. In practice this would happen only for pathological forms of $\hat{W}_0$.

We therefore *define* a primary loading path in (λ_1, λ_2)-space as a path starting from $(1, 1)$ on which $\hat{W}_0$ is *increasing*. Indeed, it can be shown that for many standard forms of strain-energy function $\hat{W}_0$ is increasing along any straight line path from $(1, 1)$. A case in point is the neo-Hookean strain-energy function, which, in terms of λ_1 and λ_2, is written

$$\hat{W}_0(\lambda_1, \lambda_2) = \frac{\mu}{2}(\lambda_1^2 + \lambda_2^2 + \lambda_1^{-2}\lambda_2^{-2} - 3), \tag{5.5}$$

where μ is the shear modulus appearing in (4.36). To illustrate this point contours of constant values of $\hat{W}_0$, which, it should be emphasized, are convex, are plotted in Figure 13 in the (λ_1, λ_2)-plane in respect of (5.5).

We emphasize that, in general, the value of η derived from (5.4) will depend on the values of the principal stretches λ_{1m} and λ_{2m} attained on a primary loading path, as well as on the specific forms of $\hat{W}_0(\lambda_1, \lambda_2)$ and $\phi(\eta)$ used. Since $\eta = 1$ at any point on the primary loading path from which unloading is initiated, it follows from equation (5.4) that

$$-\phi'(1) = \hat{W}_0(\lambda_{1m}, \lambda_{2m}) \equiv W_m, \tag{5.6}$$

wherein the notation W_m is defined. In accordance with the properties of $\hat{W}_0$, W_m increases along a primary loading path.

Simple tension For the simple tension specialization of the above equations we refer to the relevant notation in Section 4.3.1. In terms of the Biot stress t we then have

$$t = \eta \tilde{W}_0'(\lambda) = \eta t_0, \tag{5.7}$$

where t_0 is the Biot stress on the primary loading path at the same value of λ. For (5.7) to predict stress softening, with unloading governed by the energy function (5.1) and η activated at the start of unloading, we must have $\eta \leq 1$ on the unloading path, with equality only at the point where unloading begins. Here we take $\eta > 0$, so that t remains positive on unloading until $\lambda = 1$ is reached. This means that residual strains will not arise. In order to incorporate residual strains (at least for this simple model) it is necessary to relax this restriction, and this will be discussed in Section 5.4.

For simple tension (5.4) specializes to

$$\phi'(\eta) = -\tilde{W}_0(\lambda), \tag{5.8}$$

and differentiation of this with respect to λ with η regarded as a function of λ leads to

$$\phi''(\eta)\frac{d\eta}{d\lambda} = -\tilde{W}_0'(\lambda). \tag{5.9}$$

For consistency with the stress softening feature discussed above unloading must be associated with decreasing η, and since $t_0 \equiv \tilde{W}_0'(\lambda) > 0$ for $\lambda > 1$ it follows from (5.9) that

$$\phi''(\eta) < 0. \tag{5.10}$$

We assume henceforth that this inequality holds, a consequence of which is that $\phi'(\eta)$ is a monotonic decreasing function of η. Hence η is determined *uniquely* from (5.8) as a function of $\tilde{W}_0(\lambda)$, at least in principle if not explicitly. A similar conclusion may be reached in the case of biaxial deformations, i.e. η is uniquely determined as a function of the loading energy $\hat{W}_0(\lambda_1, \lambda_2)$.

Let λ_m be the principal stretch achieved in simple tension on a primary loading path. Then, by specializing the discussion in the paragraph containing equation (5.6), we see that the value of η derived from (5.8) depends on the value of λ_m attained on such a path as well as on the forms of $\tilde{W}_0(\lambda)$ and $\phi(\eta)$ employed. Moreover, on specializing (5.6) we obtain

$$\phi'(1) = -\tilde{W}_0(\lambda_m) \equiv W_m, \tag{5.11}$$

and we note that W_m is the current maximum value of the energy achieved on the primary loading path. In accordance with the properties of $\tilde{W}_0$, W_m increases along a primary loading path. In view of (5.11), the function ϕ depends (implicitly) on the point from which unloading begins through the energy expended on the loading path up to that point.

For the idealized Mullins effect, when the material is fully unloaded, with $\lambda = 1$, η attains its minimum value, η_m say. This is determined by inserting these values into equation (5.8) to give

$$\phi'(\eta_m) = -\tilde{W}_0(1) = 0, \tag{5.12}$$

where the first equation in (4.41) has been used. Since the function ϕ depends (implicitly) on the point where unloading begins then so must η_m. Thus, η_m depends on W_m and hence on λ_m. In this fully unloaded state the pseudo-energy function (5.1) has the residual value

$$\hat{W}(1, 1, \eta_m) = \phi(\eta_m). \tag{5.13}$$

This applies for general equibiaxial deformations as well as for the simple tension specialization. Thus, the residual (non-recoverable) energy $\phi(\eta_m)$ is associated with the damage in the material and is a measure of departure from pure elasticity. In simple tension $\phi(\eta_m)$ is the area between the primary loading curve and the relevant unloading curve in Figure 6.

Discussion As observed above for simple tension, unloading may be initiated from any point on a primary loading path, and the start of unloading acts as a signal for η to be activated. Recalling the notation $(\lambda_{1m}, \lambda_{2m})$ used in (5.6), then, in the notation of (4.31), $\eta_e(\lambda_{1m}, \lambda_{2m}) = 1$, which implies that the function η_e, and hence $\hat{w}$ in (4.32), depends on the point from which unloading starts. In simple tension, as discussed above, the point at which unloading began was identified by the value of the stretch (λ_m) at that point, or, equivalently, by the associated energy on the primary path. For biaxial deformations it is the value of the energy maximum on the primary loading path, again denoted W_m and given by (5.6), rather than the specific $(\lambda_{1m}, \lambda_{2m})$ pair that governs the unloading response. Thus, any other pair of (λ_1, λ_2) values corresponding to the same value of W_m could equally be taken as the starting point for unloading. The collection of all such pairs satisfies the equation

$$\hat{W}_0(\lambda_1, \lambda_2) = W_m, \tag{5.14}$$

and, for each given fixed value of W_m, forms a closed contour in (λ_1, λ_2)-space, as exemplified for the neo-Hookean strain-energy function (5.5) in Figure 13.

The contour defined by the current maximum value W_m represents the current damage threshold, and $\eta = 1$ at any point on this contour. For any deformation path within this contour $\eta < 1$ and no further damage occurs. The energy required on the initial loading path to cause the damage is not required on subsequent loading up to the contour boundary. However, if the deformation path crosses the contour primary loading is again activated, $\eta = 1$ and further damage will occur. The value of $\hat{W}_0(\lambda_1, \lambda_2)$ will increase until the next maximum value W_m is reached, at which point loading terminates and a new threshold contour is established (which encloses the previous one). It should be emphasized at this point that here we are modelling the *idealized* Mullins effect on the basis of *isotropic* theory, which, necessarily, oversimplifies the actual material behaviour. Some discussion of the limitations of the isotropy assumption is provided at the end of Section 5.4 following some analysis of the influence of residual strains.

Several other models of the Mullins effect, some of which have features in common with the theory of pseudo-elasticity discussed here, have been developed recently in the literature. For example, we mention Johnson and Beatty (1993a, b), who considered uniaxial deformations, and Beatty and Krishnaswamy (2000) and Krishnaswamy and Beatty (2000), who analyzed general deformations for incompressible and compressible materials respectively. Other contributions are those of Govindjee and Simo (1991, 1992a, b), DeSimone *et al.* (2001) and Marckmann *et al.* (2002).

5.2 The damage function $\phi(\eta)$

The function ϕ serves to determine the damage parameter in terms of the state of deformation through equation (5.4). The choice of $\phi(\eta)$ is arbitrary subject to (5.2), (5.6) and the inequality (5.10), with η satisfying $0 < \eta \leq 1$.

Let λ_{1m} and λ_{2m} be the values of the principal stretches corresponding to the point at which unloading has most recently been initiated from a primary loading path and $\hat{W}_0(\lambda_{1m}, \lambda_{2m}) = W_m$. Then, we choose $\phi(\eta)$ to be such that

$$-\phi'(\eta) = \mu m \,\mathrm{erf}^{-1}\big(r(\eta - 1)\big) + W_m, \tag{5.15}$$

where m and r are positive dimensionless parameters (material constants), μ is the shear modulus in $\mathcal{B}_r$, and erf^{-1} is the inverse of the error function, which, as pointed out by Ogden and Roxburgh (1999a), has properties particularly well suited to the task at hand. It is not, however, the only possible function for this purpose — for example, the inverse hyperbolic tangent $\tanh^{-1}$ is equally effective, and was used by Dorfmann and Ogden (2003). Note that (5.15) differs from the formula in Ogden and Roxburgh (1999a) in that the factor μ has been introduced to effect the non-dimensionalization.

It is not necessary to integrate equation (5.15) in order to satisfy (5.2) explicitly, but we note that (5.6) follows immediately from (5.15). The properties of the error function ensure that the inequality (5.10) holds. On use of (5.15) equation (5.4) may be rearranged to give an expression for η, namely

$$\eta = 1 - \frac{1}{r}\mathrm{erf}\left[\frac{1}{\mu m}\big(W_m - \hat{W}_0(\lambda_1, \lambda_2)\big)\right]. \tag{5.16}$$

Since m and r were defined to be positive and, by definition, $\hat{W}_0(\lambda_1, \lambda_2) \leq W_m$, it follows that $\eta \leq 1$, with equality holding only when $\hat{W}_0(\lambda_1, \lambda_2) = W_m$. Moreover, if we set $r > 1$ then η is necessarily positive. The minimum value η_m of η, corresponding to the undeformed configuration, is given by

$$\eta_m = 1 - \frac{1}{r}\mathrm{erf}\,(W_m/\mu m). \tag{5.17}$$

The parameters r and m have the following physical interpretations. First, we note that r is a measure of the extent of the damage relative to the virgin state. In particular, from (5.16) it follows that the larger the value of r the less the damage parameter η can depart from unity and the less damage occurs. On the other hand, the parameter m controls the distribution of the damage through the range of deformation. For small values of m, significant damage is caused for relatively small strains, and the material response in the small strain region is not markedly affected by further primary loading. For larger values of m there is relatively little damage for small strains but the material response changes significantly in the small strain region after subsequent primary loading.

Figure 14 illustrates the above observations in respect of a neo-Hookean material undergoing pure shear with r and m varied independently. In each of the six figures unloading is initiated from three separate values of the stretch λ, namely $\lambda = 2, 3, 4$. The dashed curve in each of the figures in the right-hand column is, for fixed r, the asymptotic unloading/reloading curve for large primary deformation. The equations of pure shear on which Figure 14 is based are given below in Section 5.3.

Note that if $r < 1$ were allowed then η, which enters as a multiplicative factor in the pseudo-energy function (5.1), and hence the stress would vanish before the undeformed configuration is recovered. This possibility has been ruled out in our considerations thus far but will be discussed in Section 5.4.

5.3 Numerical results

In this section we consider first the example of pure shear to highlight the theory developed above, in particular to show the relative behaviour of the main and side stresses with respect to stress softening. The theory is then used to fit the (somewhat limited) experimental data obtained by Mullins and Tobin (1957) from simple tension tests.

Pure shear When $\sigma_3 = 0$ equations (5.3) reduce to

$$\sigma_1 = \eta\lambda_1 \frac{\partial \hat{W}_0}{\partial \lambda_1}, \qquad \sigma_2 = \eta\lambda_2 \frac{\partial \hat{W}_0}{\partial \lambda_2}. \tag{5.18}$$

In pure shear equations (5.18) are specialized by setting $\lambda_1 = \lambda$, $\lambda_2 = 1$ and $\lambda_3 = \lambda^{-1}$. For a neo-Hookean material, defined by (5.5), the pure shear specialization of (5.18) yields

$$\sigma_1 = \lambda t_1 = \eta\mu(\lambda^2 - \lambda^{-2}), \qquad \sigma_2 = t_2 = \eta\mu(1 - \lambda^{-2}), \tag{5.19}$$

t_1 and t_2 being the relevant principal Biot stresses, the latter referred to as the side stress.

Corresponding expressions for the strain-energy function

$$\hat{W}_0(\lambda_1, \lambda_2) = \mu \sum_{m=1}^{N} \frac{\mu_m^*}{\alpha_m} \left(\lambda_1^{\alpha_m} + \lambda_2^{\alpha_m} + \lambda_1^{-\alpha_m}\lambda_2^{-\alpha_m} - 3\right) \tag{5.20}$$

(Ogden, 1972) are

$$\sigma_1 = \lambda t_1 = \eta\mu \sum_{m=1}^{N} \mu_m^*(\lambda^{\alpha_m} - \lambda^{-\alpha_m}), \qquad \sigma_2 = t_2 = \eta\mu \sum_{m=1}^{N} \mu_m^*(1 - \lambda^{-\alpha_m}), \tag{5.21}$$

where $\mu_m^* = \mu_m/\mu$ are dimensionless material constants satisfying $\sum_{m=1}^{N} \mu_m^* \alpha_m = 2$, the α_m themselves being dimensionless material constants.

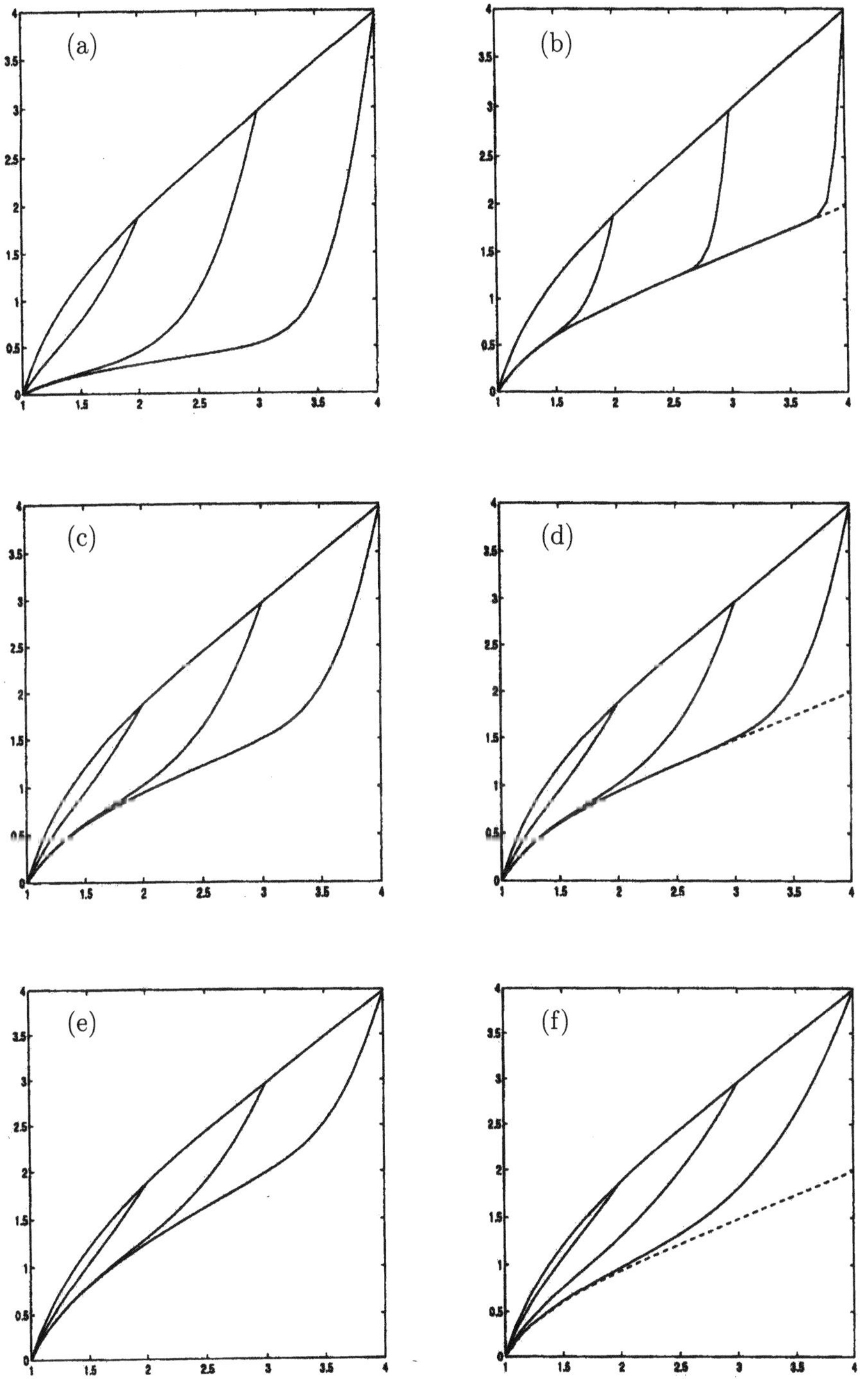

Figure 14. Plots of the main Biot stress t_1 in pure shear against the corresponding stretch λ for the neo-Hookean material, showing the effect of changing the parameters r (left column) and m (right column): (a) $r = 1.2, m = 2.0$; (b) $r = 2.0, m = 0.5$; (c), (d) $r = 2.0, m = 2.0$; (e) $r = 3.0, m = 2.0$; (f) $r = 2.0, m = 4.0$.

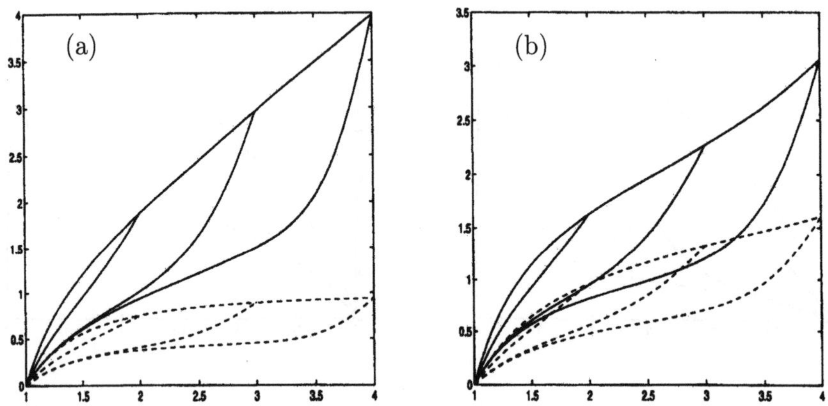

Figure 15. Pure shear loading/unloading curves for (a) the neo-Hookean material and (b) Ogden material. The main Biot stresses t_1 (continuous curves) and the side stresses t_2 (broken curves) are plotted against λ.

For illustration we consider (5.20) with $N = 3$ and the material constants given by

$$\alpha_1 = 1.3, \quad \alpha_2 = 5.0, \quad \alpha_3 = -2.0,$$
$$\mu_1^* = 1.491, \quad \mu_2^* = 0.003, \quad \mu_3^* = -0.0237,$$

(see, for example, Ogden 1972, 1997).

We now simulate a pure shear experiment in which the material is subjected to a series of loadings with the sample returned to the undeformed configuration after each loading and then reloaded beyond the previous maximum. We use the damage model discussed in Section 5.1 together with (5.16) and $m = r = 2.0$. The resulting values of the Biot stresses t_1 (main stress) and t_2 (side stress) for the constitutive equations (5.5) and (5.20) are displayed in Figures 15(a) and 15(b) respectively. In each case the continuous curves correspond to the main stress t_1, and the dashed curves to the side stresses t_2 required to maintain the pure shear deformation.

One serious problem that restricts modelling attempts in respect of the Mullins effect and other phenomena, such as stress-strain cycling hysteresis, is that there is no comprehensive set of data available for pure shear with which to compare the theory. In general there is a shortage of good data, particularly in respect of biaxial deformations. Even for simple tension, which we discuss below, the available data are limited.

Simple tension In order to compare the theory with available experimental results, we consider the special case of simple tension, for which $\lambda_1 = \lambda$ and $\lambda_2 = \lambda_3 = \lambda^{-1/2}$. With this specialization, $\sigma_2 = 0$ and the remaining non-vanishing principal stress is given by the first equation in (5.18). For the neo-Hookean material, we therefore have

$$\sigma_1 = \lambda t_1 = \eta\mu(\lambda^2 - \lambda^{-1}), \tag{5.22}$$

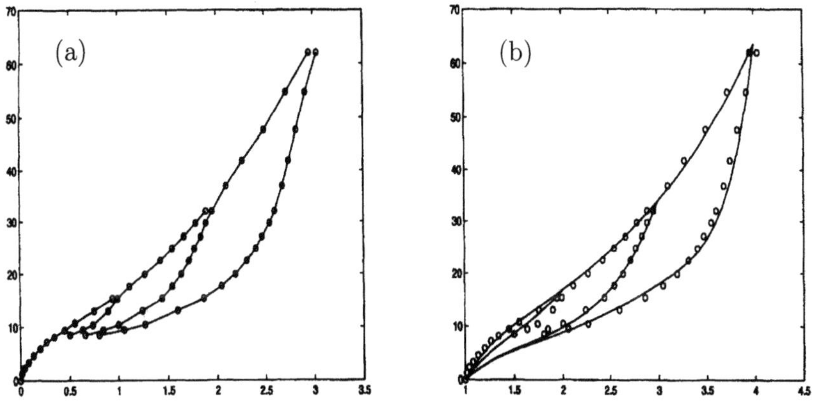

Figure 16. (a) Simple tension data (circles) from Mullins and Tobin (1957); (b) best fit curves for the pseudo-elastic model discussed in Sections 5.1 and 5.2, after Ogden and Roxburgh (1999a).

and similarly for (5.20):

$$\sigma_1 = \lambda t_1 = \eta\mu \sum_{m=1}^{N} \mu_m^*(\lambda^{\alpha_m} - \lambda^{-\alpha_m/2}). \qquad (5.23)$$

In Figure 16(a) we show data taken from Mullins and Tobin (1957, Figure 4) for a simple tension experiment in which the stress $t = t_1$ is plotted against the strain measure $\epsilon = \lambda - 1$ (the units of t are kg cm^{-2}). All the measurements were taken during loading and no unloading data points were provided. It should be noted that the material used by Mullins and Tobin (1957) exhibited permanent set on removal of the stress, so that a given sample was longer after the initial stretch than it was in its virgin state. This meant that after the initial loading the raw data curves did not pass through the origin, but intercepted the horizontal axis at some positive value. In their original paper Mullins and Tobin (1957) adjusted their data by subtracting the permanent set from the length measurements to ensure that the stress-strain curves passed through the origin. Consequently, the data points on the reloading curves have been shifted to the left from their true positions by the amount of the 'set' strain (which is different for each reloading). There must therefore be some uncertainty about the precise interpretation of the data.

Figure 16(b) shows the result of a nonlinear least squares best fit solution to the data based on a strain-energy function of the form (5.20). In this case t is plotted against λ rather than ϵ. First, a best fit was carried out for the primary loading curve to obtain the appropriate values for the material constants in (5.20). These values are

$$\mu_1 = 8.00, \quad \mu_2 = 0.76, \quad \mu_3 = -4.50 \quad (\text{kg cm}^{-2}),$$
$$\alpha_1 = 1.25, \quad \alpha_2 = 4.0, \quad \alpha_3 = -2.0.$$

With the primary loading curve fitted it is then a simple matter to find suitable values of the parameters r and m in (5.16) to fit the reloading curves. The values obtained are $r = 2.104$ and $m = 2.037$. The results are shown as continuous curves in Figure 16(b). The comparison between the theory and data is reasonable bearing in mind the reservations expressed above concerning the reliability of the data. It is clear, however, that better quality data are needed.

5.4 Residual strain

Thus far we have assumed that the damage variable η is strictly positive and we have noted that this condition excludes the possibility of residual strains arising when the stress vanishes. We now examine the consequences of removing this restriction in respect of the model (5.1). This is merely to illustrate that residual strains can be predicted with this simple model and should not be taken to imply that the model is appropriate for detailed representation of residual stresses. A more general model within the framework discussed in Section 4 may be needed.

If we consider biaxial deformations on the basis of equations (5.3) with $\sigma_3 = 0$ then it is clear that if $\eta = 0$ then $\sigma_1 = \sigma_2 = 0$. In general the stretches λ_1 and λ_2 will not both be 1 at this point so a residual strain remains after removal of the stresses. We denote the values of λ_1 and λ_2 where $\eta = 0$ by λ_{1r} and λ_{2r} respectively and set W_r to be the associated value of $\hat{W}_0$. Equation (5.4) then gives

$$\phi'(0) = -\hat{W}_0(\lambda_{1r}, \lambda_{2r}) = -W_r. \tag{5.24}$$

Since the function ϕ depends on W_m then, for a given function ϕ, equation (5.24) defines W_r associated with the residually strained state in terms of the energy W_m associated with primary loading. Hence, the residual deformation depends on the deformation at the end of the primary loading process. Note, however, that while, in view of the monotonicity of $\phi'(\eta)$, W_r is uniquely defined, the associated residual strain depends on the unloading path followed to reach this value of W_r.

In order to make explicit the dependence of ϕ on W_m we write $\phi(\eta) = \Phi(\eta, W_m)$ so that (5.24) becomes

$$\Phi_\eta(0, W_m) = -W_r, \tag{5.25}$$

where the subscript η indicates partial differentiation with respect to η. Equation (5.4) becomes

$$\Phi_\eta(\eta, W_m) = -\hat{W}_0, \qquad \hat{W}_0 \le W_m, \tag{5.26}$$

and we must have

$$\Phi_\eta(1, W_m) = -W_m. \tag{5.27}$$

When $\eta = 0$ the pseudo-energy function (5.1) has the residual value

$$\phi(0) \equiv \Phi(0, W_m). \tag{5.28}$$

This is the non-recoverable part of the energy used on loading and may be interpreted as a measure of the energy required to cause damage in the material during loading. This is analogous to the interpretation of $\phi(\eta_m)$ in Section 5.1. In a uniaxial test such as simple tension $\phi(0)$ is the area between the primary loading curve and the relevant unloading curve and above the axis $t = 0$.

With reference to the discussion in Section 5.5.1 of the contours in Figure 13 we remark that in the present context, for a given value of W_m, $\eta = 0$ defines a contour $\hat{W}_0(\lambda_1, \lambda_2) = W_r < W_m$, and points on this contour define possible values $(\lambda_{1r}, \lambda_{2r})$ of the residual stretches.

Limitations of the isotropy assumption It should be emphasized that we have assumed here that the material response is *isotropic* relative to $\mathcal{B}_r$ in both loading and unloading. This is reflected, in particular, in the symmetry about the line $\lambda_1 = \lambda_2$ of the energy contours in Figure 13. In general, however, it cannot be expected that the material response remains isotropic when damage occurs. For example, for the same value of W_m different deformations (for example, simple tension and pure shear) produce different maximum stretches and are therefore associated with different degrees of damage, i.e. there is damage-induced anisotropy. This should be mirrored in the material response on unloading. Such anisotropy is not incorporated in the theory discussed here. In principle, it is not difficult to accommodate the anisotropy in the theory but at present the available data are not adequate to assess the value of doing this.

Furthermore, residual strain may itself be associated with anisotropy since the natural (unstressed reference) configuration is no longer the initial (virgin) configuration and the material symmetry must be characterized relative to this new reference configuration. And since, in general, this is not related to the initial reference configuration by a pure dilatation the subsequent response (and hence the unloading response) will be anisotropic. Thus, the unstressed (reference) configuration is being continually modified during primary loading and a more general model will need to account for this and for the anisotropy. Some discussion of the modelling of the anisotropy is contained in the paper by Horgan *et al.* (2003), while characterization of the evolving anisotropic response due to damage (or other mechanisms for material property changes) is addressed in a forthcoming paper (Ogden, 2003).

6 Viscoelasticity

In this section we summarize some of the basic concepts in viscoelasticity and then examine a range of constitutive laws that have been proposed in the literature.

6.1 Creep, relaxation and recovery

The term *creep* is applied to the situation when (in a simple tension test, for example) the strain increases with time (to some asymptotic value) while the stress is held fixed. This is depicted in Figure 17.

Creep *recovery* (in time) occurs after creep at constant stress after which there is instantaneous elastic strain recovery following instantaneous removal of the stress and then viscoelastic recovery, as illustrated in Figure 18.

Relaxation is the term applied to the time decay of stress when a specimen is held at constant strain, in simple tension for example. The stress tends asymptotically to a

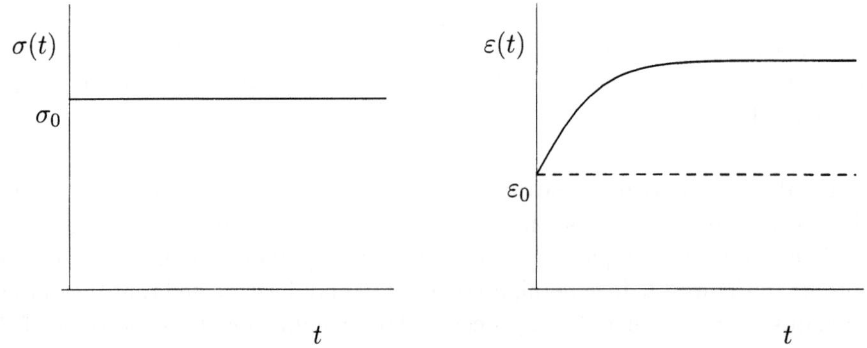

Figure 17. Constant stress σ_0 (left-hand figure) and the resulting viscoelastic creep strain $\varepsilon(t)$ with time t on the horizontal axis. The dashed line shows the corresponding elastic response with constant strain ε_0.

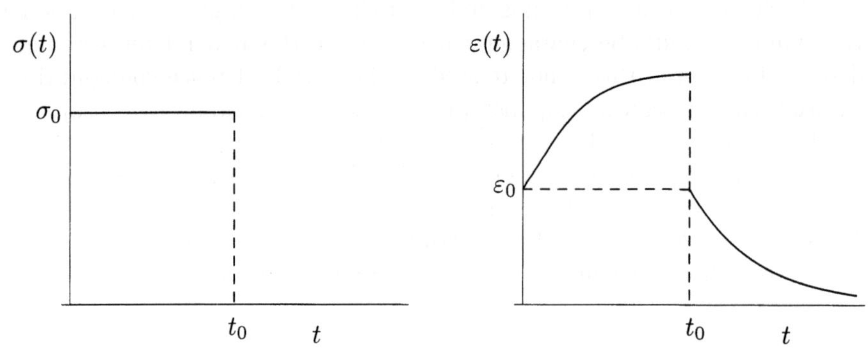

Figure 18. Creep recovery: constant stress σ_0 removed at time t_0 (left-hand figure). The right-hand figure shows the associated creep up to time t_0, instantaneous elastic recovery to strain ε_0 and then viscoelastic recovery.

non-zero value, as depicted in Figure 19.

Stress *recovery* (in time) occurs when the constant strain is removed instantaneously (which requires the instantaneous application of a reverse stress). In this case the asymptotic stress is zero. This behaviour is illustrated in Figure 20.

For a typical rubberlike material one can think in terms of the *long-term response* and the *instantaneous response.* The long-term response is appropriate for very low strain rates or frequencies (the quasi-static limit) and is modelled as elastic. The instantaneous response is the limiting behaviour at very high strain rates (or frequencies). Stress-strain

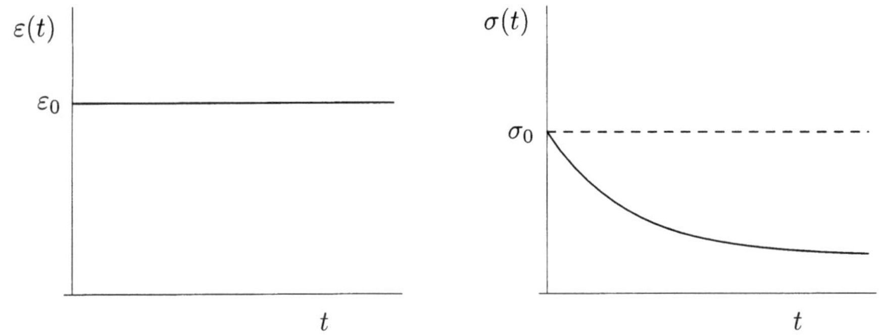

Figure 19. Constant strain ε_0 (left-hand figure) and the resulting stress relaxation from the initial value σ_0 to an asymptotic value (in general non-zero) as $t \to \infty$. The dashed line shows the corresponding elastic response with constant stress σ_0.

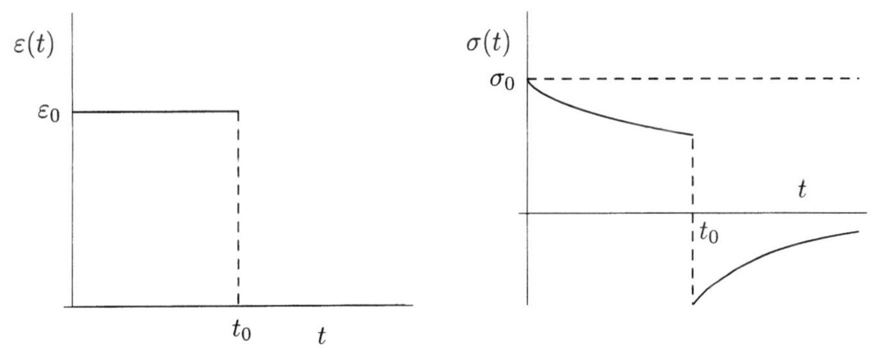

Figure 20. Stress recovery: constant strain ε_0 removed at time t_0 (left-hand figure) and the resulting stress relaxation to time t_0, application of reverse stress and the subsequent stress recovery.

curves for simple tension associated with these two limits are depicted in Figure 21 for both linear and nonlinear behaviour. Notice that the material is significantly stiffer at the higher strain rates. For intermediate rates (or frequencies) the stress-stretch response lies between these two limits. In the linear case the gradients of the lines are the Young's moduli E_0 and E_∞ for the two limits, where these notations are defined in Section 6.2 below.

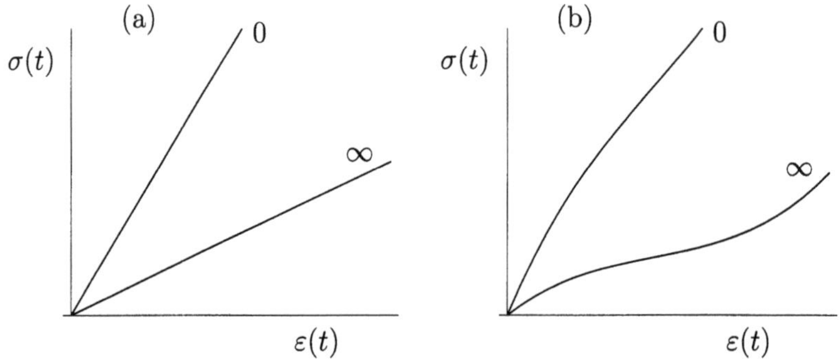

Figure 21. Limiting stress-strain response in simple tension for very low and very high strain rates: (a) linear response; (b) nonlinear response. The labels 0 and ∞ are associated with the instantaneous and long-term responses respectively.

6.2 Linear viscoelasticity

To set the scene for a more general discussion of constitutive laws we examine briefly some aspects of the classical linear theory of viscoelasticity, and we restrict attention to uniaxial behaviour. For a linear incompressible isotropic *elastic* solid the uniaxial stress σ is given in terms of the uniaxial strain ε by $\sigma = 3G\varepsilon$, where the constant G is the shear modulus ($3G$ is Young's modulus, denoted E).

Frequency response Suppose now that in a linear viscoelastic material the strain is given by

$$\varepsilon(t) = \varepsilon_0 \exp(\mathrm{i}\omega t), \tag{6.1}$$

where ε_0 is the strain *amplitude*, ω (real) is the frequency of harmonic oscillations and t is time. Then, in general, the stress is out of phase with the strain and given by

$$\sigma(t) = \sigma_0 \exp(\mathrm{i}\omega t + \delta), \tag{6.2}$$

where σ_0 is the stress amplitude and δ is called the *loss angle.* We write the relationship between stress and strain in the complex form

$$\sigma(t) = 3G^* \varepsilon(t), \tag{6.3}$$

where G^* is the complex shear modulus. From (6.1) and (6.2) it is given by

$$3G^* = \frac{\sigma_0}{\varepsilon_0} \exp(\mathrm{i}\delta) = \sigma_0(\cos\delta + \mathrm{i}\sin\delta)/\varepsilon_0 \equiv 3G_s^* + 3\mathrm{i}G_l^*, \tag{6.4}$$

wherein the *storage modulus* G_s^* and the *loss modulus* G_l^* are defined. Respectively, G_s^* and G_l^* characterize the *in-phase* and *out-of-phase* response of the material. Note that for the simple situation considered here G^* is independent of the frequency ω. However, for actual materials (in particular, rubberlike materials) G^* depends on ω, so the simple relationship (6.3) is inappropriate.

From (6.4) we have $G_l^*/G_s^* = \tan\delta$, which is called the *loss tangent.* In general, for rubber (particularly filled rubber) G_s^* and G_l^* depend on the strain amplitude as well as on the frequency and the dependence varies significantly with the percentage of filler. In other words, the viscoelastic properties of rubber are strongly influenced by the proportions of filler used. For further discussion of the viscoelastic properties of rubber we refer to the chapter by Buckley in this volume.

The linear isotropic viscoelastic constitutive law In one dimension the classical constitutive law of a linear isotropic viscoelastic solid may be written in the form

$$\sigma(t) = \int_{-\infty}^{t} E(t-\tau)\dot{\varepsilon}(\tau)\,\mathrm{d}\tau, \tag{6.5}$$

where $E(t)$ is called the *relaxation modulus*, $\varepsilon(-\infty) = 0$ and the superposed dot represents differentiation with respect to τ. Typical behaviour of $E(t)$ is represented in Figure 22. The value $E_0 = E(0)$ is the instantaneous (Young's) modulus, while $E_\infty = E(\infty)$ is the long term (Young's) modulus. As mentioned above these correspond to the gradients of the straight lines in Figure 21(a). Clearly, the relaxation modulus satisfies $E_\infty < E(t) \leq E_0$.

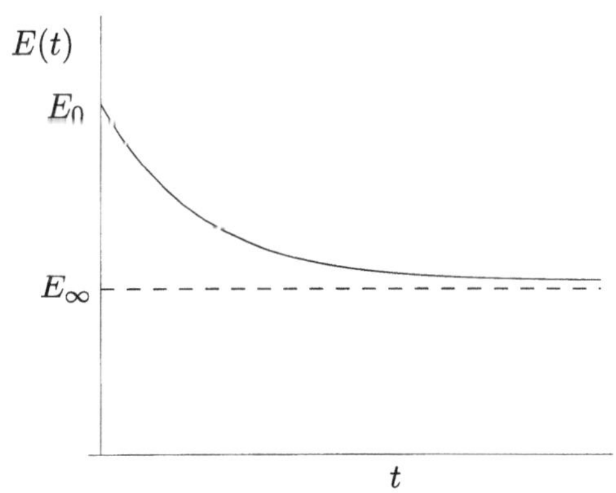

Figure 22. Dependence of the relaxation modulus $E(t)$ on time t. The dashed line corresponds to pure elastic response with Young's modulus E_∞.

On integrating (6.5) by parts and changing variables, we obtain

$$\sigma(t) = \sigma_0(t) + \frac{1}{E_0}\int_0^\infty \dot{E}(\tau)\sigma_0(t-\tau)\mathrm{d}\tau, \tag{6.6}$$

where $\sigma_0(t) = E_0\varepsilon(t)$ is the stress associated with the instantaneous elastic response for the given strain $\varepsilon(t)$.

Two further equivalent forms of (6.6) are obtained by defining the relaxation function

$$e(t) = E(t)/E_\infty - 1, \tag{6.7}$$

to give

$$\sigma(t) = \sigma_\infty(t) + \int_{-\infty}^{t} e(t-\tau)\dot{\sigma}_\infty(\tau)\,\mathrm{d}\tau, \tag{6.8}$$

or

$$\sigma(t) = \sigma_\infty(t) - \int_{0}^{\infty} e(\tau)\frac{\mathrm{d}\sigma_\infty}{\mathrm{d}\tau}(t-\tau)\mathrm{d}\tau, \tag{6.9}$$

where $\sigma_\infty(t) = E_\infty\varepsilon(t)$ is the stress associated with the long-term response at the strain $\varepsilon(t)$. Equations such as these and their three-dimensional (tensor) counterparts form the basis for many extensions of the theory to the finite strain domain.

6.3 Finite strain viscoelasticity

An example A recent example, which relates closely in structure to (6.5), is the model introduced by Yang *et al.* (2000) for the high strain-rate regime. They take the Cauchy stress tensor to have the form

$$\boldsymbol{\sigma} = \boldsymbol{\sigma}^{(e)} + \boldsymbol{\sigma}^{(v)}, \tag{6.10}$$

representing the sum of elastic and viscoelastic contributions, with $\boldsymbol{\sigma}^{(e)}$ having a particular form of the general representation

$$\boldsymbol{\sigma}^{(e)} = -p\mathbf{I} + \alpha_1\mathbf{B} + \alpha_2\mathbf{B}^2 \tag{6.11}$$

for an incompressible isotropic elastic solid (recall (2.33)), where α_1 and α_2 are function of the invariants I_1 and I_2 of the Cauchy-Green tensor $\mathbf{B}$.

The viscoelastic part is given by

$$\boldsymbol{\sigma}^{(v)} = -p^{(v)}\mathbf{I} + \mathbf{F}(t)\left[\int_{-\infty}^{t} \phi(I_1, I_2)m(t-\tau)\dot{\mathbf{E}}(\tau)\mathrm{d}\tau\right]\mathbf{F}^T(t), \tag{6.12}$$

with

$$m(t) = \sum_{i=1}^{N} \exp(-t/\tau_i), \tag{6.13}$$

where $\mathbf{F}(t)$ is the deformation gradient at time t, $\mathbf{E}(t)$ is the Green strain tensor defined by

$$\mathbf{E} = \frac{1}{2}(\mathbf{C} - \mathbf{I}) = \frac{1}{2}(\mathbf{F}^T\mathbf{F} - \mathbf{I}), \tag{6.14}$$

and τ_i $(i = 1, 2, \ldots, N)$ are relaxation times. The authors used a specific form of the function ϕ and $N = 1$ to fit data from high strain-rate tests. Their data show that the material stiffens significantly at very high strain rates.

A similar basis for such nonlinear viscoelastic models has been used by other authors, as exemplified by Johnson *et al.* (1995).

The BKZ model An early success in finite strain viscoelastic modelling was due to Bernstein *et al.* (1963). Their model was based on incompressible isotropic theory and may be expressed compactly as

$$\boldsymbol{\sigma}(t) = -p(t)\mathbf{I} + \int_0^t \hat{\boldsymbol{\sigma}}(t,\tau)\mathrm{d}\tau, \tag{6.15}$$

where

$$\hat{\boldsymbol{\sigma}}(t,\tau) = \alpha_1(t,\tau)\mathbf{B}(t,\tau) + \alpha_2(t,\tau)\mathbf{B}(t,\tau)^2, \tag{6.16}$$

i.e. the same as the latter two terms in (6.11) except that the deformation is now a *relative* deformation, $\mathbf{F}(t,\tau)$ being the deformation gradient at time t relative to the configuration at time τ. Actually, Bernstein *et al.* (1963) expressed their $\hat{\boldsymbol{\sigma}}$ in terms of an 'energy' function. Many special cases of this model have been used in the literature.

The constitutive laws just discussed are embraced within a general framework in which the stress is a *functional* of the deformation and there are many examples that are cast within such a framework. We do not discuss any other specific models of this kind but refer to a guide to the literature in the paper by Schapery (2000). See also Holzapfel (2000) and the volumes of proceedings of the three conferences on Constitutive Models for Rubber (Dorfmann and Muhr, 1999; Besdo *et al.*, 2001; Busfield and Muhr, 2003) for further references.

6.4 Internal variable models

An alternative and widely-used approach to modelling viscoelasticity is based on the concept of *internal variables* and is set within a thermodynamic framework in which the internal variables account for dissipation. Internal variables may, for example, be strain-like or stress-like variables that reflect changes in the material microstructure due macroscopic stresses and strains. Viscoelastic rate effects are introduced through evolution equations governing the internal variables. Each internal variable is associated with a time scale through a relaxation time constant and many such variables may be needed to capture the viscoelastic behaviour over a range of time scales.

A typical internal variable theory might have the following structure. Let $\mathbf{q}_\alpha$, $\alpha = 1, 2, \ldots, N$, be N internal variables of (second-order) tensor type. For example, they could represent internal strains. The Helmholtz free energy function is decomposed as

$$\Psi(\mathbf{E}, \mathbf{q}_1, \ldots, \mathbf{q}_N) = \Psi^{(\infty)}(\mathbf{E}) + \Psi^{(v)}(\mathbf{E}, \mathbf{q}_1, \ldots, \mathbf{q}_N), \tag{6.17}$$

where $\mathbf{E}$ is the Green strain tensor, $\Psi^{(\infty)}$ characterizes the long-term equilibrium (elastic) response and $\Psi^{(v)}$ the viscoelastic (dissipative) response. The Green strain tensor is used so that (6.17) is objective provided the internal variables are themselves objective. The latter are taken to be objective Lagrangian tensors so that their material time derivatives are also objective.

The second Piola-Kirchhoff stress tensor, which is conjugate to $\mathbf{E}$, is denoted here by $\mathbf{P}$ and is calculated as

$$\mathbf{P} = \frac{\partial \Psi}{\partial \mathbf{E}} = \mathbf{P}^{(\infty)} + \mathbf{P}^{(v)}, \tag{6.18}$$

where

$$\mathbf{P}^{(\infty)} = \frac{\partial \Psi^{(\infty)}}{\partial \mathbf{E}}, \quad \mathbf{P}^{(v)} = \frac{\partial \Psi^{(v)}}{\partial \mathbf{E}}. \tag{6.19}$$

Note that in terms of the nominal stress tensor $\mathbf{S}$, $\mathbf{P}$ is given by $\mathbf{P} = \mathbf{S}\mathbf{F}^{-T}$.

A general evolution equation for the $\mathbf{q}_\alpha$ may have a form such as

$$\dot{\mathbf{q}}_\alpha = \mathbf{\Gamma}_\alpha(\mathbf{F}, \mathbf{q}_1, \ldots, \mathbf{q}_N), \tag{6.20}$$

where the superposed dot represents the material time derivative and $\mathbf{\Gamma}_\alpha$, $\alpha = 1, \ldots, N$, are suitable tensor functions. Together, equations (6.17)–(6.20) provide the constitutive law of a viscoelastic material.

It is also possible to define stress-like variables, denoted $\mathbf{Q}_\alpha$, $\alpha = 1, \ldots, N$, conjugate to $\mathbf{q}_\alpha$, $\alpha = 1, \ldots, N$ through

$$\mathbf{Q}_\alpha = -\frac{\partial \Psi^{(v)}}{\partial \mathbf{q}_\alpha}. \tag{6.21}$$

The rate of dissipation associated with a viscoelastic process, which must be non-negative, is then calculated as

$$\sum_{\alpha=1}^{N} \operatorname{tr}(\mathbf{Q}_\alpha \dot{\mathbf{q}}_\alpha) \geq 0. \tag{6.22}$$

The constitutive law has to be consistent with this inequality. For a discussion of the relevant thermodynamic background we refer to Holzapfel (2000), for example.

In the specific constitutive model considered by Holzapfel and Simo (1996) $\Psi^{(v)}$ was decoupled in the form

$$\Psi^{(v)}(\mathbf{E}, \mathbf{q}_1, \ldots, \mathbf{q}_N) = \sum_{\alpha=1}^{N} \Psi_\alpha^{(v)}(\mathbf{E}, \mathbf{q}_\alpha), \tag{6.23}$$

with a corresponding decoupling of the viscoelastic stress:

$$\mathbf{P}^{(v)} = \sum_{\alpha=1}^{N} \mathbf{P}_\alpha^{(v)}, \quad \mathbf{P}_\alpha^{(v)} = \frac{\partial \Psi_\alpha^{(v)}}{\partial \mathbf{E}}. \tag{6.24}$$

The identification $\mathbf{P}_\alpha^{(v)} = \mathbf{Q}_\alpha$ was then made. Note that we are using different notation from that used by Holzapfel and Simo (1996).

Instead of evolution equations for the $\mathbf{q}_\alpha$ Holzapfel and Simo (1996) used evolution equations for the associated internal stresses. These equations are expressed in the form

$$\dot{\mathbf{Q}}_\alpha + \frac{1}{\tau_\alpha}\mathbf{Q}_\alpha = \dot{\mathbf{\Sigma}}_\alpha, \tag{6.25}$$

where again the superposed dot represents the material time derivative, the right-hand side being the rate of a stress-like quantity $\mathbf{\Sigma}_\alpha$, $\alpha = 1, \ldots, N$. Initial conditions are given as

$$\mathbf{Q}_\alpha(t = 0) = \mathbf{Q}_\alpha^{(0)}. \tag{6.26}$$

With use of (6.26) equations (6.25) are integrated to give

$$\mathbf{Q}_\alpha = \exp(-t/\tau_\alpha)\mathbf{Q}_\alpha^{(0)} + \int_0^t \exp[-(t-\tau)/\tau_\alpha]\dot{\mathbf{\Sigma}}_\alpha(\tau)\,\mathrm{d}\tau, \quad \alpha = 1,\ldots,N. \tag{6.27}$$

The tensor $\mathbf{\Sigma}_\alpha$ is supposed to be independent of the internal variables and to depend only on the macroscopic strain $\mathbf{E}$. It may be written in terms of a potential function, for example, so that

$$\mathbf{\Sigma}_\alpha = \frac{\partial \Psi_\alpha^{(0)}}{\partial \mathbf{E}}, \tag{6.28}$$

where $\Psi_\alpha^{(0)} = \Psi_\alpha^{(0)}(\mathbf{E})$, which may be interpreted as the instantaneous elastic energy associated with the τ_α relaxation process. Note that $\mathbf{E} = \mathbf{E}(t)$ is the Green strain at time t.

Govindjee and Simo (1992b) set (in different notation)

$$\Psi_\alpha^{(0)}(\mathbf{E}) = \beta_\alpha^{(\infty)}\Psi^{(\infty)}(\mathbf{E}), \quad \alpha = 1,\ldots,N, \tag{6.29}$$

where $\beta_\alpha^{(\infty)}$, $\alpha = 1,\ldots,N$, is a constant associated with the τ_α relaxation. In fact, the theory of Govindjee and Simo (1992b) used the viscoelastic stresses $\mathbf{Q}_\alpha$ directly as internal variables satisfying evolution equations of the form (6.25) without the intermediate strain-like variables.

The resulting expression for the Piola-Kirchhoff stress $\mathbf{P}$ is now given in the form (6.18) with $\mathbf{P}^{(\infty)}$ given by the first equation in (6.19). Thus, after some rearrangement,

$$\mathbf{P} = \mathbf{P}^{(\infty)} + \sum_{\alpha=1}^{N} \mathbf{Q}_\alpha = g(t)\frac{\partial \Psi^{(\infty)}}{\partial \mathbf{E}}(\mathbf{E}_0) + \int_0^t g(t-\tau)\frac{\mathrm{d}}{\mathrm{d}\tau}\left[\frac{\partial \Psi^{(\infty)}}{\partial \mathbf{E}}(\mathbf{E}(\tau))\right]\mathrm{d}\tau, \tag{6.30}$$

where

$$g(t) = 1 + \sum_{\alpha=1}^{N} \beta_\alpha^{(\infty)}\,\mathrm{e}^{-t/\tau_\alpha}. \tag{6.31}$$

In this form the structure is very similar to that in the models discussed in Sections 6.3.1 and 6.3.2 except that here isotropy and incompressibility have not been assumed. In both Govindjee and Simo (1992b) and Holzapfel and Simo (1996) the theory was specialized to isotropy and the isochoric and dilatational parts of the response were decoupled.

In a paper by Dafalias (1991) a relatively simple model with a single non-equilibrium stress treated as an internal variable was considered with an (objective) evolution equation for the stress. Many models consider the viscoelastic (or non-equilibrium) stress as an addition to the equilibrium stress. The former is sometimes referred to as the *overstress*. In some other models an intermediate (inelastic) reference configuration is used and treated as an internal variable requiring an evolution equation, as exemplified in the papers by Lubliner (1985), Reese and Govindjee (1998) and Huber and Tsakmakis (2000). Some relevant experimental findings are contained in the papers by Lion (1996) and Bergström and Boyce (2000).

6.5 Differential models

Finally, we mention briefly differential models of the form

$$\boldsymbol{\sigma} = \bar{\boldsymbol{\sigma}}(\mathbf{F}, \mathbf{A}_1, \mathbf{A}_2, \ldots, \mathbf{A}_N), \tag{6.32}$$

where $\mathbf{A}_n$, $n = 1, \ldots, N$, are the Rivlin-Ericksen tensors defined by

$$\mathbf{F}^T \mathbf{A}_n \mathbf{F} = \frac{\partial^n}{\partial t^n} \mathbf{C}, \tag{6.33}$$

$\mathbf{C}$ being the right Cauchy-Green deformation tensor and $\partial/\partial t$ being the material time derivative. Note that $\mathbf{A}_1 = \mathbf{D}$, which is the Eulerian strain-rate tensor.

For an incompressible isotropic material it is known (see, for example, Truesdell and Noll, 1965) that the most general form of this representation when restricted to $N = 1$ can be expressed in the form

$$\begin{aligned} \boldsymbol{\sigma} = & -p\mathbf{I} + \alpha_1 \mathbf{B} + \alpha_2 \mathbf{B}^2 + \alpha_3 \mathbf{D} \\ & + \alpha_4 (\mathbf{DB} + \mathbf{BD}) + \alpha_5 (\mathbf{DB}^2 + \mathbf{B}^2\mathbf{D}) + \alpha_6 \mathbf{D}^2 \\ & + \alpha_7 (\mathbf{D}^2\mathbf{B} + \mathbf{BD}^2) + \alpha_8 (\mathbf{D}^2\mathbf{B}^2 + \mathbf{B}^2\mathbf{D}^2), \end{aligned} \tag{6.34}$$

where $\mathbf{I}$ is the identity tensor, p is the arbitrary hydrostatic stress (a Lagrange multiplier) associated with the incompressibility constraint and α_i, $i \in \{1, 2, \ldots 8\}$ are material functions that depend on the invariants $I_1, I_2, I_4, \ldots I_9$. We write

$$\alpha_i = \alpha_i(I_1, I_2, I_4, \ldots I_9) \tag{6.35}$$

and note that, since for an incompressible material $\operatorname{tr}\mathbf{D} = 0$, the independent invariants of $\mathbf{B}$ and $\mathbf{D}$ are

$$I_1 = \operatorname{tr}\mathbf{B}, \quad I_2 = \operatorname{tr}(\mathbf{B}^{-1}) \equiv \frac{1}{2}\left((\operatorname{tr}\mathbf{B})^2 - \operatorname{tr}(\mathbf{B}^2)\right), \tag{6.36}$$

$$I_4 = \operatorname{tr}(\mathbf{DB}), \quad I_5 = \operatorname{tr}(\mathbf{DB}^2), \quad I_6 = \operatorname{tr}(\mathbf{D}^2), \tag{6.37}$$

$$I_7 = \operatorname{tr}(\mathbf{D}^2\mathbf{B}), \quad I_8 = \operatorname{tr}(\mathbf{D}^2\mathbf{B}^2), \quad I_9 = \operatorname{tr}(\mathbf{D}^3). \tag{6.38}$$

Models of this kind have not been utilized to any great extent in the context of rubber viscoelasticity but may be worth exploring in some detail. However, without modification they are not able to model stress relaxation, which would require terms involving objective stress rates.

In conclusion, it should be mentioned that there is a pressing need for more extensive data on the inelastic behaviour of rubberlike solids, particularly time- and rate-dependent behaviour, in both uniaxial and multiaxial tests. Such data is essential as a basis for discriminating between competing models or classes of models and for leading to the development of a model that can be regarded as reliable for computational implementation over a range of behaviours.

References

Barenblatt, G.I., and Joseph, D.D. (1996). *Collected Papers of R.S. Rivlin.* Springer, New York.

Beatty, M.F., and Krishnaswamy, S. (2000). A theory of stress softening in incompressible isotropic elastic materials. *J. Mech. Phys. Solids* 48:1931–1965.

Bergström, J.S., and Boyce, M.C. (2000). Large strain time-dependent behaviour of filled elastomers. *Mech. Materials* 32:627–644.

Bernstein, B., Kearsley, A., and Zapas, L.J. (1963). A study of stress relaxation with finite strain. *Trans. Soc. Rheol.* 7:391–410.

Besdo, D., Schuster, R.H., and Ihlemann, J. (2001). *Constitutive Models for Rubber II.* Rotterdam: Balkema.

Bouasse, H., and Carrière, Z. (1903). Courbes de traction du caoutchouc vulcanisé. *Ann. Fac. Sciences de Toulouse* 5:257–283.

Boyce, M.C., and Arruda, E.M. (2000). Constitutive models of rubber elasticity: a review. *Rubber Chem. Technol.* 73:504–523.

Busfield, J.J.C., and Muhr, A.H. (2003). *Constitutive Models for Rubber III.* Rotterdam: Balkema.

Dafalias, Y.F. (1991). Constitutive model for large viscoelastic deformations of elastomeric materials. *Mech. Res. Comm.* 18:61–66.

DeSimone, A., Marigo, J.J., and Teresi, L. (2001). A damage mechanics approach to stress softening and its application to rubber. *Eur. J. Mech. A/Solids* 20:873–892.

Dorfmann, A., and Muhr, A. (1999). *Constitutive Models for Rubber.* Rotterdam: Balkema.

Dorfmann, A., and Ogden, R.W. (2003). A pseudo-elastic model for loading, partial unloading and reloading of particle-reinforced rubber. *Int. J. Solids Structures* 40:2699–2714.

Fu, Y.B., and Ogden, R.W. (2001). *Nonlinear Elasticity: Theory and Applications.* Cambridge University Press.

Govindjee, S., and Simo, J. C. (1991). A micro-mechanically based continuum damage model for carbon black-filled rubbers incorporating the Mullins' effect. *J. Mech. Phys. Solids* 39:87–112.

Govindjee, S., and Simo, J. C. (1992a). Transition from micro-mechanics to computationally efficient phenomenology: carbon black filled rubbers incorporating Mullins' effect. *J. Mech. Phys. Solids* 40:213–233.

Govindjee, S., and Simo, J. C. (1992b). Mullins' effect and the strain amplitude dependence of the storage modulus. *Int. J. Solids Structures* 29:1737–1751.

Hayes, M.A., and Saccomandi, G. (2001). *Topics in Finite Elasticity*, CISM Courses and Lectures, vol. 424. Wien: Springer.

Holzapfel, G.A. (2000). *Nonlinear Solid Mechanics.* Chichester: John Wiley.

Holzapfel, G.A., and Simo, J.C. (1996). A new viscoelastic constitutive model for continuous media at finite thermodynamical changes. *Int. J. Solids Structures* 33:3019–3034.

Horgan, C.O., Ogden, R.W., and Saccomandi, G. (2003). A theory of stress softening of elastomers based on finite chain extensibility. *Proc. R. Soc. Lond. A*, in press.

Huber, N., and Tsakmakis, C. (2000). Finite deformation viscoelasticity laws. *Mech. Materials* 32:1–18.

Johnson, A.R., Quigley, C.J., and Freese, C.E. (1995). A viscohyperelastic finite element model for rubber. *Comp. Methods Appl. Mech. Eng.* 127:163–180.

Johnson, M. A., and Beatty, M. F. (1993a). The Mullins effect in uniaxial extension and its influence on the transverse vibration of a rubber string. *Continuum Mech. Thermodyn.* 5:83–115.

Johnson, M. A., and Beatty, M. F. (1993b). A constitutive equation for the Mullins effect in stress controlled uniaxial extension experiments. *Continuum Mech. Thermodyn.* 5:301–318.

Jones, D.F., and Treloar, L.R.G. (1975). The properties of rubber in pure homogeneous strain. *J. Phys. D, Appl. Phys.* 8:1285–1304.

Krishnaswamy, S., and Beatty, M.F. (2000). The Mullins effect in compressible solids. *Int. J. Engng Sci.* 38:1397–1414.

Lion, A. (1996). A constitutive model for carbon black filled rubber: experimental investigation and mathematical representation. *Cont. Mech. Thermodyn.* 8:153–169.

Lubliner, J. (1985). A model of rubber viscoelasticity. *Mech. Res. Comm.* 12:93–99.

Marckmann, G., Verron, E., Gornet, L., Chagnon, G., Charrier, P., and Fort, P. (2002). A theory of network alteration for the Mullins effect. *J. Mech. Phys. Solids* 50:2011–2028.

Muhr, A.H., Gough, J., and Gregory, I.H. (1999). Experimental determination of model for liquid silicone rubber: Hyperelasticity and Mullins' effect. In Dorfmann, A., and Muhr, A., eds., *Proceedings of the First European Conference on Constitutive Models for Rubber.* Rotterdam: Balkema. 181–187.

Mullins, L. (1947). Effect of stretching on the properties of rubber. *J. Rubber Research* 16:275–289.

Mullins, L. (1969). Softening of rubber by deformation. *Rubber Chem. Technol.* 42:339–362.

Mullins, L., and Tobin, N. R. (1957). Theoretical model for the elastic behaviour of filler-reinforced vulcanized rubbers. *Rubber Chem. Technol.* 30:551–571.

Ogden, R.W. (1972). Large deformation isotropic elasticity: on the correlation of theory and experiment for incompressible rubberlike solids. *Proc. R. Soc. Lond. A* 326:565–584.

Ogden, R.W. (1982). Elastic deformation of rubberlike solids. In Hopkins, H.G., and Sewell, M.J., eds., *Mechanics of Solids*, The Rodney Hill 60th Anniversary Volume. Oxford: Pergamon Press. 499–537.

Ogden, R.W. (1986). Recent advances in the phenomenological theory of rubber elasticity. *Rubber Chemistry and Technology* 59:361–383.

Ogden, R.W. (1997). *Non-linear Elastic Deformations.* New York: Dover Publications.

Ogden, R.W. (2001). Pseudo-elasticity and stress softening. In Fu, Y.B., and Ogden, R.W., eds., *Nonlinear Elasticity: Theory and Applications.* Cambridge University Press. 491–522.

Ogden, R.W. (2003). On an anisotropic theory of pseudo-elasticity. Manuscript in preparation.

Ogden, R.W., and Roxburgh, D.G. (1999a). A pseudo-elastic model for the Mullins effect in filled rubber. *Proc. R. Soc. Lond. A* 455:2861–2877.

Ogden, R.W., and Roxburgh, D.G. (1999b). An energy-based model of the Mullins effect. In Dorfmann, A., and Muhr, A., eds., *Proceedings of the First European Conference on Constitutive Models for Rubber.* Rotterdam: Balkema. 23–28.

Reese, S., and Govindjee, S. (1998). A theory of finite visocelasticity and numerical aspects. *Int. J. Solids Structures* 35:3455–3482.

Schapery, R.A. (2000). Nonlinear viscoelastic solids. *Int. J. Solids Structures* 37:359–366.

Treloar, L.R.G. (1944). Stress-strain data for vulcanized rubber under various types of deformation. *Trans. Faraday Soc.* 40:59–70.

Treloar, L.R.G. (1975). *The Physics of Rubber Elasticity*, 3rd edition. Oxford University Press.

Truesdell, C.A., and Noll, W. (1965). In Flügge, S., ed., *The Nonlinear Field Theories of Mechanics: Handbuch der Physik Vol. III/3.* Berlin: Springer.

Valanis, K.C., and Landel, R.F. (1967). The strain-energy function of a hyperelastic material in terms of the extension ratios. *J. Appl. Phys.* 38:2997–3002.

Vangerko, H., and Treloar, L.R.G. (1978). The inflation and extension of rubber tube for biaxial strain studies. *J. Phys. D, Appl. Phys.* 11:1969–1978.

Yang, L.M., Shim, V.P.W., and Lim, C.T. (2000) A visco-hyperelastic approach to modelling the constitutive behaviour of rubber. *Int. J. Impact Eng.* 24:545–560.

Numerical Issues in Finite Elasticity and Viscoelasticity*

Sanjay Govindjee

Structural Engineering, Mechanics and Materials, Department of Civil Engineering, University of California, Berkeley, CA 74720, USA
govindjee@ce.berkeley.edu

Abstract. This chapter starts with an introduction to the notion of finite element methods in solid mechanics. After an overview of basic concepts in 1-D, the multi-D problem is investigated within the context of finite elasticity. Particular attention is paid to the so-called mixed variational principles that are needed for the solution of nearly incompressible solids. Special forms for elasticity in principal stretches are also given. Details of commonly used notations for actual implementations are covered. The chapter closes with several sections covering important special cases that arise in the modeling of elastomeric solids. First, numerical issues in the solution linear and nonlinear finite deformation viscoelasticity are discussed. This includes both convolution form models and multiplicative split type models. This is followed by the application of finite element methods to the solution of inverse design problems of the type that arise in mold design. Next, the general numerical methods are applied to the case of steady state spinning which is important in the analysis of automotive tires. The last two sections deal with the special cases of damage models and energy release rate computations in elastomers.

1 Introduction

1.1 Quick refresher on linear weak and strong statements

The finite element method is a way to approximately solve boundary value problems (BVP). The strong form of a boundary value problem is made up of an ordinary differential equation (ODE) or a partial differential equation (PDE) and a set of boundary conditions. The strong form of a BVP is the most common form; however, the finite element method is based on what is called the weak form of the BVP. A third form of a BVP is a minimization form. Not all BVP's have minimization forms, but all BVP's have weak forms. BVP's dealing with elastic bodies in mechanics have minimization forms.

To make this quick review more concrete let us consider the problem of a 1-D elastic rod with both force and displacement boundary conditions. As shown in Fig. 1, the rod is subject to a prescribed displacement of $\bar{u}$ at the left end, $x = 0$, a prescribed force $\bar{F}$ at the right end, $x = L$, and a prescribed body force $b(x)$ per unit length of the rod. With these prescribed loads we want to find the displacement at each point in the rod. Knowing that, we will also know the strains and hence the stresses. Assume for simplicity that the cross-sectional area of the rod, A, and the Young's modulus, E, are both constants.

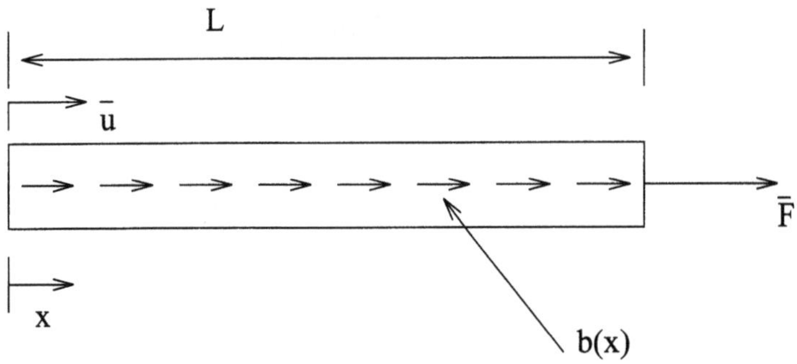

Figure 1. 1-D bar under body and end-loads.

The strong form of the BVP says:

$$(S)\begin{cases}\text{Find the function } u(x) \text{ such that} \\ AEu_{,xx}+b = 0 & \text{for } x\in(0,L) \\ u(0) = \bar{u} \\ AEu_{,x}(L) = \bar{F}\end{cases} \tag{1}$$

Certain approximation methods such as finite differences work directly with the strong form of the BVP. However, we will need the weak form of the BVP. The weak form of the BVP is found via a simple algebraic manipulation of the strong form plus an integration by parts (the divergence theorem in multi-dimensions). In mechanics, the weak form corresponds to the virtual work equations.

To obtain the weak form, it proves useful to first define two different spaces or sets of functions. The first set of functions is denoted by the symbol $\mathcal{S}$. It is defined as the set of all possible functions that still satisfy the displacement boundary conditions (the essential or Dirichlet boundary conditions) and the differentiability requirements. In this situation, it implies that the displacement needs to have square-integrable first derivatives. This space of functions is the Sobolev space $W^{1,2}(\Omega)$ also known as $H^1(\Omega)$; it is a Hilbert space. $\Omega=(0,L)$ is the open-set representing the material points of the rod. Thus we write

$$\mathcal{S}=\left\{u\in W^{1,2}(\Omega)\mid u(0)=\bar{u}\right\}. \tag{2}$$

The vertical bar, $|$, is read as "such that". So the statement above is read as: "The set of all functions in $W^{1,2}(\Omega)$ such that they are equal to $\bar{u}$ at x equals zero". Note that there are an infinite number of functions in the set $\mathcal{S}$. It will also prove useful to define the set of functions that are equal to zero at the essential boundary conditions. This set is denoted by the symbol $\mathcal{V}$ and is called the set of weighting functions, virtual displacements, variations, test functions, or trial functions. It is written as

$$\mathcal{V}=\left\{v\in W^{1,2}(\Omega)\mid v(0)=0\right\}. \tag{3}$$

The statement above would be read as: "The set of all functions in $W^{1,2}(\Omega)$ such that they are zero when evaluated at $x = 0$". Note that this corresponds to the concept in mechanics of an admissible virtual displacement.

Remark 1. The choice of making the test functions zero at the essential boundaries is simply a convenient choice (which leads to classical finite element methods). □

Remark 2. The choice concerning the differentiability conditions is also somewhat of an arbitrary choice. We could use even weaker differentiability requirements and end up with alternative finite element methods such as the discontinuous Galerkin method (DG). □

The weak form statement can now be constructed. Pick an arbitrary weighting function $v(x)$ from the set $\mathcal{V}$, i.e. $v \in \mathcal{V}$. Multiply both sides of the ODE in the strong form by this function, integrate both sides from $x = 0$ to $x = L$, then integrate by parts so that the number of derivatives on $v(x)$ is the same as the number of derivatives on $u(x)$. The weak form of the BVP then says:

$$(W) \begin{cases} \text{Find the function } u \in \mathcal{S} \text{ such that for all } v \in \mathcal{V} \\ \displaystyle\int_0^L v_{,x} AE u_{,x}\, dx - \int_0^L vb\, dx + v(L)\bar{F}\,. \end{cases} \tag{4}$$

This form of the BVP is sometimes termed a variational form.

Remark 3. One should stop for a moment and consider the "unusual" form the BVP takes in the weak statement. An interpretation of it is that the solution to the BVP is found by choosing a function in the set $\mathcal{S}$, plugging it into the integral expression, and then testing whether the integral expression is true for each and every function in the set $\mathcal{V}$. Note that there are an infinite number of functions to choose in the set $\mathcal{S}$. Also, there are an infinite number of functions in the set $\mathcal{V}$. All the choices from $\mathcal{S}$ have to be tested against all the choices from $\mathcal{V}$ in the integral expression in (W). The process continues until the element $u \in \mathcal{S}$ is found that satisfies the integral expression in (W) for all the elements $v \in \mathcal{V}$. Clearly, this is an intractable task. The finite element method gives us a prescription for reducing this horribly large search to a "finite dimensional" search. □

Remark 4. The LHS of (4) represents a coercive continuous bi-linear form in terms of the displacement field and the field of variations. The RHS is a continuous linear operator on the space of variations. Together these statements imply stability of the displacement field in the $H^1(\Omega)$ norm. □

Also note that by direct manipulation, it can be shown that the weak form (W) and the strong form (S) are equivalent statements of the BVP (under appropriate smoothness assumptions); ie. the solution to (S) is also the solution to (W) and the solution to (W) is also the solution to (S). By working from the strong form to derive the weak form we have already shown that solutions of (S) satisfy (W). To show that a solution of (W) satisfies (S), we need to show that the equations that comprise the strong form can be derived as consequences of the weak form. Starting with the left-hand side of (4)

$$\int_0^L v_{,x} AEu_{,x}\, dx = \int_0^L (vAEu_{,x})_{,x} - vAEu_{,xx}\, dx \tag{5}$$

$$= vAEu_{,x}\Big|_0^L - \int_0^L vAEu_{,xx}\, dx \tag{6}$$

$$= v(L)AEu_{,x}(L) - \int_0^L vAEu_{,xx}\, dx \tag{7}$$

In the above manipulations, use has been made of the fact that $v(0) = 0$. Now take (7) and combine it with the right-hand side of (4):

$$\int_0^L v[AEu_{,xx} + b]\, dx + v(L)\bar{F} = v(L)AEu_{,x}(L) \tag{8}$$

Since (8) must hold for any function in $\mathcal{V}$, make the following choice:

$$v(x) = x(L-x)[AEu_{,xx} + b] \tag{9}$$

With this choice (8) becomes

$$\int_0^L x(L-x)[AEu_{,xx} + b]^2\, dx = 0 \tag{10}$$

Since $x(L-x)$ is always greater than zero in the open interval $(0, L)$ and $[AEu_{,xx} + b]^2 \geq 0$, it follows that

$$AEu_{,xx} + b = 0\,. \tag{11}$$

Thus the ODE of the strong form problem statement is a "consequence" of the weak form problem statement. The boundary conditions can also be shown to hold. First, the Dirichlet boundary condition $u(0) = \bar{u}$ is automatically satisfied by construction of $\mathcal{S}$. The Neumann (or natural) boundary condition is seen to hold true by substituting (11) back into (8):

$$v(L)\bar{F} = v(L)AEu_{,x}(L) \tag{12}$$

After dividing both sides by $v(L)$ the desired Neumann boundary condition is seen to hold.

1.2 Quick refresher on linear FEA approximation

The basic idea of finite elements is to replace the infinitely large spaces $\mathcal{S}$ and $\mathcal{V}$ by finite dimensional ones. The approximation to $\mathcal{S}$ is denoted as $\mathcal{S}^h$. $\mathcal{S}^h$ is a subspace of $\mathcal{S}$; ie. $\mathcal{S}^h \subset \mathcal{S}$. This is done so that elements of $\mathcal{S}^h$ are always elements of the original space that we were using to find the solution to the BVP. The approximation to the space $\mathcal{V}$ is denoted by $\mathcal{V}^h$. Just as for $\mathcal{S}^h$, $\mathcal{V}^h$ is chosen to be a subspace of the space it is approximating; ie. $\mathcal{V}^h \subset \mathcal{V}$. In the particular kind of finite element methods we will use, elements $u^h \in \mathcal{S}^h$ are expressed in the particular form

$$u^h = v^h + g^h \tag{13}$$

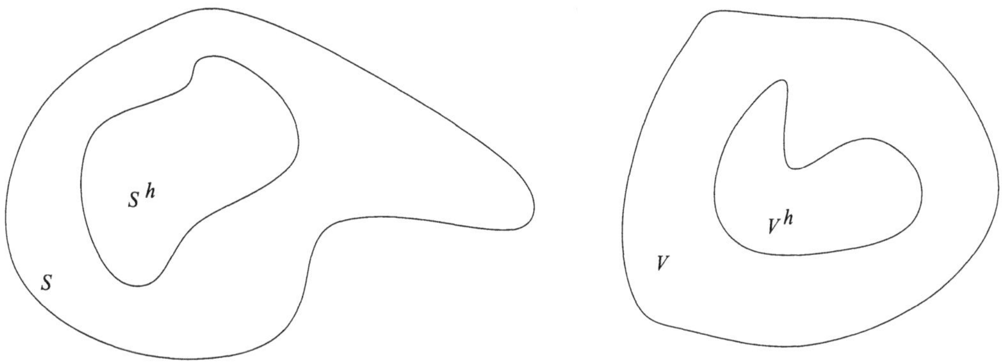

Figure 2. Function space relations

where $v^h \in \mathcal{V}^h$ and g^h is a function equal to $\bar{u}$ at $x = 0$. Note that this form satisfies all of our requirements for elements of the space $\mathcal{S}^h$. The superscript h's are to remind us that our solution is associated with a discretization of our domain into elements of length h. Pictorially, the relation between the real spaces and the approximate spaces is as shown in Fig. 2.

The finite element prescription continues by stating that any element of $\mathcal{V}^h$ is given by a linear combination of shape functions, such as the hat functions. Hence, we may write

$$v^h = \sum_{A=1}^{n} N_A(x) v_A \tag{14}$$

where the $N_A(x)$'s are called shape functions and the v_A's are constants. n is the dimensionality of the finite dimensional space $\mathcal{V}^h$. $N_1(x)$ is the shape function associated with node 1; $N_2(x)$ is the shape function associated with node 2; etc. It is a required property (by standard assumption) that the shape functions are equal to unity at the node to which they are associated and zero at all other nodes; ie.

$$N_A(x_B) = 1 \text{ for } A = B \tag{15}$$

$$N_A(x_B) = 0 \text{ for } A \neq B \tag{16}$$

Or more simply, $N_A(x_B) = \delta_{AB}$, where δ_{AB} is the Kronecker delta function. For the hat functions this appears as shown in Fig. 3.

The elements of $\mathcal{S}^h$ are fully specified by the definition for the elements of $\mathcal{V}^h$ and the definition that

$$g^h = N^g(x)\bar{u} \tag{17}$$

where $N^g(x)$ is the shape function associated with the node at which the essential boundary condition is specified. Note that $N^g(0) = 1$ and is zero at all other nodes; see the picture above. Thus an element of $\mathcal{S}^h$ is expressed as

$$u^h = \sum_{A=1}^{n} N_A(x) u_A + \bar{u} N^g(x) \tag{18}$$

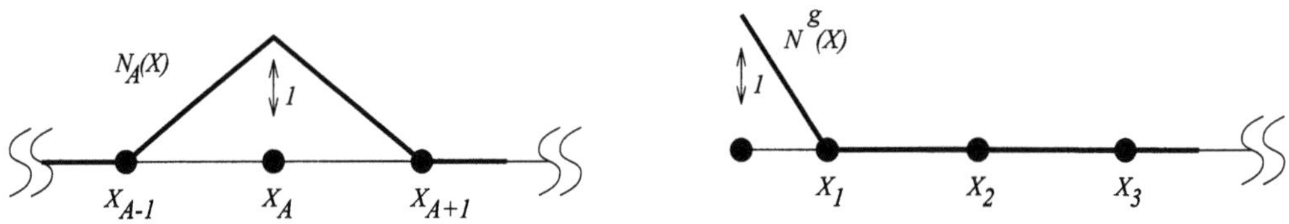

Figure 3. Linear hat functions for the 1-D problem.

Let us now examine the approximate statement of the weak form that the FEM tells us to solve:

$$(W\ approximate)\left\{\begin{array}{l}\text{Find the function } u^h \in \mathcal{S}^h \text{ such that for all } v^h \in \mathcal{V}^h \\ \displaystyle\int_0^L v^h_{,x} AE u^h_{,x}\, dx = \int_0^L v^h b\, dx + v^h(L)\bar{F}\end{array}\right. \tag{19}$$

Remark 5. Using the abstract properties of the bi-linear form on the LHS one can easily show that the finite element solution error is "a best approximation" in terms of the $H^1(\Omega)$ semi-norm – ie. the energy norm. □

If we plug the expressions for u^h and v^h into the approximate weak form above, we get the result that

$$\begin{aligned}\sum_{A=1}^{n} v_A \Bigg[\sum_{B=1}^{n} \left(\int_0^L N_{A,x} AE N_{B,x}\, dx \right) u_B - \int_0^L N_A b\, dx - N_A(L)\bar{F} \\ + \int_0^L N_{A,x} AE N^g_{,x} \bar{u}\, dx \Bigg] = 0\,.\end{aligned} \tag{20}$$

Since this equation is to hold for all elements of $\mathcal{V}^h$, it must hold for arbitrary choices of the constants v_A. Hence the expression in the square brackets must be zero; i.e.

$$\sum_{B=1}^{n} \left(\int_0^L N_{A,x} AE N_{B,x}\, dx \right) u_B - \int_0^L N_A b\, dx - N_A(L)\bar{F} + \int_0^L N_{A,x} AE N^g_{,x} \bar{u}\, dx = 0 \tag{21}$$

We now define an $n \times n$ matrix $\boldsymbol{K}$ with components

$$K_{AB} = \int_0^L N_{A,x} AE N_{B,x}\, dx \tag{22}$$

a vector $\boldsymbol{u}$ with components u_B and a vector $\boldsymbol{F}$ with components given by

$$F_A = \int_0^L N_A b\, dx + N_A(L)\bar{F} - \int_0^L N_{A,x} AE N^g_{,x} \bar{u}\, dx \tag{23}$$

This allows us to rewrite our expression as

$$\boldsymbol{K}\boldsymbol{u} = \boldsymbol{F} \tag{24}$$

All the elements of $\boldsymbol{K}$ and $\boldsymbol{F}$ are known. $\boldsymbol{u}$ is called the vector of unknown nodal displacements; note that $u^h(x_A) = u_A$ by our definition above. In structural mechanics, $\boldsymbol{K}$ is called the stiffness matrix and $\boldsymbol{F}$ is the load vector. Also observe that all we need to do is solve this system of matrix equations to obtain the unknown values of the vector $\boldsymbol{u}$. Once these are known, we then have an expression for the solution to the approximate weak form of the BVP.

There are several important properties of the matrix $\boldsymbol{K}$ in typical linear problems.

1. $\boldsymbol{K}$ is positive definite. This means that we have a unique solution to our approximate problem. Mathematically, positive definiteness is the condition that

$$\boldsymbol{u} \cdot \boldsymbol{K}\boldsymbol{u} > 0 \text{ for } \boldsymbol{u} \neq \boldsymbol{o} \tag{25}$$

$$\boldsymbol{u} \cdot \boldsymbol{K}\boldsymbol{u} = 0 \text{ for } \boldsymbol{u} = \boldsymbol{o} \tag{26}$$

   Its importance here is that it assures us a unique solution for the vector $\boldsymbol{u}$.
2. $\boldsymbol{K}$ is symmetric; i.e. $K_{AB} = K_{BA}$. This is also expressed by saying that the transpose of $\boldsymbol{K}$ is equal to $\boldsymbol{K}$; i.e. $\boldsymbol{K}^T = \boldsymbol{K}$. One important consequence of this is that one only needs to compute half of the matrix to know all the values. And one can use streamlined linear equation solvers.
3. For functions like the hat functions, $\boldsymbol{K}$ has a banded structure. In particular for the hat functions shown above, $K_{AB} = 0$ for $|A - B| > 1$. The importance here is that this allows us to use special storage and solution techniques when working with $\boldsymbol{K}$ in the computer.

2 Strong Form Finite Elasticity

The strong form problem for finite elasticity can be stated as follows: Find $\varphi : \mathcal{B} \to \mathcal{S} \subset \mathbb{R}^3$ such that

$$\text{DIV}[\boldsymbol{P}] + \boldsymbol{b} = 0 \qquad \forall \boldsymbol{X} \in \mathcal{B}, \tag{27}$$

where $\boldsymbol{b}(\boldsymbol{X})$ is a given body force per unit reference volume, $\boldsymbol{P}(\boldsymbol{F})$ is the 1st Piola-Kirchhoff stress tensor, and $\boldsymbol{F} = \text{GRAD}[\varphi]$ is the deformation gradient. The body $\mathcal{B}$ is subject to the boundary conditions that $\varphi = \overline{\varphi}$ for points $\boldsymbol{X} \in \partial\mathcal{B}_\phi$ and imposed tractions $\boldsymbol{P}\boldsymbol{N} = \bar{\boldsymbol{t}}$ for points $\boldsymbol{X} \in \partial\mathcal{B}_t$. In the above $\boldsymbol{N}$ is the unit outward normal to the reference placement of the body $\mathcal{B}$. Further for reasons of existence of solutions we assume $\partial\mathcal{B} = \overline{\partial\mathcal{B}_\phi \cup \partial\mathcal{B}_t}$ and that $\partial\mathcal{B}_\phi \cap \partial\mathcal{B}_t = \emptyset$. For convenience, the inertial terms have been omitted.

The finite elasticity constitutive relation can be given strictly as a "stress-strain" relation (i.e. Cauchy elasticity) or it can be assumed to emanate from a potential (i.e. hyperelasticity also known as Green elasticity). For our purposes we will mainly consider the case of hyperelasticity. Thus we will assume $\boldsymbol{P} = \partial W(\boldsymbol{F})/\partial \boldsymbol{F}$. Note that for a well defined problem the strain energy density $W(\cdot)$ can not simply be arbitrary. It must conform to the principles of frame indifference and thus actually be a function of $\boldsymbol{C} = \boldsymbol{F}^T\boldsymbol{F}$ the right Cauchy-Green deformation tensor. Also it should, at the minimum, be quasi-convex and more likely poly-convex in terms of $\boldsymbol{F}$. A complete discussion of the required properties is well beyond the scope of these introductory notes. Interested readers are directed to Ball (1977b), Ball (1977a), Ciarlet (1988), and Dacorogna (1989) for a thorough introduction to these topics.

Remark 6. Even though we will not discuss these issues in any depth it is important to realize that they have very important implications relative to finite element analysis in finite deformation. Constitutive models must have all the correct behavior as outlined by these theories otherwise the FEA models have the possibility of failing. It is also important to realize that sometimes the models have the correct properties for a limited range of deformation gradients but not over the entire range required by the existence theories. In such situations one is often tempted to claim that the finite element analysis will be limited to the "valid" range of the model and thus be "ok". It should be carefully noted, however, that often during the solution of the nonlinear FEA equations it is quite possible for a trial solution(iterate) to reach out into the "invalid" range of the model and subsequently "crash" the computation. □

3 Weak Form Finite Elasticity

To determine the weak equations of equilibrium we start by defining the space of variations as maps $\delta\varphi : \mathcal{B} \to \mathcal{S} \subset \mathbb{R}^3$ in the space

$$\mathcal{V} = \left\{ \delta\varphi \in [W^{1,p}(\mathcal{B})]^3 \mid \delta\varphi = \mathbf{0} \quad \forall \boldsymbol{X} \in \mathcal{B}_\phi \right\} , \tag{28}$$

where the choice of p depends upon the material model at hand. We can now take a dot product of an arbitrary test function with the equilibrium equation.

$$G(\varphi; \delta\varphi) = \int_{\mathcal{B}} \delta\varphi \cdot \mathrm{DIV}[\boldsymbol{P}] + \delta\varphi \cdot \boldsymbol{b} = 0. \tag{29}$$

Integration by parts (i.e. application of the divergence theorem) gives

$$\begin{aligned} G(\varphi; \delta\varphi) &= \int_{\mathcal{B}} -\mathrm{DIV}\left[\delta\varphi \cdot [\boldsymbol{P}]\right] + \boldsymbol{P} : \mathrm{GRAD}[\delta\varphi] - \delta\varphi \cdot \boldsymbol{b} \\ &= \int_{\partial\mathcal{B}} -\delta\varphi \cdot \boldsymbol{P}\boldsymbol{N} + \int_{\mathcal{B}} \boldsymbol{P} : \mathrm{GRAD}[\delta\varphi] - \delta\varphi \cdot \boldsymbol{b} \\ &= \int_{\mathcal{B}} \boldsymbol{P} : \mathrm{GRAD}[\delta\varphi] - \int_{\mathcal{B}} \delta\varphi \cdot \boldsymbol{b} - \int_{\partial\mathcal{B}_t} \delta\varphi \cdot \bar{\boldsymbol{t}} = 0 , \end{aligned} \tag{30}$$

which must hold for all $\delta\varphi \in \mathcal{V}$.

The weak statement of the problem is: find $\varphi \in \mathcal{S}$ such that (30) holds for all $\delta\varphi \in \mathcal{V}$. Here we take the space of solutions to be

$$\mathcal{S} = \left\{ \varphi \in [W^{1,p}]^3 \mid J = \det[\boldsymbol{F}] > 0 \quad \forall \boldsymbol{X} \in \mathcal{B} \quad \text{and} \quad \varphi = \overline{\varphi} \quad \forall \boldsymbol{X} \in \mathcal{B}_\phi \right\} , \tag{31}$$

where $\boldsymbol{F} = \mathrm{GRAD}[\varphi]$ and $\boldsymbol{P} = \widehat{\boldsymbol{P}}(\boldsymbol{F})$.

Remark 7. Note that in this "traditional" weak form setting the equilibrium equation is expressed in weak form but the kinematic relation and the constitutive relation are still expressed in strong form. □

Remark 8. Note that even though we use the traditional variational notation of $\delta\varphi$, we in no way are assuming that $\delta\varphi$ is small. Nowhere have we invoked any assumptions of smallness – only arbitrariness. □

3.1 Weak enforcement of constitution and kinematics

As noted in the last remark, in the traditional weak expression of the boundary value problem, the constitution and governing kinematic relation are given in strong form. This obviously need not be the case. We can in fact introduce the space of deformation gradients and the space of 1st Piola-Kirchhoff stress tensors and suitable corresponding spaces of test functions. With these two sets of additional spaces we can, if we desire, rewrite the constitution and kinematic relations in weak form. Doing so will give what is often termed a *mixed formulation*. The term *mixed* is associated with the fact that the types of variables to be directly solved for are of differing type.

In the present setting we would create two new weak forms:

$$G_P(\varphi, \boldsymbol{F}, \boldsymbol{P}; \delta\boldsymbol{P}) = 0 = \int_{\mathcal{B}} \delta\boldsymbol{P} : (\boldsymbol{F} - \mathrm{GRAD}[\varphi]) \tag{32}$$

for the weak statement of the kinematics and

$$G_F(\varphi, \boldsymbol{F}, \boldsymbol{P}; \delta\boldsymbol{F}) = 0 = \int_{\mathcal{B}} \delta\boldsymbol{F} : \left(\frac{\partial W}{\partial \boldsymbol{F}} - \boldsymbol{P}\right) \tag{33}$$

for the weak statement of the constitution. In the above we have introduced the conventional variational notation of $\delta\boldsymbol{P}$ and $\delta\boldsymbol{F}$ to represent "variations" of the stresses and deformation gradients. The general functional spaces for $\boldsymbol{P}, \delta\boldsymbol{P}, \boldsymbol{F}, \delta\boldsymbol{F}$ are such that the integrals above are well defined. Note in particular that the above relations do not involve gradients of the stress or the deformation gradient. Thus the components of these variables will in general lie in $W^{0,q}(\Omega) = L^q(\Omega)$, where q needs to be chosen so that the integrals are computable.

Remark 9. This form of the weak statement, while perfectly mathematically valid is not particularly common in practice. More common in practice is a set of weak relations that only "mixes" part of the deformation gradient and part of the stresses – viz. the Jacobian determinant and the pressure. □

4 Minimization Form of Finite Elasticity

A third alternative way of writing the boundary value problem of finite elasticity is to phrase it in terms of a variational minimization problem. This way of writing the governing equations sometimes gives advantage in understanding and/or deriving a variety of finite element formulations. In particular, let us consider the potential energy of a body as

$$\Pi(\varphi) = \int_{\mathcal{B}} W(\boldsymbol{F}) - \int_{\mathcal{B}} \boldsymbol{b} \cdot \boldsymbol{u} - \int_{\partial\mathcal{B}_t} \bar{\boldsymbol{t}} \cdot \boldsymbol{u}\,, \tag{34}$$

where $\boldsymbol{u} = \boldsymbol{x} - \boldsymbol{X}$, $\boldsymbol{x} = \varphi(\boldsymbol{X})$, and $\boldsymbol{F} = \mathrm{GRAD}[\varphi]$. If we ask that the variation of the potential be stationary, then we wish to find the fields for which the variational derivative of Π with respect to φ are zero. Note that in varying φ we require that the essential boundary conditions be strictly satisfied; i.e. that the variations $\delta\varphi \in \mathcal{V}$ as defined in the previous section.

This gives:

$$G(\varphi;\delta\varphi) = \delta_\phi \Pi = \frac{d}{d\epsilon}\bigg|_{\epsilon=0} \Pi(\varphi+\epsilon\delta\varphi)$$
$$= \int_{\mathcal{B}} \boldsymbol{P} : \mathrm{GRAD}[\delta\varphi] - \int_{\mathcal{B}} \delta\varphi \cdot \boldsymbol{b} - \int_{\partial\mathcal{B}_t} \delta\varphi \cdot \bar{\boldsymbol{t}} = 0\,, \tag{35}$$

where we have written (strongly) $\boldsymbol{P} = \partial W/\partial \boldsymbol{F}$ and $\boldsymbol{F} = \mathrm{GRAD}[\varphi]$. This is simply the weak form of the equilibrium equation and it must hold true for all $\delta\varphi \in \mathcal{V}$.

4.1 3-field mixed minimization forms

It is also possible to write a minimization form that generates the weak enforcement of the constitution and kinematics. Consider for instance the problem of finding the stationary point of the following functional

$$\Pi(\varphi, \boldsymbol{F}, \boldsymbol{P}) = \int_{\mathcal{B}} W(\boldsymbol{F}) - \boldsymbol{b}\cdot\boldsymbol{u} - \boldsymbol{P} : (\boldsymbol{F} - \mathrm{GRAD}[\varphi]) - \int_{\partial\mathcal{B}_t} \bar{\boldsymbol{t}} \cdot \boldsymbol{u}\,. \tag{36}$$

Taking the variation with respect to the first argument give the weak equilibrium equation

$$G_\phi(\varphi, \boldsymbol{F}, \boldsymbol{P}; \delta\varphi) = \delta_\phi \Pi = \frac{d}{d\epsilon}\bigg|_{\epsilon=0} \Pi(\varphi+\epsilon\delta\varphi, \boldsymbol{F}, \boldsymbol{P})$$
$$= \int_{\mathcal{B}} \boldsymbol{P} : \mathrm{GRAD}[\delta\varphi] - \int_{\mathcal{B}} \delta\varphi \cdot \boldsymbol{b} - \int_{\partial\mathcal{B}_t} \delta\varphi \cdot \bar{\boldsymbol{t}} = 0\,. \tag{37}$$

Taking the variation with respect to the deformation gradient gives

$$G_F(\varphi, \boldsymbol{F}, \boldsymbol{P}; \delta\boldsymbol{F}) = \delta_F \Pi = \frac{d}{d\epsilon}\bigg|_{\epsilon=0} \Pi(\varphi, \boldsymbol{F}+\epsilon\delta\boldsymbol{F}, \boldsymbol{P})$$
$$= \int_{\mathcal{B}} \delta\boldsymbol{F} : \left(\frac{\partial W}{\partial \boldsymbol{F}} - \boldsymbol{P}\right) = 0\,. \tag{38}$$

And lastly taking a variation with respect to the stresses gives

$$G_P(\varphi, \boldsymbol{F}, \boldsymbol{P}; \delta\boldsymbol{P}) = \delta_P \Pi = \frac{d}{d\epsilon}\bigg|_{\epsilon=0} \Pi(\varphi, \boldsymbol{F}, \boldsymbol{P}+\epsilon\delta\boldsymbol{P})$$
$$= -\int_{\mathcal{B}} \delta\boldsymbol{P} : (\boldsymbol{F} - \mathrm{GRAD}[\varphi]) = 0\,. \tag{39}$$

4.2 3-field, u, p, θ formulation

A very important variation of the 3-field minimization principle stated in the last section is one in which only the volumetric parts of the stress and deformation gradient are "mixed". This stationary problem is given in terms of finding the stationary equations for the following principle

$$\Pi(\varphi, p, \theta) = \int_{\mathcal{B}} W(\overline{\boldsymbol{F}}) - \boldsymbol{b}\cdot\boldsymbol{u} - p\,(\theta - \det[\mathrm{GRAD}(\varphi)]) - \int_{\partial\mathcal{B}_t} \bar{\boldsymbol{t}} \cdot \boldsymbol{u}\,. \tag{40}$$

In the above $\overline{\boldsymbol{F}}$ is the so called mixed deformation gradient and is given by $\overline{\boldsymbol{F}} = (\theta/J)^{1/3}\boldsymbol{F}$. θ is known as the mixed Jacobian and p is the mixed pressure. The variation with respect to the motion gives:

$$\begin{aligned} G_\phi(\varphi, p, \theta; \delta\varphi) = \delta_\phi \Pi &= \left.\frac{d}{d\epsilon}\right|_{\epsilon=0} \Pi(\varphi + \epsilon\delta\varphi, p, \theta) \\ &= \int_{\mathcal{B}} \left\{ (\theta/J)^{1/3} \boldsymbol{P}(\overline{\boldsymbol{F}}) : [\mathbb{I} - \frac{1}{3}\boldsymbol{F} \otimes \boldsymbol{F}^{-T}] + pJ\boldsymbol{F}^{-T} \right\} : \mathrm{GRAD}[\delta\varphi] \\ &- \int_{\mathcal{B}} \delta\varphi \cdot \boldsymbol{b} - \int_{\partial\mathcal{B}_t} \delta\varphi \cdot \bar{\boldsymbol{t}} = 0\,. \end{aligned} \tag{41}$$

Taking the variation with respect to the mixed Jacobian gives

$$\begin{aligned} G_\theta(\varphi, p, \theta; \delta\theta) = \delta_\theta \Pi &= \left.\frac{d}{d\epsilon}\right|_{\epsilon=0} \Pi(\varphi, p, \theta + \epsilon\delta\theta) \\ &= \int_{\mathcal{B}} \delta\theta : \left(\frac{1}{3} \left[\frac{1}{\theta} \boldsymbol{P}(\overline{\boldsymbol{F}}) \overline{\boldsymbol{F}}^T \right] : \mathbf{1} - p \right) = 0\,. \end{aligned} \tag{42}$$

And lastly taking a variation with respect to the mixed pressure gives

$$\begin{aligned} G_P(\varphi, p, \theta; \delta p) = \delta_p \Pi &= \left.\frac{d}{d\epsilon}\right|_{\epsilon=0} \Pi(\varphi, p + \epsilon\delta p, \theta) \\ &= -\int_{\mathcal{B}} \delta p \,(\theta - \det[\mathrm{GRAD}[\varphi]]) = 0\,. \end{aligned} \tag{43}$$

Remark 10. Often in this setting it is assumed that the strain energy density splits additively into a deviatoric and volumetric form; i.e. that

$$W(\boldsymbol{F}) = \widetilde{W}(\widetilde{\boldsymbol{F}}) + U(J)\,, \tag{44}$$

where $\widetilde{\boldsymbol{F}} = J^{-1/3}\boldsymbol{F}$ is the volume free part of the deformation gradient. While this is a popular assumption, it is not necessary. □

5 Nonlinear finite elements

We consider first the situation where the boundary value problem to be solved is given in terms of the so-called displacement form. This is represented by the weak form expression (35). The spaces to be approximated are the space of solutions $\mathcal{S}$ which we represent in terms of the displacement field $\boldsymbol{u} = \boldsymbol{x} - \boldsymbol{X} = \varphi(\boldsymbol{X}) - \boldsymbol{X}$ and the space of variations $\mathcal{V}$. For simplicity we will assume throughout that the essential boundary condition are zero. In doing so $\mathcal{S} = \mathcal{V}$ and there is a small simplification of notation.

5.1 Notational conventions

We will assume that the geometry is meshed isoparametrically so that

$$\boldsymbol{X} = \sum_\alpha N^\alpha(\boldsymbol{\xi})\boldsymbol{X}^\alpha\,, \tag{45}$$

where α ranges over the number of nodes in the mesh, $N^{\alpha}(\cdot)$ are the element shape functions in terms of the isoparametric coordinates $\boldsymbol{\xi}$, and $\boldsymbol{X}^{\alpha}$ are the nodal coordinate vectors. It is convenient to introduce the following matrix-vector notation. First, define

$$\boldsymbol{N} = \left[\boldsymbol{N}^1 \, \boldsymbol{N}^2 \cdots\right]$$
$$\boldsymbol{N}^{\alpha} = \begin{bmatrix} N^{\alpha} & 0 & 0 \\ 0 & N^{\alpha} & 0 \\ 0 & 0 & N^{\alpha} \end{bmatrix} \tag{46}$$

and further define the mapping of the components X_i^{α} into a column vector so that

$$\boldsymbol{X} = \begin{pmatrix} X_1^1 \\ X_2^1 \\ X_3^1 \\ X_1^2 \\ \vdots \end{pmatrix} . \tag{47}$$

With this notation at hand we can write the isoparametric map (45) as

$$\boldsymbol{X} = \boldsymbol{N}\boldsymbol{X} , \tag{48}$$

where there is an obvious abuse of notation involved that must be understood from context. Using this set up we can write the displacements and the variations as

$$\boldsymbol{u}(\boldsymbol{X}) = \sum_{\alpha} N^{\alpha} \boldsymbol{u}^{\alpha} = \boldsymbol{N}\boldsymbol{u} , \tag{49}$$

$$\delta\boldsymbol{\varphi}(\boldsymbol{X}) = \sum_{\alpha} N^{\alpha} \delta\boldsymbol{\varphi}^{\alpha} = \boldsymbol{N}\delta\boldsymbol{\varphi} . \tag{50}$$

It will also prove to be useful to introduce the following notational convention for symmetric gradients with respect to the spatial parameterization of the body.

$$\mathrm{sym}(\mathrm{grad}[\boldsymbol{u}]) = \mathrm{sym}\left(\sum_{\alpha} \mathrm{grad}[N^{a}]\boldsymbol{u}^{\alpha}\right) = \boldsymbol{B}\boldsymbol{u} , \tag{51}$$

where

$$\boldsymbol{B} = \left[\boldsymbol{B}^1 \, \boldsymbol{B}^2 \cdots\right]$$
$$\boldsymbol{B}^{\alpha} = \begin{bmatrix} N^{\alpha}_{,1} & 0 & 0 \\ 0 & N^{\alpha}_{,2} & 0 \\ 0 & 0 & N^{\alpha}_{,3} \\ N^{\alpha}_{,2} & N^{\alpha}_{,1} & 0 \\ 0 & N^{\alpha}_{,3} & N^{\alpha}_{,2} \\ N^{\alpha}_{,3} & 0 & N^{\alpha}_{,1} \end{bmatrix} . \tag{52}$$

In this last equation the indicated derivatives are to be taken with respect to x_1, x_2, and x_3. It is also convenient to map symmetric rank-2 tensors to column vectors and doubly symmetric rank-4 tensors to 6×6 matrices. Suppose $\boldsymbol{\tau}$ is symmetric rank-2, then we define

$$\boldsymbol{\tau} \rightarrow \begin{bmatrix} \tau_{11} & \tau_{12} & \tau_{13} \\ \tau_{21} & \tau_{22} & \tau_{23} \\ \tau_{31} & \tau_{32} & \tau_{33} \end{bmatrix} \rightarrow \begin{pmatrix} \tau_{11} \\ \tau_{22} \\ \tau_{33} \\ \tau_{12} \\ \tau_{23} \\ \tau_{31} \end{pmatrix} . \tag{53}$$

If $\mathbb{C}$ is a rank-4 tensor with major and minor symmetries then its entries are mapped to matrix with the following ordering

$$\mathbb{C} \rightarrow \begin{bmatrix} \mathbb{C}_{1111} & \mathbb{C}_{1122} & \mathbb{C}_{1133} & \mathbb{C}_{1112} & \mathbb{C}_{1123} & \mathbb{C}_{1131} \\ \mathbb{C}_{2211} & \mathbb{C}_{2222} & \mathbb{C}_{2233} & \mathbb{C}_{2212} & \mathbb{C}_{2223} & \mathbb{C}_{2231} \\ \mathbb{C}_{3311} & \mathbb{C}_{3322} & \mathbb{C}_{3333} & \mathbb{C}_{3312} & \mathbb{C}_{3323} & \mathbb{C}_{3331} \\ \mathbb{C}_{1211} & \mathbb{C}_{1222} & \mathbb{C}_{1233} & \mathbb{C}_{1212} & \mathbb{C}_{1223} & \mathbb{C}_{1231} \\ \mathbb{C}_{2311} & \mathbb{C}_{2322} & \mathbb{C}_{2333} & \mathbb{C}_{2312} & \mathbb{C}_{2323} & \mathbb{C}_{2331} \\ \mathbb{C}_{3111} & \mathbb{C}_{3122} & \mathbb{C}_{3133} & \mathbb{C}_{3112} & \mathbb{C}_{3123} & \mathbb{C}_{3131} \end{bmatrix} \tag{54}$$

Remark 11. The convention of using vectors for symmetric rank-2 tensors and 6×6 matrices for rank-4 tensors is known as Voigt notation. When symmetric rank-2 kinematic tensors are mapped to vectors, it is convention to multiply the last 3 terms by 2.

5.2 Displacement form

We now return to (35). First let us manipulate it to remove the 1st Piola-Kirchhoff stress tensor and replace it by the Kirchhoff stress tensor $\boldsymbol{\tau} = \boldsymbol{P}\boldsymbol{F}^T$. Doing so gives the weak form

$$G(\boldsymbol{\varphi}; \delta\boldsymbol{\varphi}) = \int_{\mathcal{B}} \boldsymbol{\tau} : \mathrm{sym}(\mathrm{grad}[\delta\boldsymbol{\varphi}]) - \int_{\mathcal{B}} \delta\boldsymbol{\varphi} \cdot \boldsymbol{b} - \int_{\partial\mathcal{B}_t} \delta\boldsymbol{\varphi} \cdot \bar{\boldsymbol{t}} = 0 . \tag{55}$$

If we now introduce our special notational conventions we will have that

$$\boldsymbol{\varphi}^T \left[\int_{\mathcal{B}} \boldsymbol{B}^T \boldsymbol{\tau} - \int_{\mathcal{B}} \boldsymbol{N}^T \boldsymbol{b} - \int_{\partial\mathcal{B}_t} \boldsymbol{N}^T \bar{\boldsymbol{t}} \right] = 0 . \tag{56}$$

Since this expression is to hold for all variations we have the following "residual" equations:

$$\boldsymbol{R}(\boldsymbol{u}) = \underbrace{\int_{\mathcal{B}} \boldsymbol{B}^T \boldsymbol{\tau}(\mathbf{1} + \mathrm{GRAD}\boldsymbol{u})}_{\boldsymbol{F}^{\mathrm{int}}(\boldsymbol{u})} - \underbrace{\left(\int_{\mathcal{B}} \boldsymbol{N}^T \boldsymbol{b} + \int_{\partial\mathcal{B}_t} \boldsymbol{N}^T \bar{\boldsymbol{t}} \right)}_{\boldsymbol{F}^{\mathrm{ext}}} = 0 . \tag{57}$$

Thus the objective is to solve $\boldsymbol{R}(\boldsymbol{u}) = \boldsymbol{F}^{\mathrm{int}}(\boldsymbol{u}) - \boldsymbol{F}^{ext} = 0$ for the (number of free nodes times 3) nodal unknowns $\boldsymbol{u}$.

5.3 Equation solving

The residual equations (57) constitute (typically) a large system of nonlinear equations in terms of the unknowns $\boldsymbol{u}$. There are a large number of solution methods that one can employ in solving these equations. For the most, however, they are solved using variants of Newton's method. Thus we will briefly review Newton's methods. This scheme attempts to solve the nonlinear equations iteratively by solving a sequence of linear problems. Assume that we have a starting guess $\boldsymbol{u}^{(0)}$ for the solution vector. Then we desire a generic rule for finding $\boldsymbol{u}^{(k+1)} = \boldsymbol{u}^{(k)} + \Delta\boldsymbol{u}$ when $\boldsymbol{u}^{(k)}$ is known. It is further desired that in general $\|\boldsymbol{R}(\boldsymbol{u}^{(k+1)})\| < \|\boldsymbol{R}(\boldsymbol{u}^{(k)})\|$ – a situation that is not always guaranteed by Newton's method. To find the needed rule we write $\boldsymbol{R}(\boldsymbol{u}^{(k+1)})$ as a linear expansion about the known state $\boldsymbol{u}^{(k)}$ and set it equal to zero (the desired result); i.e.

$$0 = \boldsymbol{R}(\boldsymbol{u}^{(k+1)}) \approx \boldsymbol{R}(\boldsymbol{u}^{(k)}) + \frac{\partial \boldsymbol{R}}{\partial \boldsymbol{u}}(\boldsymbol{u}^{(k)}) \cdot \Delta\boldsymbol{u} \,. \tag{58}$$

Solving for $\Delta\boldsymbol{u}$ gives

$$\Delta\boldsymbol{u} = -\left[\frac{\partial \boldsymbol{R}}{\partial \boldsymbol{u}}(\boldsymbol{u}^{(k)})\right]^{-1} \boldsymbol{R}(\boldsymbol{u}^{(k)}) \,. \tag{59}$$

One can then update the displacement field using $\boldsymbol{u}^{(k+1)} = \boldsymbol{u}^{(k)} + \Delta\boldsymbol{u}$ and continue with the process – terminating when the norm of the residual equations is below some specified tolerance.

Remark 12. This is the most basic version of Newton's iteration scheme. Its ability to converge is strongly influence by the severity of the imposed load. For large loads one normally applies a fraction of the load and then converges the residual. Then the load is further increased toward the desired load and the residual is converged again. The process is repeated until the load is at its target value. In the description given, quantities are always computed relative to the undeformed configuration and thus there is no build up of residual errors during the process. During the iteration process at each load step the iterations converge quadratically (asymptotically). □

Remark 13. The matrix $\left[\frac{\partial \boldsymbol{R}}{\partial \boldsymbol{u}}(\boldsymbol{u}^{(k)})\right]$ is known as the finite element tangent matrix. For the nonlinear problem this matrix is composed of two parts – one material and one geometric. □

5.4 Tangent for the displacement form

To find a useful expression for the tangent one typically rewrites (35) in terms of the 2nd Piola-Kirchhoff stress tensor $\boldsymbol{S} = 2\partial W/\partial \boldsymbol{C} = \boldsymbol{F}^{-1}\boldsymbol{P}$. Thus we can write

$$G(\varphi;\delta\varphi) = \int_{\mathcal{B}} (\boldsymbol{F}\boldsymbol{S}) : \mathrm{GRAD}[\delta\varphi] - \int_{\mathcal{B}} \delta\varphi \cdot \boldsymbol{b} - \int_{\partial\mathcal{B}_t} \delta\varphi \cdot \bar{\boldsymbol{t}} = 0 \,. \tag{60}$$

One now computes the second variation of this expression with respect to the motion:

$$\begin{aligned} DG(\varphi;\delta\varphi)[\Delta\varphi] &= \left.\frac{d}{d\epsilon}\right| G(\varphi + \epsilon\Delta\varphi;\delta\varphi) \\ &= \int_{\mathcal{B}} \underbrace{(\mathrm{GRAD}[\Delta\varphi]\boldsymbol{S}) : \mathrm{GRAD}[\delta\varphi]}_{\text{geometric tangent}} + \underbrace{J\mathrm{grad}[\Delta\varphi] : \varphi_*(\mathbb{C}) : \mathrm{grad}[\delta\varphi]}_{\text{material tangent}} \,, \end{aligned} \tag{61}$$

where $\varphi_*(\mathbb{C}) = \frac{1}{J}\boldsymbol{F}\boldsymbol{F}\partial^2 W/\partial \boldsymbol{C}^2 \boldsymbol{F}^T \boldsymbol{F}^T$ is known as the push forward of the material tangent.

Remark 14. In matrix form, the material tangent has the usual form $\int_{\mathcal{B}} J\boldsymbol{B}^T \varphi_*(\mathbb{C})\boldsymbol{B}$, where it is understood that the rank-4 tensor is mapped to the 6×6 (Voigt) notation.

Remark 15. A major drawback of this finite element procedure is that if the material is nearly incompressible it will perform rather poorly in finite precision arithmetic – regardless of the interpolation order utilized. A very loose qualitative reason for this is that in the expression for the material tangent the terms associated with the volumetric response will dominate the deviatoric response terms. In finite precision the deviatoric terms will be completely lost and the formulation will result in near zero displacements. This argument is a paraphrase of Fried (1974). There are a large number of methods for treating such problems – some rigorous, some ad hoc. Here we shall only treat one common method for circumventing this problem – mixed formulations. For a fuller discussion of other treatments the reader is referred to Zienkiewicz and Taylor (2000) and references therein. □

5.5 Mixed form

Perhaps the most popular method for treating locking due to incompressible material behavior is to use the 3-field $\boldsymbol{u}, p, \theta$ formulation. In the finite element setting one takes advantage of the fact that the pressure and mixed Jacobian fields are in $W^{0,p} = L^p$ and thus can be discontinuous across element boundaries. The practical meaning of this point is that one does not have to globally solve the weak expressions for kinematics and constitution; rather, one can do this on an element by element basis. This is an important consideration since one does not want to introduce extra unknowns at the global level. Let us start with a typical decoupled ansatz for the strain energy

$$\Pi(\varphi, p, \theta) = \int_{\mathcal{B}} \widetilde{W}(\widetilde{\boldsymbol{C}}) + U(\theta) - \boldsymbol{b}\cdot\boldsymbol{u} - p\,(\theta - \det[\mathrm{GRAD}(\varphi)]) - \int_{\partial\mathcal{B}_t} \bar{\boldsymbol{t}}\cdot\boldsymbol{u}\,. \tag{62}$$

In the above $\widetilde{\boldsymbol{C}} = J^{-2/3}\boldsymbol{C}$ is the volume preserving part of the right Cauchy-Green deformation tensor. The variation with respect to the motion gives:

$$\begin{aligned} G_\phi(\varphi, p, \theta; \delta\varphi) = &\int_{\mathcal{B}} \left\{ J^{-2/3}\left(2\frac{\partial \widetilde{W}}{\partial \widetilde{\boldsymbol{C}}}\right) : [\mathbb{I} - \frac{1}{3}\boldsymbol{C}\otimes\boldsymbol{C}^{-1}] + pJ\boldsymbol{C}^{-1}\right\} : \boldsymbol{F}^T\mathrm{GRAD}[\delta\varphi] \\ &- \int_{\mathcal{B}} \delta\varphi\cdot\boldsymbol{b} - \int_{\partial\mathcal{B}_t} \delta\varphi\cdot\bar{\boldsymbol{t}} = 0\,. \end{aligned} \tag{63}$$

Taking the variation with respect to the mixed Jacobian gives

$$G_\theta(\varphi, p, \theta; \delta\theta) = \int_{\mathcal{B}} \delta\theta \left[\frac{\partial U}{\partial \theta} - p\right] = 0\,. \tag{64}$$

And lastly taking a variation with respect to the mixed pressure gives

$$G_p(\varphi, p, \theta; \delta p) = -\int_{\mathcal{B}} \delta p\,(\theta - \det[\mathrm{GRAD}[\varphi]]) = 0\,. \tag{65}$$

In the finite element setting assume that the pressures and mixed Jacobian are given in terms of shape functions over individual elements which are discontinuous across element boundaries. Arranging these shape functions in a column vector $\boldsymbol{\Gamma}$ and the element parameters into column vectors $\boldsymbol{\theta}$, $\delta\boldsymbol{\theta}$, $\boldsymbol{p}$, and $\delta\boldsymbol{p}$, allows us to compute the element level mixed pressure and Jacobian from (64) and (65) as:

$$p = \boldsymbol{\Gamma}^T \boldsymbol{H}^{-1} \int_{\mathcal{B}_e} \boldsymbol{\Gamma} \frac{\partial U}{\partial \theta} \tag{66}$$

$$\theta = \boldsymbol{\Gamma}^T \boldsymbol{H}^{-1} \int_{\mathcal{B}_e} \boldsymbol{\Gamma} J \,, \tag{67}$$

where $\mathcal{B}_e$ is the reference domain of an element e and $\boldsymbol{H} = \int_{\mathcal{B}_e} \boldsymbol{\Gamma}\boldsymbol{\Gamma}^T$.

Remark 16. To find the tangent expression one can back substitute these two expression on an element by element level. Then the weak equilibrium equation can be expressed solely in terms of the motion (elementwise). The second variation can then be taken. The algebra is somewhat complex and we refer the interested reader to Simo and Taylor (1991) and Zienkiewicz and Taylor (2000) for the detailed expressions. □

Mixed Element Technology The primary element utilized in finite elasticity is a mixed element in the form outlined above. The most common choice is the so-called Q1/P0 element. It uses bi-linear (or tri-linear in 3-D) displacement interpolations and constant pressure interpolations. While this element is used in the majority of production finite element analyses is does not satisfy the BB inf-sup condition which is needed for showing the stability of the finite element solution (in the linear case). A second popular element is the so-called Q2/P1 element. This element utilizes bi-quadratic displacement interpolations and linear pressure interpolations. In 2-D, for the linear problem, it is known to satisfy the BB inf-sup conditions; see e.g. Johnson (1987) and Brezzi and Fortin (1991). There are a great many other element types that have been proposed but none are as popular as these two for finite elasticity/viscoelasticity. In general, it could be said that one should prefer the Q2/P1 element due to its theoretical underpinnings and its generally superior behavior. However, the element does suffer from difficulties associated with converging the Newton iterates with large time steps. For this reason one often employs the Q1/P0 element.

Remark 17. Another modestly popular element class is the enhanced class of elements. These are more recent and, while popular with some researchers, do not yet enjoy complete acceptance due to a number of theoretical difficulties associated with element stability.

5.6 Principal stretches

One very important class of finite elasticity models are those expressed in terms of the eigenvalues of the deformation gradient. In particular consider a strain energy density of the form $W = \widehat{W}(\lambda_1, \lambda_2, \lambda_3)$. Where the spectral representation of the deformation gradient is given as

$$\boldsymbol{F} = \sum_A \lambda_A \boldsymbol{n}_A \otimes \boldsymbol{N}_A \,, \tag{68}$$

where $\boldsymbol{n}_A$ and $\boldsymbol{N}_A$ are the eigenvectors of $\boldsymbol{b}$ and $\boldsymbol{C}$ the left and right Cauchy-Green deformation tensors, respectively. In such a situation the material is necessarily isotropic and the various stress tensors have the following forms:

$$\boldsymbol{S} = \sum_A S_A \boldsymbol{N}_A \otimes \boldsymbol{N}_A = \sum_A \frac{1}{\lambda_A} \frac{\partial W}{\partial \lambda_A} \boldsymbol{N}_A \otimes \boldsymbol{N}_A \tag{69}$$

$$\boldsymbol{P} = \sum_A P_A \boldsymbol{n}_A \otimes \boldsymbol{N}_A = \sum_A \frac{\partial W}{\partial \lambda_A} \boldsymbol{n}_A \otimes \boldsymbol{N}_A \tag{70}$$

$$\boldsymbol{\tau} = \sum_A \tau_A \boldsymbol{n}_A \otimes \boldsymbol{n}_A = \sum_A \lambda_A \frac{\partial W}{\partial \lambda_A} \boldsymbol{n}_A \otimes \boldsymbol{n}_A \,. \tag{71}$$

Remark 18. There relations are easy to derive if one notes that

$$\frac{\partial \lambda_A}{\partial \boldsymbol{C}} = \frac{1}{2\lambda_A} \boldsymbol{N}_A \otimes \boldsymbol{N}_A \qquad (\text{no sum})\,. \tag{72}$$

□

In terms of computation, one of the main issues is the determination of the tangent – especially when the principal stretches are repeated. In the paper of Simo and Taylor (1991) a particular scheme has been given for treating such problems and is popularly quoted. Computationally, however, this method has not proven itself to be robust. A better set of tangent relations is given in Zienkiewicz and Taylor (2000) using the formulas based on Ogden (1984). For example the material tangent defined from $\dot{\boldsymbol{S}} = \mathbb{C} : \frac{1}{2}\dot{\boldsymbol{C}}$ is expressed as

$$\begin{aligned} \mathbb{C} = & \sum_{A,B} \frac{1}{\lambda_B} \frac{\partial S_A}{\partial \lambda_B} \boldsymbol{N}_A \otimes \boldsymbol{N}_A \otimes \boldsymbol{N}_B \otimes \boldsymbol{N}_B \\ & + \sum_{A \neq B} \frac{1}{2} \frac{S_B - S_A}{\lambda_B^2 - \lambda_A^2} [\boldsymbol{N}_A \otimes \boldsymbol{N}_B \otimes \boldsymbol{N}_A \otimes \boldsymbol{N}_B + \boldsymbol{N}_A \otimes \boldsymbol{N}_B \otimes \boldsymbol{N}_B \otimes \boldsymbol{N}_A]\,, \end{aligned} \tag{73}$$

where S_A are the principal values of $\boldsymbol{S}$. This expression looks like it would have trouble in the second set of sums for the case of repeated eigenvalues. However, the term that looks like it has a divide by zero is actually well defined; i.e. in the limit as say $\lambda_A \to \lambda_B$

$$\frac{S_B - S_A}{\lambda_B^2 - \lambda_A^2} \to \frac{\partial (S_B - S_A)}{\partial \lambda_B^2} \,. \tag{74}$$

This change over is smooth and can be effected by monitoring the difference in the two expressions. The eigenvalues and eigenvectors themselves are most easily found using Jacobi iterations (Golub and Van Loan, 1996, chap. 8).

6 Viscoelasticity

Much of the material in the prior sections also holds true for viscoelasticity. In this section we review two basic models of viscoelasticity suitable for use in a finite deformation framework. The models are presented without derivation and this is followed by a brief discussion of the primary computational issues.

6.1 Finite linear viscoelastic model

The model in this section first appeared in Simo (1987). What is presented below is adapted from Govindjee and Simo (1992a). The model utilizes N-relaxation mechanisms each of which has its own non-equilibrium over-stress. For simplicity we will only consider materials that are viscoelastic in their deviatoric response and volumetrically elastic. The basic model states that the 2nd Piola-Kirchhoff stress is given by

$$\boldsymbol{S}(t) = 2\frac{\partial W^{\infty}}{\partial \boldsymbol{C}}(\boldsymbol{C}(t)) + \mathrm{DEV}(t)\left[\sum_{k=1}^{N} \boldsymbol{Q}^k(t)\right], \tag{75}$$

where W^{∞} is the "long-time" hyper-elastic stored energy function for the material, the reference deviatoric projector is $\mathrm{DEV}(\cdot) = J^{-2/3}\{(\cdot) - 1/3[(\cdot) : \boldsymbol{C}]\boldsymbol{C}^{-1}\}$, and $\boldsymbol{Q}^k$ are a set of viscoelastic second Piola-Kirchhoff over-stresses. More specifically

$$\boldsymbol{Q}^k(t) = \int_{-\infty}^{t} \exp\left(-(t-s)/\tau_k\right) \frac{d}{ds}\left\{\mathrm{DEV}(s)\left[2\frac{\partial W^k}{\partial \boldsymbol{C}}(\boldsymbol{C}(s))\right]\right\} ds, \tag{76}$$

where τ_k are the relaxation times and the $W^k(\cdot)$ are stored energy functions associated with each relaxation mechanism.

Remark 19. The appealing feature of this model is that mathematically it exactly mimics the small strain case. It is also fully objective and admits anisotropic models. Its primary downside is that there is no known proof that it satisfies the 2nd Law of Thermodynamics (contrary to several claims in the literature). □

Time integration Time integration of such a model can be done very efficiently using the method of Herrmann and Peterson (1968). The thermal extension of this model can be integrated using the related method of Taylor et al. (1970). The setting is as follows. Assume that one has computed the complete solution of an IBVP up to some time t_n including the stresses and the over-stresses. It is now desired to increment the loads and find the new solution for the displacement of the body, the stresses, and over-stresses. In the process of finding the displacement field one will need to solve the residual equations at time $t_{n+1} = t_n + \Delta t$, $\Delta t > 0$. To do this one will need to compute the stresses at t_{n+1} and the over-stresses. Since this will be done with a set of Newton iterations one can presume that the displacement field at t_{n+1} is known and that the main issue at hand is finding the over-stresses at t_{n+1}. One can write

$$
\begin{aligned}
\boldsymbol{Q}_{n+1}^{k} &= \int_{-\infty}^{t_{n+1}} \exp\left(-(t_{n+1}-s)/\tau_k\right) \frac{d}{ds}\left\{ \mathrm{DEV}(s)\left[2\frac{\partial W^k}{\partial \boldsymbol{C}}(\boldsymbol{C}(s))\right]\right\} ds \\
&= \int_{t_n}^{t_{n+1}} \exp\left(-(t_{n+1}-s)/\tau_k\right) \frac{d}{ds}\left\{ \mathrm{DEV}(s)\left[2\frac{\partial W^k}{\partial \boldsymbol{C}}(\boldsymbol{C}(s))\right]\right\} ds \\
&+ \int_{-\infty}^{t_n} \exp\left(-(t_{n+1}-s)/\tau_k\right) \frac{d}{ds}\left\{ \mathrm{DEV}(s)\left[2\frac{\partial W^k}{\partial \boldsymbol{C}}(\boldsymbol{C}(s))\right]\right\} ds \\
&= \int_{t_n}^{t_{n+1}} \exp\left(-(t_{n+1}-s)/\tau_k\right) \frac{d}{ds}\left\{ \mathrm{DEV}(s)\left[2\frac{\partial W^k}{\partial \boldsymbol{C}}(\boldsymbol{C}(s))\right]\right\} ds \\
&+ \exp\left(-\Delta t/\tau_k\right) \boldsymbol{Q}_n^k \\
&\approx \exp\left(-\Delta t/2\tau_k\right)\left\{ \mathrm{DEV}(t_{n+1})\left[2\frac{\partial W^k}{\partial \boldsymbol{C}}(\boldsymbol{C}(t_{n+1}))\right] - \mathrm{DEV}(t_n)\left[2\frac{\partial W^k}{\partial \boldsymbol{C}}(\boldsymbol{C}(t_n))\right]\right\} \\
&+ \exp\left(-\Delta t/\tau_k\right) \boldsymbol{Q}_n^k \, .
\end{aligned}
\tag{77}
$$

Note that this manipulation gives us a very efficient means to evaluate the convolution integrals as long as we have stored the prior time step's solution. The prior history storage cost here is 1 symmetric rank-2 tensor per relaxation mechanism per Gauss point.

Remark 20. If one is using a Newton type iteration method one will also need to compute the contribution to the material tangent from the viscoelastic over-stresses. Conceptually, this is straightforward and we refer the reader to Simo and Hughes (1998, chap. 10) for the details. □

Remark 21. The terminology of finite linear viscoelasticity stems from the fact that the convolution form for the over-stresses can be shown to be equivalent to a system of *linear* first order ODEs. The model in the next section is a finite nonlinear model. It has the advantage of being exactly 2nd Law satisfying by construction – at the expense of some computational costs.

6.2 Finite nonlinear viscoelastic model

This section summarizes the model of Govindjee and Reese (1997), Reese and Govindjee (1998b), and Reese and Govindjee (1998a); this model is quite similar to the model of Bergstrom and Boyce (1998). The model is a finite nonlinear model of viscoelasticity based upon a multiplicative split of the deformation gradient (somewhat akin to finite plasticity models). For simplicity we will restrict ourselves to the case of a single relaxation time. The model is based on an exploitation of the multiplicative decomposition in viscoelasticity of Sidoroff (1974). One first assumes a deformation gradient decomposition

$$\boldsymbol{F} = \boldsymbol{F}^e \boldsymbol{F}^v \, , \tag{78}$$

where $\boldsymbol{F}^v$ can be interpreted as a set of kinematical internal variables that are associated with the viscous motion of the material. The variables $\boldsymbol{F}^e$ are implicitly defined for a given motion through this last equation and may be interpreted as the "elastically relaxing" portion of the motion associated with the individual relaxation mechanisms. The "physical" picture associated with this assumption is that the deformation of each viscous mechanism is decomposed into an energetically active portion and a viscously dissipating portion the two of which, together, give the total deformation, loosely represented by $\boldsymbol{F}$. The energetic contribution of each mechanism is assumed to depend on $\boldsymbol{F}^e$ through $\boldsymbol{b}^e = [\boldsymbol{F}^e]^T \boldsymbol{F}^e$ such that the overall free energy of the material can be expressed as

$$W(\boldsymbol{b}, \boldsymbol{b}^e) = \Psi^{EQ}(\boldsymbol{b}) + \Psi^{NEQ}(\boldsymbol{b}^e), \tag{79}$$

where Ψ^{EQ} represents the long time equilibrium strain energy and Ψ^{NEQ} represents the strain energy in the part of the material that is relaxing. If one looks at the the isothermal Clausius-Duhem inequality one finds that it can be satisfied by choosing (1):

$$\boldsymbol{\tau} = 2\boldsymbol{F}(\partial W/\partial \boldsymbol{C})\boldsymbol{F}^T = \boldsymbol{\tau}^{EQ} + \boldsymbol{\tau}^{NEQ} = 2(\partial \Psi^{EQ}/\partial \boldsymbol{b})\boldsymbol{b} + 2(\partial \Psi^{NEQ}/\partial \boldsymbol{b}_e)\boldsymbol{b}^e \tag{80}$$

and (2) by adopting following evolution law for $\boldsymbol{b}^e$:

$$-\frac{1}{2}\mathcal{L}_v \boldsymbol{b}^e \boldsymbol{b}^{e-1} = \mathcal{V}^{-1} : \boldsymbol{\tau}^{NEQ}, \tag{81}$$

where $\mathcal{V}^{-1}$ is the inverse viscosity in the system and $\mathcal{L}_v$ denotes the Lie derivative along the vector field defined by the motion of the body.

Remark 22. From the rheological viewpoint, the proposed constitutive equations can easily be interpreted as a parallel-type rheological model with a Maxwell element in parallel with an equilibrium spring. Note, however, that following to the arguments of Poincaré (1929, pp. 177-183), one could just as well interpret these constitutive equations as a series-type rheological model due to the non-uniqueness associated with the rheological interpretation of viscoelastic constitutive equations. □

Remark 23. If one assumes that $\mathcal{V}$ is an isotropic linear operator, it must have the form (Gurtin, 1981, §37)

$$\mathcal{V}^{-1} = \frac{1}{2\eta_D}\left[\mathbb{I} - \frac{1}{3}\mathbf{1} \otimes \mathbf{1}\right] + \frac{1}{9\eta_V}\mathbf{1} \otimes \mathbf{1}, \tag{82}$$

where η_D and η_V are "deviatoric" and "volumetric" material viscosities which may possibly be deformation dependent. $\mathbb{I}$ is the rank 4 symmetric identity tensor and $\mathbf{1}$ is the rank 2 identity tensor. Note that even if η_D and η_V are taken to be constants the evolution equation is still highly nonlinear. □

Remark 24. For the technologically important case of elastic bulk behavior we may assume

$$\mathcal{V}^{-1} = \frac{1}{2\eta_D}\left[\mathbb{I} - \frac{1}{3}\mathbf{1} \otimes \mathbf{1}\right]. \tag{83}$$

For small strains and near the elastic equilibrium the corresponding relaxation time is easily computed to be $\tau = \eta_D/\mu$, where μ is the small strain shear moduli associated with the non-equilibrium mechanism. □

A crucial issue in the use of the nonlinear evolution law is the development of an efficient integration methodology so that the constitution may be evaluated for a given loading history. In what follows, we assume the motion $\varphi(\boldsymbol{X}, t)$ is given and that one desires to evaluate the stresses. An examination of the expression for the stresses shows that to evaluate the stresses one needs to know $\boldsymbol{b}$, $\boldsymbol{b}^e$. Note that if the motion is known then $\boldsymbol{b}$ is known immediately. Thus it is seen that the evaluation of the viscous internal variables really reduces to the evaluation of $\boldsymbol{b}^e$ as a function of time. This can be viewed as the need to integrate $\dot{\boldsymbol{b}}^e$ in time.

The integration of $\dot{\boldsymbol{b}}_e$ is effected by use of the identity

$$\dot{\boldsymbol{b}}^e = \mathcal{L}_v \boldsymbol{b}^e + \boldsymbol{l}\boldsymbol{b}^e + \boldsymbol{b}^e \boldsymbol{l}^T = \dot{\boldsymbol{F}}[\boldsymbol{C}^v]^{-1}\boldsymbol{F}^T + \boldsymbol{F}[\boldsymbol{C}^v]^{-1}\dot{\boldsymbol{F}}^T + \boldsymbol{F}\overline{[\boldsymbol{C}^v]^{-1}}^{\bullet}\boldsymbol{F}^T , \tag{84}$$

where $\boldsymbol{l}$ is the spatial velocity gradient. To integrate this expression we use an operator split integration method similar to that of finite deformation elastoplasticity where an equation of similar form appears; see e.g. Weber and Anand (1990), Eterovic and Bathe (1990), or Simo (1992). In particular, we can consider (84) together with $\dot{\boldsymbol{F}} = \partial^2\varphi/\partial\boldsymbol{X}\partial t$ as a system of equations for the time evolution of $\boldsymbol{F}$ and $\boldsymbol{F}^v$. The operator split, which can be interpreted as a total motion predictor and viscous corrector algorithm, first assumes that $\dot{\boldsymbol{F}}^v = \boldsymbol{0}$ and that $\boldsymbol{F}$ evolves. This provides the predicted values which are the initial conditions to the corrector phase of the solution where $\dot{\boldsymbol{F}} = \boldsymbol{0}$ and $\boldsymbol{F}^v$ evolves. In the following presentation the predicted quantities at time t_{n+1} will be denoted with a superposed hat. Note that $\boldsymbol{b}^e$ evolves in both the predictor and corrector parts of the solution algorithm. In the predictor phase, the integration of $\dot{\boldsymbol{F}} = \partial^2\varphi/\partial\boldsymbol{X}\partial t$ is considered to be specified by the global finite element solution scheme; so $\hat{\boldsymbol{F}}_{n+1}$ is assumed to be given on the constitutive level and from (84) we have that

$$\text{[Predictor]} \qquad \dot{\boldsymbol{b}}^e = \dot{\boldsymbol{F}}[\boldsymbol{C}^v]^{-1}\boldsymbol{F}^T + \boldsymbol{F}[\boldsymbol{C}^v]^{-1}\dot{\boldsymbol{F}}^T . \tag{85}$$

This expression has the exact solution of $\hat{\boldsymbol{b}}^e_{(n+1)} = \hat{\boldsymbol{F}}_{n+1}[\boldsymbol{C}^v_{(n)}]^{-1}\hat{\boldsymbol{F}}^T_{n+1}$. In the viscous corrector phase, since $\dot{\boldsymbol{F}} = \boldsymbol{0}$ we find that $\boldsymbol{F}_{n+1} = \hat{\boldsymbol{F}}_{n+1}$ and, further, that

$$\text{[Corrector]} \qquad \dot{\boldsymbol{b}}^e = \boldsymbol{F}\overline{[\boldsymbol{C}^v]^{-1}}^{\bullet}\boldsymbol{F}^T = \mathcal{L}_v \boldsymbol{b}^e = \left[-2[\mathcal{V}]^{-1} : \boldsymbol{\tau}^{NEQ}\right] \boldsymbol{b}^e , \tag{86}$$

subject to the initial conditions of $\hat{\boldsymbol{b}}^e_{(n+1)}$ from the predictor phase. This expression can be integrated to first order using the exponential map and an Euler method to give (in the case at hand)

$$\boldsymbol{b}^e_{(n+1)} = \exp\left(-\frac{\Delta t}{\eta_D}\operatorname{dev}[\boldsymbol{\tau}^{NEQ}_{(n+1)}]\right) \hat{\boldsymbol{b}}^e_{(n+1)} . \tag{87}$$

This expression gives an implicit relation for $\boldsymbol{b}^e_{(n+1)}$. By a careful exploitation of isotropy it is seen to really represent 3 nonlinear equations for the 3 eigenvalues of $\boldsymbol{b}^e_{(n+1)}$. They are easily solvable using standard iterative methods. Note that the eigenvectors of $\boldsymbol{b}^e_{(n+1)}$ are the same as those of $\hat{\boldsymbol{b}}^e_{(n+1)}$ by the isotropy assumptions.

Remark 25. This algorithmic procedure involves more computational expense than the recursion relations of the convolution model. For this extra effort, however, one gains added modeling flexibility – viz., a model for finite nonlinear viscoelasticity as opposed to finite linear viscoelasticity.

Note, further, that if one linearizes the evolution equation in the finite nonlinear model about the elastic equilibrium ($\boldsymbol{b}^e = \mathbf{1}$), then an iterative procedure for the determination of the eigenvalues of $\boldsymbol{b}^e$ is not needed, since the equations become linear. □

Remark 26. Further details on appropriate tangent expressions and specialization to the case elastic bulk response can be found in Reese and Govindjee (1998b).

Remark 27. The algorithm presented is incrementally objective and thus suitable for large deformations; see e.g. Weber and Anand (1990). □

Remark 28. This model like the convolution model also has an extension to the thermoviscoelastic range. The details of this extension can be found in Reese and Govindjee (1998a).

7 Inverse Motion Design Problems

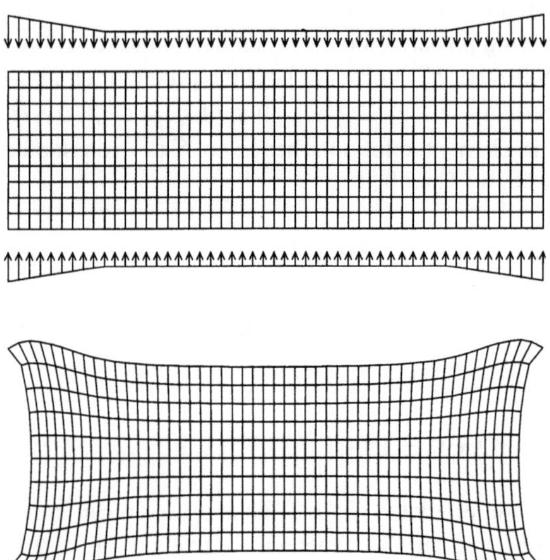

Figure 4. Top: Deformed gasket cross-section with loading, Bottom: Undeformed gasket cross-section.

A problem encountered in the design of finitely deformed elastomeric parts is one in which the initial undeformed shape of a body is unknown and the final deformed shape, applied Cauchy tractions, and displacement boundary conditions are known. The problem being to compute the undeformed shape. For an illustration, consider the design of the "manufactured shape" of a gasket. To prevent leakage, the gasket is required to have an increased clamping force along the edges, and fit into a rectangular region; a cross section of the known deformed gasket is shown in the top of Fig. 4. Computational aspects aside, the gasket to be manufactured must have a cross-sectional shape as shown in the bottom of Fig. 4 (where the top and bottom surfaces have been constrained from lateral motion).

Euler (1744) first examined a problem of this type for a tip-loaded cantilevered elastica; Truesdell (1953) considered Euler's problem but generalized the allowable loads on the system. For higher dimensional continua, this class of problems was studied by Shield (1967) who posed the "inverse deformation" problem as a set of balance equations written in terms of the inverse deformation and standard boundary conditions. Later, Chadwick (1975) showed the existence of various duality relations between the inverse and forward problem. In particular, Chadwick noted a duality between the Cauchy stress tensor and Eshelby's Energy Momentum tensor; see Eshelby (1956) and Eshelby (1975). Using this result, Chadwick recognized that under certain restrictions Shield's equilibrium equations could be formulated in terms of Eshelby's tensor. In contrast to other inverse problems, the inverse deformation problem at hand can be shown to be "well-posed" in accordance with Hadamard's definition.

Recently a new numerical method has been proposed for this class of problems by Govindjee and Mihalic (1996) and Govindjee and Mihalic (1998a). In these papers, the authors present two formulations – one based on Eshelby's Energy Momentum tensor and a second based on a re-parameterization of the equilibrium equations. The energy momentum formulation was shown to be deficient in several regards. In particular, the Energy Momentum formulation places strong continuity requirements on the motion and Eshelby's tensor lacks direct physical connection to the stated problem creating difficulties with the boundary conditions. The re-parameterization approach was shown to require only C^0 continuity and it had a direct physical connection to the problem at hand, eliminating boundary condition difficulties. The resulting numerical formulation was easily implemented using standard (forward) numerical methods. In the next sections we review the basic numerical setup for solving such problems. Further elaboration on such methods can be found in Govindjee (1999) and Koishi and Govindjee (2001) where the such methods are extended to include the case of inverse modeling in the presence of axis-symmetric geometries and temperature changes – a point important when using high temperature molding operations. For conciseness we will only treat here the nearly incompressible case and thus will pose a mixed formulation.

7.1 Forward Problem

First let us review the standard forward motion problem which measures the motion from the reference placement of the body. Let the open set $\mathcal{B} \subset \mathbb{R}^3$ be the reference placement of a continuum body containing the material points $\boldsymbol{X} \in \mathcal{B}$. Points in the reference placement are mapped to the deformed configuration $\mathcal{S} \subset \mathbb{R}^3$ by the motion $\boldsymbol{x} = \boldsymbol{\phi}(\boldsymbol{X})$ where $\mathcal{S} = \boldsymbol{\phi}(\mathcal{B})$ and points in the deformed configuration are denoted by $\boldsymbol{x} \in \mathcal{S}$.

We will assume a hyper-elastic material with a strain energy function, $W : \mathrm{Lin}_+ \longrightarrow \mathbb{R}$ per unit reference volume where Lin_+ is the space of second order tensors with positive determinant. Thus the Cauchy stress will be given as

$$\boldsymbol{\sigma} = \frac{1}{J}\frac{\partial W(\boldsymbol{F})}{\partial \boldsymbol{F}}\boldsymbol{F}^T = \frac{1}{J}\boldsymbol{P}(\boldsymbol{F})\boldsymbol{F}^T \tag{88}$$

where $\boldsymbol{P} = \partial W(\boldsymbol{F})/\partial \boldsymbol{F}$ is the first Piola-Kirchhoff stress tensor and $J = \det[\boldsymbol{F}]$ as defined in the earlier sections. The standard forward boundary value problem for the unknown motion $\boldsymbol{\phi}$ is defined by the following equilibrium equations and boundary conditions: for all $\boldsymbol{x} \in \mathcal{S}$

$$\operatorname{div}[\boldsymbol{\sigma}] + \boldsymbol{b} = \boldsymbol{o} \quad \text{and} \quad \boldsymbol{\sigma} = \boldsymbol{\sigma}^T; \tag{89}$$

for all $\boldsymbol{x} \in \partial\mathcal{S}_t$

$$\boldsymbol{\sigma}\boldsymbol{n} = \bar{\boldsymbol{t}}; \tag{90}$$

and for all $\boldsymbol{x} \in \partial\mathcal{S}_\phi$

$$\boldsymbol{\phi} = \bar{\boldsymbol{\phi}}, \tag{91}$$

where div[·] is the divergence operator with respect to $\boldsymbol{x}$, $\boldsymbol{b}$ a given body force per unit spatial volume, $\bar{\boldsymbol{t}}$ a given traction function per unit deformed area, $\boldsymbol{n}$ the boundary outward normal, $\bar{\boldsymbol{\phi}}$ a given surface motion, $\partial\mathcal{S}_t \cap \partial\mathcal{S}_\phi = \emptyset$, and $\overline{\partial\mathcal{S}_t \cup \partial\mathcal{S}_\phi} = \partial\mathcal{S}$ the boundary of $\mathcal{S}$.

7.2 Inverse Problem

Let $\boldsymbol{\varphi} = \boldsymbol{\phi}^{-1}$ be the inverse motion. In the inverse problem the primary unknown is the inverse motion $\boldsymbol{\varphi}(\boldsymbol{x})$. First let us define a set of duality relations: (1) a strain energy function $w : \text{Lin}_+ \longrightarrow \mathbb{R}$ as $w = W/J$ and (2) an inverse deformation gradient $\boldsymbol{f} = \text{grad}(\boldsymbol{\varphi})$ where $\text{grad}(\cdot)$ is the gradient operator with respect to $\boldsymbol{x}$. The inverse deformation gradient is related to its forward dual through the relation $\boldsymbol{f} = \boldsymbol{F}^{-1} \circ \boldsymbol{\varphi}$, where $\circ$ is the composition symbol. The inverse Jacobian is similarly defined as $j = det(\boldsymbol{f}) = 1/J \circ \boldsymbol{\varphi}$. The boundary value problem for the unknown motion $\boldsymbol{\varphi}$ is found through the trivial observation that (88) through (90) can be re-parameterized in terms of the inverse motion. Also note the displacement boundary condition (91) can be prescribed with reference to the inverse motion. Thus, the equilibrium equations and boundary conditions for the inverse problem become: for all $\boldsymbol{x} \in \mathcal{S}$

$$\operatorname{div}[\boldsymbol{\sigma}] + \boldsymbol{b} = \boldsymbol{o} \quad \text{and} \quad \boldsymbol{\sigma} = \boldsymbol{\sigma}^T, \tag{92}$$

for all $\boldsymbol{x} \in \partial\mathcal{S}_t$

$$\boldsymbol{\sigma}\boldsymbol{n} = \bar{\boldsymbol{t}}; \tag{93}$$

and for all $\boldsymbol{x} \in \partial\mathcal{S}_\varphi$

$$\boldsymbol{\varphi} = \bar{\boldsymbol{\varphi}}. \tag{94}$$

The constitutive relation may be expressed in terms of the inverse motion as

$$\boldsymbol{\sigma} = j\boldsymbol{P}(\boldsymbol{f}^{-1})\boldsymbol{f}^{-T}. \tag{95}$$

7.3 Incompressibility

Consider now the standard forward problem of finite elasticity defined by (88) - (91) and the added constraint

$$J - 1 = 0. \tag{96}$$

The addition of the constraint insures that only isochoric motions occur. Equation (96) is referred to as an internal constraint on the material behavior (see for example Ogden, 1984, p. 198). Numerical approaches for adding a constraint to a boundary value problem include: penalty methods, Lagrange multipliers, or a combinations of both methods. These methods, all developed for forward problems, are equally effective in the inverse setting. In the formulation presented below, a penalty parameter κ will be used in combination with a two-field variational principle.

7.4 Strong Form Forward Quasi-Incompressible Problem

A convenient assumption in quasi-incompressible elasticity is that the deviatoric stresses are caused by purely deviatoric strains. This is achieved in finite deformation elasticity through the use of a multiplicative split of the deformation gradient (Flory, 1961). into purely volumetric and purely deviatoric parts. The deviatoric and volumetric parts of the deformation gradient are, respectively, defined as

$$\tilde{\boldsymbol{F}} = J^{-1/3}\boldsymbol{F} \tag{97}$$

and

$$\boldsymbol{F}_{vol} = J^{1/3}\mathbf{1}. \tag{98}$$

Note that $\boldsymbol{F} = \tilde{\boldsymbol{F}}\boldsymbol{F}_{vol}$, $\det(\tilde{\boldsymbol{F}}) = 1$, and $\det(\boldsymbol{F}_{vol}) = J$. These definitions allow us to state the strong form equations of the forward quasi-incompressible problem as: for all $\boldsymbol{x} \in \mathcal{S}$

$$p = \kappa(J-1), \tag{99}$$

and

$$\operatorname{div}[\tilde{\boldsymbol{\sigma}} + p\mathbf{1}] + \boldsymbol{b} = 0 \quad \text{and} \quad \boldsymbol{\sigma} = \boldsymbol{\sigma}^T; \tag{100}$$

for all $\boldsymbol{x} \in \partial\mathcal{S}_t$

$$\boldsymbol{\sigma}\boldsymbol{n} = \bar{\boldsymbol{t}}, \tag{101}$$

and for all $\boldsymbol{x} \in \partial\mathcal{S}_\phi$

$$\boldsymbol{\phi} = \bar{\boldsymbol{\phi}}, \tag{102}$$

where $\boldsymbol{\sigma} = \tilde{\boldsymbol{\sigma}} + p\mathbf{1}$ is the total stress, $p \in \mathbb{R}$ denotes the "pressure", and $\kappa \in \mathbb{R}_+$ is a penalty parameter chosen large. The constitutive relation for the "deviatoric" portion of the stress is defined over the purely deviatoric motions as:

$$\tilde{\boldsymbol{\sigma}} = \frac{\partial W(\tilde{\boldsymbol{F}})}{\partial \tilde{\boldsymbol{F}}}\tilde{\boldsymbol{F}}^T. \tag{103}$$

Remark 29. Note that (99) is the multiplication of a penalty parameter (usually chosen large) with the constraint which is approaching zero. The result is a finite value for the "pressure" p (Lagrange multiplier). Note that p is the pressure and $\tilde{\boldsymbol{\sigma}}$ the deviatoric stress only in the limit $\kappa \to \infty$. □

7.5 Strong Form Inverse Quasi-incompressible Problem

It is again a trivial observation that we can re-parameterize the (forward) strong form equations (99) - (103) as a function of the inverse motion $\boldsymbol{\varphi} = \boldsymbol{\phi}^{-1}$. This gives the strong form equations for the inverse incompressible problem as: for all $\boldsymbol{x} \in \mathcal{S}$

$$p = \kappa(1/j - 1), \tag{104}$$

and

$$\operatorname{div}[\tilde{\boldsymbol{\sigma}} + p\mathbf{1}] + \boldsymbol{b} = 0 \quad \text{and} \quad \boldsymbol{\sigma} = \boldsymbol{\sigma}^T; \tag{105}$$

for all $\boldsymbol{x} \in \partial\mathcal{S}_t$

$$\boldsymbol{\sigma n} = \bar{\boldsymbol{t}}; \tag{106}$$

for all $\boldsymbol{x} \in \partial\mathcal{S}_\varphi$

$$\boldsymbol{\varphi} = \bar{\boldsymbol{\varphi}}. \tag{107}$$

The constitutive relation is given by:

$$\tilde{\boldsymbol{\sigma}} = \boldsymbol{P}(\tilde{\boldsymbol{f}}^{-1})\tilde{\boldsymbol{f}}^{-T}, \tag{108}$$

where $\tilde{\boldsymbol{f}} = \tilde{\boldsymbol{F}}^{-1} = j^{-1/3}\boldsymbol{f}$.

7.6 Weak Form Inverse Quasi-incompressible Problem

The weak form equations for the inverse incompressible problem are obtained by multiplying (104)–(105) by arbitrary admissible weighting functions, integrating over the domain, and performing integration by parts on the result. The resulting weak form expressions are:

$$\tilde{G}_1(\boldsymbol{\varphi}, p; \boldsymbol{\eta}) = \int_{\mathcal{S}} [\tilde{\boldsymbol{\sigma}} : \text{grad}(\boldsymbol{\eta}) + p\,\text{div}(\boldsymbol{\eta})] + G_{ext} = 0 \tag{109}$$

and

$$\tilde{G}_2(\boldsymbol{\varphi}, p; \beta) = \int_{\mathcal{S}} [\kappa(\frac{1}{j} - 1) - p]\beta = 0. \tag{110}$$

where $\boldsymbol{\eta} : \mathcal{S} \longrightarrow \mathbb{R}^3$ and $\boldsymbol{\eta} = 0$ on $\partial\mathcal{S}_\varphi$, $\beta : \mathcal{S} \longrightarrow \mathbb{R}$, and G_{ext} contains the contribution of the tractions $\bar{\boldsymbol{t}}$ and body forces $\boldsymbol{b}$. Note that the pressure-volume expression is given its own variational equation in the inverse motion problem just as it is in the forward motion problem.

7.7 Inverse Finite Element Formulation

The finite element formulation of the weak form problem defined by (109) and (110) can be solved using suitable approximations to $\boldsymbol{\varphi}$, $\boldsymbol{\eta}$, p, and β. As a concrete example consider the Q1/P0 element class which assumes a constant approximation per element for β and p. With this, just as in the forward case, we can solve (110) explicitly over an individual element e for the pressure p_e as,

$$p_e = \frac{1}{v_e}\int_{\mathcal{S}_e} \kappa(\frac{1}{j} - 1), \tag{111}$$

where $\mathcal{S}_e$ refers to an individual element domain and v_e is the "spatial element volume." We can then substitute this result back into (109) to arrive at a single weak form expression,

$$G(\boldsymbol{\varphi}; \boldsymbol{\eta}) = \tilde{G}_1(\boldsymbol{\varphi}, p_e(\boldsymbol{\varphi}); \boldsymbol{\eta}) = 0. \tag{112}$$

Equation (112) represents a system of non-linear equations which can be solved for the motion $\boldsymbol{\varphi}$, given suitable element subspaces for $\boldsymbol{\eta}$ and $\boldsymbol{\varphi}$.

7.8 Tangent Linearization

In typical implicit codes a Newton-Raphson method is used to solve (112) for the unknown motion. This is based upon the linearization of (112) about a current iterate $\boldsymbol{\varphi}^{(k)}$ in the arbitrary direction $\boldsymbol{\nu} : \mathcal{S} \longrightarrow \mathbb{R}^3$, where $\boldsymbol{\nu} = 0$ on $\partial\mathcal{S}_\varphi$. This gives

$$\mathcal{L}G(\boldsymbol{\varphi}^{(k)};\boldsymbol{\eta})[\boldsymbol{\nu}] = G(\boldsymbol{\varphi}^{(k)};\boldsymbol{\eta}) + DG(\boldsymbol{\varphi}^{(k)};\boldsymbol{\eta})[\boldsymbol{\nu}]. \tag{113}$$

The "element tangent" is given by:

$$\begin{aligned} DG(\boldsymbol{\varphi}^{(k)};\boldsymbol{\eta})[\boldsymbol{\nu}] = & \int_{\mathcal{S}_e} 2\,\mathrm{sym}[\mathrm{grad}(\boldsymbol{\eta})] : \mathrm{DEV}\left[\frac{\partial\tilde{\boldsymbol{\sigma}}}{\partial\tilde{\boldsymbol{c}}}\right] : \mathrm{sym}[\boldsymbol{f}^T\mathrm{grad}(\boldsymbol{\nu})] \\ & - \left(\int_{\mathcal{S}_e} \mathrm{div}(\boldsymbol{\eta})\right)\frac{\kappa}{v_e}\left(\int_{\mathcal{B}_e} J^2\mathrm{DIV}(\boldsymbol{\nu})\right), \end{aligned} \tag{114}$$

where

$$\mathrm{DEV}[\cdot] = j^{-2/3}\left([\cdot] - \frac{1}{3}\left([\cdot]:\boldsymbol{c}\right)\boldsymbol{c}^{-1}\right), \tag{115}$$

$\boldsymbol{c} = \boldsymbol{f}^T\boldsymbol{f}$, $\tilde{\boldsymbol{c}} = \tilde{\boldsymbol{f}}^T\tilde{\boldsymbol{f}}$, and $\mathrm{sym}[\cdot] = 1/2([\cdot] + [\cdot]^T)$. Equation (113) may be set to zero and solved for $\boldsymbol{\nu}$ and the iterate updated via the Newton-Raphson formula:

$$\boldsymbol{\varphi}^{(k+1)} = \boldsymbol{\varphi}^{(k)} + \boldsymbol{\nu}\,. \tag{116}$$

Remark 30. The lack of symmetry of the second term is consistent with the first term and characteristic of this problem class. The linearization represents, in a way, one variation of the potential energy of the body with respect to the forward motion and one variation with respect to the inverse motion. Thus there should be no expectation of symmetry of the formulation. □

Remark 31. The procedure provided above gives a general approach for developing the inverse problem for other methods of enforcing incompressibility. For example, we may have instead considered the three field formulation of Simo and Taylor (1991). While some of the details change, the general procedure is the same. □

Remark 32. In the case of exact incompressibility, we could have instead chosen to use $1-1/J = 0$ as the constraint in contrast to (96). This choice yields a simpler linearizations as seen by (111) which would become

$$p_e = \frac{1}{v_e}\int_{\mathcal{S}_e} \kappa(1-j). \tag{117}$$

Note the resulting tangent would no longer contain the J^2 in the last integral of (114). In the quasi-incompressible case, this change would alter the pressure-volume relation. □

7.9 Matrix Formulation: 3D case

The terms in the tangent (114) can be easily converted to a matrix formulation. Consider the following approximations for the arbitrary variations and solution field:

$$\boldsymbol{\eta} = \sum_{A=1}^{nen} N_A\boldsymbol{\eta}_A, \quad \boldsymbol{\nu} = \sum_{A=1}^{nen} N_A\boldsymbol{\nu}_A, \quad \text{and} \quad \boldsymbol{\varphi} = \sum_{A=1}^{nen} N_A\boldsymbol{\varphi}_A, \tag{118}$$

where N_A are the shape functions and $\boldsymbol{\eta}_A \in \mathbb{R}^3$, $\boldsymbol{\nu}_A \in \mathbb{R}^3$, and $\boldsymbol{\varphi}_A \in \mathbb{R}^3$ are discrete nodal values with nen being the number of element nodes. The discrete nodal values can be arranged in a compact vector form as,

$$\underline{\boldsymbol{\nu}} = [\nu_1^1, \nu_2^1, \nu_3^1, \ldots, \nu_1^{nen}, \nu_2^{nen}, \nu_3^{nen}]^T. \tag{119}$$

Define the following block matrices:

$$\underline{\boldsymbol{B}} = [\boldsymbol{B}_1, \boldsymbol{B}_2, \cdots, \boldsymbol{B}_{nen}], \tag{120}$$
$$\underline{\boldsymbol{b}} = [\boldsymbol{b}_1, \boldsymbol{b}_2, \cdots, \boldsymbol{b}_{nen}], \tag{121}$$
$$\underline{\boldsymbol{f}} = \mathrm{diag}[\boldsymbol{f}, \boldsymbol{f}, \cdots, \boldsymbol{f}]_{3nen \times 3nen}, \tag{122}$$
$$\underline{\boldsymbol{F}} = \mathrm{diag}[\boldsymbol{F}, \boldsymbol{F}, \cdots, \boldsymbol{F}]_{3nen \times 3nen}, \tag{123}$$
$$\underline{\boldsymbol{c}} = [c_{11}, c_{22}, c_{33}, 2c_{12}, 2c_{23}, 2c_{13}]^T, \text{ and} \tag{124}$$
$$\underline{\boldsymbol{c}}^{-1} = [c_{11}^{-1}, c_{22}^{-1}, c_{33}^{-1}, c_{12}^{-1}, c_{23}^{-1}, c_{13}^{-1}]^T, \tag{125}$$

where

$$\boldsymbol{B}_A = \begin{bmatrix} N_{A,1} & 0 & 0 \\ 0 & N_{A,2} & 0 \\ 0 & 0 & N_{A,3} \\ N_{A,2} & N_{A,1} & 0 \\ 0 & N_{A,3} & N_{A,2} \\ N_{A,3} & 0 & N_{A,1} \end{bmatrix} \quad \text{and} \quad \boldsymbol{b}_A = \begin{bmatrix} N_{A,1} \; N_{A,2} \; N_{A,3} \end{bmatrix}. \tag{126}$$

Using this notation we arrive at the following relations:

$$\mathrm{sym}[\mathrm{grad}(\boldsymbol{\eta})] = \underline{\boldsymbol{B}}\, \underline{\boldsymbol{\eta}}, \tag{127}$$
$$\mathrm{sym}[\boldsymbol{f}^T \mathrm{grad}(\boldsymbol{\nu})] = \underline{\boldsymbol{B}}\, \underline{\boldsymbol{f}}^T\, \underline{\boldsymbol{\nu}}, \tag{128}$$
$$\mathrm{div}[\boldsymbol{\eta}] = \underline{\boldsymbol{b}}\, \underline{\boldsymbol{\eta}}, \text{ and} \tag{129}$$
$$\mathrm{DIV}[\boldsymbol{\nu}] = \underline{\boldsymbol{b}}\, \underline{\boldsymbol{F}}\, \underline{\boldsymbol{\nu}}. \tag{130}$$

The material stiffness is mapped to a 6×6 matrix $\underline{\tilde{\boldsymbol{D}}}$ as:

$$\underline{\tilde{\boldsymbol{D}}} = \begin{bmatrix} \partial\tilde{\sigma}_{11}/\partial\tilde{c}_{11} & \partial\tilde{\sigma}_{11}/\partial\tilde{c}_{22} & \partial\tilde{\sigma}_{11}/\partial\tilde{c}_{33} & \partial\tilde{\sigma}_{11}/\partial\tilde{c}_{12} & \partial\tilde{\sigma}_{11}/\partial\tilde{c}_{23} & \partial\tilde{\sigma}_{11}/\partial\tilde{c}_{13} \\ \partial\tilde{\sigma}_{22}/\partial\tilde{c}_{11} & \partial\tilde{\sigma}_{22}/\partial\tilde{c}_{22} & \partial\tilde{\sigma}_{22}/\partial\tilde{c}_{33} & \partial\tilde{\sigma}_{22}/\partial\tilde{c}_{12} & \partial\tilde{\sigma}_{22}/\partial\tilde{c}_{23} & \partial\tilde{\sigma}_{22}/\partial\tilde{c}_{13} \\ \partial\tilde{\sigma}_{33}/\partial\tilde{c}_{11} & \partial\tilde{\sigma}_{33}/\partial\tilde{c}_{22} & \partial\tilde{\sigma}_{33}/\partial\tilde{c}_{33} & \partial\tilde{\sigma}_{33}/\partial\tilde{c}_{12} & \partial\tilde{\sigma}_{33}/\partial\tilde{c}_{23} & \partial\tilde{\sigma}_{33}/\partial\tilde{c}_{13} \\ \partial\tilde{\sigma}_{12}/\partial\tilde{c}_{11} & \partial\tilde{\sigma}_{12}/\partial\tilde{c}_{22} & \partial\tilde{\sigma}_{12}/\partial\tilde{c}_{33} & \partial\tilde{\sigma}_{12}/\partial\tilde{c}_{12} & \partial\tilde{\sigma}_{12}/\partial\tilde{c}_{23} & \partial\tilde{\sigma}_{12}/\partial\tilde{c}_{13} \\ \partial\tilde{\sigma}_{23}/\partial\tilde{c}_{11} & \partial\tilde{\sigma}_{23}/\partial\tilde{c}_{22} & \partial\tilde{\sigma}_{23}/\partial\tilde{c}_{33} & \partial\tilde{\sigma}_{23}/\partial\tilde{c}_{12} & \partial\tilde{\sigma}_{23}/\partial\tilde{c}_{23} & \partial\tilde{\sigma}_{23}/\partial\tilde{c}_{13} \\ \partial\tilde{\sigma}_{13}/\partial\tilde{c}_{11} & \partial\tilde{\sigma}_{13}/\partial\tilde{c}_{22} & \partial\tilde{\sigma}_{13}/\partial\tilde{c}_{33} & \partial\tilde{\sigma}_{13}/\partial\tilde{c}_{12} & \partial\tilde{\sigma}_{13}/\partial\tilde{c}_{23} & \partial\tilde{\sigma}_{13}/\partial\tilde{c}_{13} \end{bmatrix}. \tag{131}$$

This definition permits us to express the element tangent matrix as:

$$\underline{k}^e = \int_{\mathcal{S}_e} 2\, \underline{\boldsymbol{B}}^T\, \underline{\hat{\boldsymbol{D}}}\, \underline{\boldsymbol{B}}\, \underline{\boldsymbol{f}}^T - \left(\int_{\mathcal{S}_e} \underline{\boldsymbol{b}}^T \right) \frac{\kappa}{v_e} \left(\int_{\mathcal{B}_e} J^2\, \underline{\boldsymbol{b}}\, \underline{\boldsymbol{F}} \right), \tag{132}$$

where

$$\underline{\hat{\boldsymbol{D}}} = j^{-2/3} \left(\underline{\tilde{\boldsymbol{D}}} - \frac{1}{3}\, \underline{\tilde{\boldsymbol{D}}}\, \underline{\boldsymbol{c}}\, \underline{\boldsymbol{c}}^{-T} \right) \tag{133}$$

and $\mathcal{S}_e$ and $\mathcal{B}_e$ represent the spatial and reference element domain.

Remark 33. To convert a standard forward element to an inverse element one merely needs to replace the tangent matrix by (132) and evaluate the internal force vector using the stresses in terms of the inverse motion. □

7.10 Thermoelastic Case

The point of departure for hyperelastic thermoelasticity is the seminal work of Chadwick (1974). If one assumes for simplicity a constant heat capacity, then the Helmholtz free energy function can be expressed as

$$\Psi(\boldsymbol{F},T) = \frac{T}{T_o}\Psi_o(\boldsymbol{F}) + \left(1 - \frac{T}{T_o}\right) e_o(\boldsymbol{F}) + \bar{c}\left(T - T_o - T\ln\frac{T}{T_o}\right), \tag{134}$$

where T is the current absolute temperature, T_o is a reference absolute temperature, Ψ_o is the free energy at T_o, e_o is the internal energy at T_o, and $\bar{c}$ is the heat capacity. Note that in what follows the heat capacity does not play a role and thus assuming it to be constant provides no restrictions on the presented theory.

In the special case where one has a separable free energy function for volumetric and deviatoric behavior then Eq. (134) is specialized to

$$\begin{aligned}\Psi(\boldsymbol{F},T) = \frac{T}{T_o}W_o(\widetilde{\boldsymbol{F}}) + \frac{T}{T_o}U_o(J) + \left(1 - \frac{T}{T_o}\right) e_o(J) \\ +\bar{c}\left(T - T_o - T\ln\frac{T}{T_o}\right),\end{aligned} \tag{135}$$

where $\widetilde{\boldsymbol{F}} = J^{-1/3}\boldsymbol{F}$ is the volume preserving part of the deformation gradient, W_o is the deviatoric free energy at T_o, and U_o is the volumetric free energy at T_o.

As a particular example of the general setting, consider the case of a regularized neo-Hookean material. Appropriate choices for the functions in Eq. (135) are

$$W_o(\widetilde{\boldsymbol{F}}) = \frac{\mu}{2}(\widetilde{\boldsymbol{F}} : \widetilde{\boldsymbol{F}} - 3) \tag{136}$$

$$U_o(J) = \frac{\kappa}{4}(J^2 - 2\ln(J) - 1) \tag{137}$$

$$e_o(J) = 3\kappa\alpha T_o \ln(J), \tag{138}$$

where μ is the shear modulus, κ is the bulk modulus, and α is the coefficient of thermal expansion. Note that the above choices properly reduce to their small strain counterparts when the model is linearized about the reference configuration. The corresponding Cauchy stress tensor is given as

$$\boldsymbol{\sigma} = \frac{T}{T_o}\frac{\mu}{J}\text{dev}[\widetilde{\boldsymbol{b}}] + \left[\frac{T}{T_o}\frac{\kappa}{2}\left(J - \frac{1}{J}\right) + (T_o - T)3\kappa\alpha\frac{1}{J}\right]\boldsymbol{1}, \tag{139}$$

where $\text{dev}[\cdot] = (\cdot) - \frac{1}{3}(\boldsymbol{1} : (\cdot))\boldsymbol{1}$, and $\widetilde{\boldsymbol{b}} = \widetilde{\boldsymbol{F}}\widetilde{\boldsymbol{F}}^T$. If we express Eq. (139) in terms of the inverse motion then one has

$$\boldsymbol{\sigma} = \frac{T}{T_o}\mu j\text{dev}[\widetilde{\boldsymbol{c}}^{-1}] + \left[\frac{T}{T_o}\frac{\kappa}{2}\left(\frac{1}{j} - j\right) + (T_o - T)3\kappa\alpha j\right]\boldsymbol{1}, \tag{140}$$

where $j = 1/J$, $\boldsymbol{c} = \boldsymbol{f}^T\boldsymbol{f}$, and $\boldsymbol{f} = \boldsymbol{F}^{-1}$. For convenience split the stress as $\boldsymbol{\sigma} = \boldsymbol{s} + p\mathbf{1}$ so that the deviatoric stress

$$\boldsymbol{s} = \frac{T}{T_o}\mu j \operatorname{dev}[\tilde{\boldsymbol{c}}^{-1}] \tag{141}$$

and the pressure

$$p = \hat{p}(j) = \frac{T}{T_o}\frac{\kappa}{2}\left(\frac{1}{j} - j\right) + (T_o - T)3\kappa\alpha j\,. \tag{142}$$

To further elaborate on the method shown in the prior section we will consider here the three field formulation of Simo and Taylor (1991). Following from the appendix of Govindjee and Mihalic (1998a), the Jacobian of the inverse motion and the pressure are introduced as second and third field variables in the weak form. If the deformed body is denoted as $\mathcal{S}$ with boundary $\partial\mathcal{S} = \overline{\partial\mathcal{S}_t \cup \partial\mathcal{S}_\varphi}$ such that $\partial\mathcal{S}_t \cap \partial\mathcal{S}_\varphi = \emptyset$ and points in this configuration are denoted by $\boldsymbol{x}$, then the strong form equations for the three field inverse (penalized) incompressible problem are given by: for all $\boldsymbol{x} \in \mathcal{S}$

$$j = \theta, \tag{143}$$

$$p = \hat{p}(\theta), \tag{144}$$

and

$$\operatorname{div}[\boldsymbol{s} + p\mathbf{1}] + \boldsymbol{b} = 0 \quad \text{and} \quad \boldsymbol{\sigma} = \boldsymbol{\sigma}^T; \tag{145}$$

for all $\boldsymbol{x} \in \partial\mathcal{S}_t$

$$\boldsymbol{\sigma}\boldsymbol{n} = \bar{\boldsymbol{t}}; \tag{146}$$

for all $\boldsymbol{x} \in \partial\mathcal{S}_\varphi$

$$\boldsymbol{\varphi} = \bar{\boldsymbol{\varphi}}, \tag{147}$$

with $\boldsymbol{b}$, $\bar{\boldsymbol{t}}$ and $\bar{\boldsymbol{\varphi}}$ as given data.

The corresponding weak form equations are given by

$$\tilde{G}_\theta(\boldsymbol{\varphi}, \theta; \beta) = \int_{\mathcal{S}} [j - \theta]\beta = 0 \tag{148}$$

$$\tilde{G}_p(\theta, p; \alpha) = \int_{\mathcal{S}} [\hat{p}(\theta) - p]\,\alpha = 0, \tag{149}$$

and

$$\tilde{G}_\varphi(\boldsymbol{\varphi}, p; \boldsymbol{\eta}) = \int_{\mathcal{S}} [\boldsymbol{s} : \operatorname{grad}(\boldsymbol{\eta}) + p\operatorname{div}(\boldsymbol{\eta})] + G_{ext} = 0 \tag{150}$$

where the first variations are such that $\boldsymbol{\eta} : \mathcal{S} \longrightarrow \mathbb{R}^3$ and $\boldsymbol{\eta} = 0$ on $\partial\mathcal{S}_\varphi$, $\beta : \mathcal{S} \longrightarrow \mathbb{R}$, $\alpha : \mathcal{S} \longrightarrow \mathbb{R}$, and G_{ext} contains the contribution of the tractions $\bar{\boldsymbol{t}}$ and body forces $\boldsymbol{b}$.

For the Q1/P0 approximation one assumes constant approximations per element for β, α, θ, and p. This results in explicit expressions on the element level for the mixed inverse Jacobian and pressure as:

$$\theta_e(\boldsymbol{\varphi}) = \frac{1}{v_e}\int_{\mathcal{S}_e} j \tag{151}$$

and

$$p_e(\boldsymbol{\varphi}) = \hat{p}(\theta_e)\,, \tag{152}$$

where $\mathcal{S}_e$ refers to an individual element domain and v_e is the "spatial element volume." We can then substitute (152) back into (150) to arrive at a single weak form expression,

$$G(\boldsymbol{\varphi};\boldsymbol{\eta}) = \tilde{G}_{\varphi}(\boldsymbol{\varphi}, p_e(\boldsymbol{\varphi});\boldsymbol{\eta}) = 0. \tag{153}$$

Equation (153) represents a set of non-linear equations which can be solved for the inverse motion $\boldsymbol{\varphi}$.

The Newton-Raphson method can be applied to (153) to solve for the unknown motion. The needed tangent operator for using this technique in terms of an admissible second variation $\boldsymbol{\nu} : \mathcal{B} \longrightarrow \mathbb{R}^3$ ($\boldsymbol{\nu} = 0$ on $\partial\mathcal{S}_{\varphi}$) is given by

$$\begin{aligned} DG(\boldsymbol{\varphi};\boldsymbol{\eta})[\boldsymbol{\nu}] = & \int_{\mathcal{S}_e} \mathrm{sym}[\mathrm{grad}(\boldsymbol{\eta})] : \left[2\frac{\partial \boldsymbol{s}}{\partial \boldsymbol{c}}\right] : \mathrm{sym}[\boldsymbol{f}^T \mathrm{grad}(\boldsymbol{\nu})] \\ & + \left(\int_{\mathcal{S}_e} \mathrm{div}(\boldsymbol{\eta})\right) \frac{d\hat{p}(\theta)/d\theta}{v_e} \left(\int_{\mathcal{B}_e} \mathrm{DIV}(\boldsymbol{\nu})\right). \end{aligned} \tag{154}$$

For the model at hand we have that

$$\frac{d\hat{p}}{d\theta}(\theta) = \frac{T}{T_o}\frac{\kappa}{2}\left(-\frac{1}{\theta^2} - 1\right) + (T_o - T)3\kappa\alpha\,. \tag{155}$$

And

$$\frac{\partial \boldsymbol{s}}{\partial \boldsymbol{c}} = \mu\frac{T}{T_o} j^{1/3}\left[\frac{5}{6}\mathrm{dev}\widetilde{\boldsymbol{c}}^{-1} \otimes \widetilde{\boldsymbol{c}}^{-1} - \mathbb{I}_{\widetilde{\boldsymbol{c}}^{-1}} + \frac{1}{3}\mathbf{1} \otimes \widetilde{\boldsymbol{c}}^{-2}\right], \tag{156}$$

where $\widetilde{\boldsymbol{c}} = \widetilde{\boldsymbol{f}}^T \widetilde{\boldsymbol{f}}$, $\widetilde{\boldsymbol{f}} = j^{-1/3}\boldsymbol{f}$, and $[\mathbb{I}_{\widetilde{\boldsymbol{c}}^{-1}}]_{ijkl} = \frac{1}{2}(\widetilde{c}_{ik}^{-1}\widetilde{c}_{jl}^{-1} + \widetilde{c}_{il}^{-1}\widetilde{c}_{jk}^{-1})$. The implementation of the model at this stage follows directly as shown in the previous section.

7.11 Closure

This section has presented a two-field displacement-pressure and a three-field displacement-pressure-volume formulation for the calculation of quasi-incompressible inverse motion problems. In particular, elements designed for computing forward motion problems in quasi-incompressible hyper-elasticity can be easily converted to inverse motion elements with small changes to the tangent matrices. The presented methods supersede elastomeric design methodologies based upon "trial and error correction" methods of optimization theory. The methodology reviewed here allows one to solve such problems directly.

8 Steady State Spinning

In this section we will be concerned with the efficient analysis of rotating bodies in contact with loading devices. Our canonical example will be a tire in contact with a roadway. Thus, the problem of interest consists of a deformable circular cylinder attached to a rigid circular core spinning with motion $\theta(t)$; see Fig. 5. The outer cylinder is then brought into contact with a flat frictional surface a distance h from the center of spin. The material in the cylinder is assumed to be either elastic or even viscoelastic. For the purposes of efficiency one can not simply

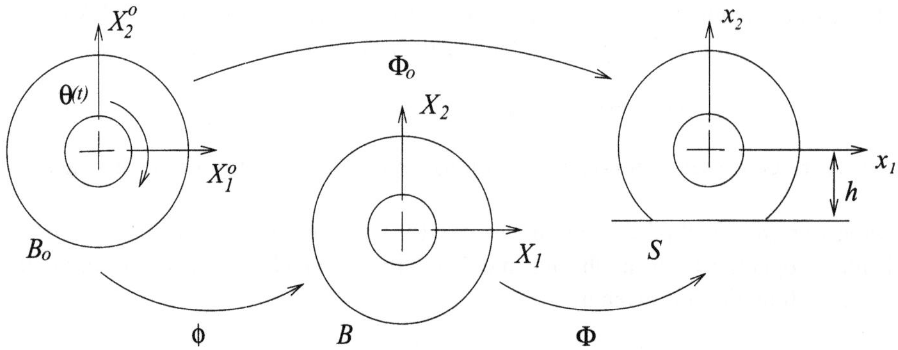

Figure 5. Inertial and non-inertial spinning reference frames; spatial frames; spatial frame.

directly model such a system for any appreciable time frame of interest if one desires any type of reasonable spatial accuracy from the simulation. An alternative approach with a great deal of useful engineering information is to model the system as being in a state of steady rotation. While this obviously leaves out certain types of problems, it covers a wide range of important cases. The essence of the procedure is to cast the problem in terms of a spinning reference configuration. A technique of this nature in a finite element setting was proposed by Lynch (1969) for the rolling of viscoelastic plates. For the spinning cylinder, formulations of varying degrees of sophistication have been proposed by many authors; see e.g. Oden and Lin (1986), Le Tallec and Rahier (1994), Govindjee and Mihalic (1998b), Padovan and Paramodilok (1985), Padovan (1987), Bass (1987), Kennedy and Padovan (1987), and other references therein.

A crucial issue to note when looking at such problems is that the spinning cylinder problem affects the constitutive relations when expressed in the moving frame. This issue is not relevant to Lynch's original work on flat plates but is important for all the subsequent work on the circular geometry. The formalism of Le Tallec and Rahier (1994) makes this issue nearly transparent and provides an avenue for the easy specification of the correct constitutive relations on the moving reference frame; see also Govindjee and Mihalic (1998b).

8.1 Review of Kinematics and Equilibrium

Consider a point on a deformable cylinder spinning about a fixed axis. This point will naturally appear differently to an observer on the fixed axis as compared to an observer spinning with the cylinder. Specifically, the spinning observer will only see the strain of the point and not its rigid rotation. In solving the steady state spinning of a cylinder, we formulate the problem on a moving reference frame which is analogous to the rotating observer. For the spinning cylinder, the moving reference frame is a time dependent frame whose rotation about the axis of the cylinder is given by $\theta(t)$. For the steady state case, we can derive equilibrium equations with respect to the moving frame which are time independent. The final equations contain additional inertial terms associated with the moving frame. In choosing the presentation of the moving reference frame and the subsequent derivation of the weak form expressions, the notational simplifications of Le Tallec and Rahier (1994) are used as they most clearly illustrate the needed constitutive relations.

A notational convention is first introduced for the moving frame. Next, we consider an application of this approach to the steady state spinning of a cylinder. The equilibrium equations and boundary conditions are stated on the fixed reference configuration and a set of weak form equations are derived. A recasting of the weak form equations onto the moving reference frame yields a complete statement of the balance equations for the steady state problem.

Moving Reference Frame Let the open set $\mathcal{B}_o \subset \mathbb{R}^3$ be the fixed reference placement of a continuum body containing the material points $\boldsymbol{X}_o \in \mathcal{B}_o$. Points in the fixed reference placement are mapped to the deformed configuration $\mathcal{S} \subset \mathbb{R}^3$ by the motion $\boldsymbol{x} = \boldsymbol{\Phi}_o(\boldsymbol{X}_o)$ where $\mathcal{S} = \boldsymbol{\Phi}_o(\mathcal{B}_o)$ and points in the deformed configuration are denoted by $\boldsymbol{x} \in \mathcal{S}$. Points in the moving reference configuration $\boldsymbol{X} \in \mathcal{B}$ are related to the fixed reference placement by the motion $\boldsymbol{X} = \boldsymbol{\phi}(\boldsymbol{X}_o)$ where $\mathcal{B} = \boldsymbol{\phi}(\mathcal{B}_o)$. The unknown motion of a particle with respect to the moving configuration is $\boldsymbol{x} = \boldsymbol{\Phi}(\boldsymbol{X}) = \boldsymbol{\Phi}_o \circ \boldsymbol{\phi}^{-1}(\boldsymbol{X})$. We define the deformation gradient with respect to the moving reference placement as $\boldsymbol{F} = \mathrm{GRAD}(\boldsymbol{\Phi})$ where $\mathrm{GRAD}(\cdot)$ denotes the gradient with respect to $\boldsymbol{X}$. The associated "right Cauchy-Green" deformation tensor is given by $\boldsymbol{C} = \boldsymbol{F}^T\boldsymbol{F}$.

Steady State Spinning of a Cylinder Define the moving reference configuration of a spinning cylinder by the motion

$$\boldsymbol{\phi}(\boldsymbol{X}_o) = \boldsymbol{R}\boldsymbol{X}_o, \tag{157}$$

where $\boldsymbol{R}$ is a time dependent rotation θ about the axis of the cylinder. For this motion, the deformation measures with respect to the fixed reference placement become

$$\boldsymbol{F}_o = \boldsymbol{F}\boldsymbol{R} \quad \text{and} \quad \boldsymbol{C}_o = \boldsymbol{R}^T\boldsymbol{C}\boldsymbol{R}, \tag{158}$$

where all quantities with a subscript o are with respect to the fixed reference placement.

The velocity and acceleration of a particle $\boldsymbol{x}$ in the current configuration for the steady spinning case can be computed as

$$\dot{\boldsymbol{x}} = \frac{\partial \boldsymbol{x}}{\partial \boldsymbol{X}}\boldsymbol{\Omega}\boldsymbol{X} \tag{159}$$

and

$$\ddot{\boldsymbol{x}} = (\frac{\partial^2 \boldsymbol{x}}{\partial \boldsymbol{X}^2}\boldsymbol{\Omega}\boldsymbol{X} + \frac{\partial \boldsymbol{x}}{\partial \boldsymbol{X}}\boldsymbol{\Omega})\boldsymbol{\Omega}\boldsymbol{X}, \tag{160}$$

where $\boldsymbol{\Omega} = \dot{\boldsymbol{R}}\boldsymbol{R}^T$ is the time independent skew-symmetric spin tensor associated with the mapping from the fixed reference configuration to the moving reference frame.

Equilibrium The boundary value problem for the unknown motion $\boldsymbol{\Phi}_o$ with respect to the fixed reference frame is defined by the following equilibrium equations and boundary conditions; for all $\boldsymbol{X}_o \in \mathcal{B}_o$,

$$\mathrm{DIV}_o[\boldsymbol{F}_o\boldsymbol{S}_o] + \rho_o\boldsymbol{b} = \rho_o\ddot{\boldsymbol{x}} \quad \text{and} \quad \boldsymbol{S}_o = \boldsymbol{S}_o^T \tag{161}$$

for all $\boldsymbol{X}_o \in \partial\mathcal{B}_{ot}$

$$\boldsymbol{F}_o\boldsymbol{S}_o\boldsymbol{N}_o = \bar{\boldsymbol{t}} \tag{162}$$

and for all $\boldsymbol{X}_o \in \partial\mathcal{B}_{o\phi}$

$$\boldsymbol{\Phi}_o = \bar{\boldsymbol{\Phi}}_o, \tag{163}$$

where $\boldsymbol{S}_o$ is the 2nd Piola-Kirchhoff stress tensor, $\text{DIV}_o[\cdot]$ is the divergence operator with respect to $\boldsymbol{X}_o$, $\boldsymbol{b}$ is a given body force per unit mass, $\bar{\boldsymbol{t}}$ is a given traction function per unit reference area, $\boldsymbol{N}_o$ is the reference surface normal, $\bar{\boldsymbol{\Phi}}_o$ is a given surface motion, $\partial\mathcal{B}_{ot} \cap \partial\mathcal{B}_{o\phi} = \emptyset$, and $\overline{\partial\mathcal{B}_{ot} \cup \partial\mathcal{B}_{o\phi}} = \partial\mathcal{B}_o$ the boundary of $\mathcal{B}_o$.

8.2 Weak Form and Linearization

The weak form equations for the equilibrium problem are formed by multiplying (161) by arbitrary admissible weighting functions, integrating over the fixed reference domain, and performing integration by parts on the result. The resulting weak form expression is,

$$G_o(\boldsymbol{\Phi}_o;\boldsymbol{\eta}) = \int_{\mathcal{B}_o} (\boldsymbol{F}_o\boldsymbol{S}_o) : \frac{\partial\boldsymbol{\eta}}{\partial\boldsymbol{X}_o} - \int_{\mathcal{B}_o} \rho_o\boldsymbol{b}\cdot\boldsymbol{\eta} - \int_{\partial\mathcal{B}_{ot}} \bar{\boldsymbol{t}}\cdot\boldsymbol{\eta} + \int_{\mathcal{B}_o} \rho_o\ddot{\boldsymbol{x}}\cdot\boldsymbol{\eta} = 0, \tag{164}$$

where the admissible weighting functions $\boldsymbol{\eta} : \mathcal{B}_o \longrightarrow \mathbb{R}^3$ and $\boldsymbol{\eta} = 0$ on $\partial\mathcal{B}_{o\phi}$. Recasting (164) onto the moving reference frame and substituting in (158a) and the steady state expression for the acceleration (160) yields,

$$\begin{aligned} G(\boldsymbol{\Phi};\boldsymbol{\eta}) &= \int_{\mathcal{B}} (\boldsymbol{F}\boldsymbol{S}) : \frac{\partial\boldsymbol{\eta}}{\partial\boldsymbol{X}} - \int_{\mathcal{B}} \rho(\boldsymbol{b} - \boldsymbol{\Omega}^2\boldsymbol{X})\cdot\boldsymbol{\eta} \\ &- \int_{\partial\mathcal{B}_t} \bar{\boldsymbol{t}}\cdot\boldsymbol{\eta} - \int_{\mathcal{B}} \rho(\frac{\partial\boldsymbol{u}}{\partial\boldsymbol{X}}\boldsymbol{\Omega}\boldsymbol{X})\cdot(\frac{\partial\boldsymbol{\eta}}{\partial\boldsymbol{X}}\boldsymbol{\Omega}\boldsymbol{X}) = 0, \end{aligned} \tag{165}$$

where $\boldsymbol{u} = \boldsymbol{x} - \boldsymbol{X}$ is the displacement field with respect to the moving frame and $\boldsymbol{S}$ is *defined* to be

$$\boldsymbol{S} = \boldsymbol{R}\boldsymbol{S}_o\boldsymbol{R}^T. \tag{166}$$

Remark 34. In arriving at this result we have assumed that $\dot{\boldsymbol{X}}\cdot\boldsymbol{N} = (\boldsymbol{\Omega}\boldsymbol{X})\cdot\boldsymbol{N} = 0$ on the outer boundary. Thus, (165) applies only when the boundary is circular. In particular when modeling tires one can not model within this framework transverse grooves in the tread. How to treat this issue effectively is still an open problem in computational tire mechanics. □

Remark 35. For completeness, we note the contribution to the linearization of (165) by the terms associated with the moving frame. Linearization of (165) in the arbitrary direction $\boldsymbol{\nu} : \mathcal{B} \longrightarrow \mathbb{R}^3$; where $\boldsymbol{\nu} = 0$ on $\partial\mathcal{B}_\phi$ yields the tangent

$$\boldsymbol{K} = DG(\boldsymbol{\Phi};\boldsymbol{\eta})[\boldsymbol{\nu}] = \tilde{\boldsymbol{K}} - \int_{\mathcal{B}} \rho(\frac{\partial\boldsymbol{\eta}}{\partial\boldsymbol{X}}\boldsymbol{\Omega}\boldsymbol{X})\cdot(\frac{\partial\boldsymbol{\nu}}{\partial\boldsymbol{X}}\boldsymbol{\Omega}\boldsymbol{X}), \tag{167}$$

where $\tilde{\boldsymbol{K}}$ is the "standard tangent." Note the last term in (167) is in a way decreasing the stiffness of the system. In fact if $\|\boldsymbol{\Omega}\|$ is large enough, this term will induce a bifurcation in the balance equations and lead to the existence of standing waves on the rotating object. □

Remark 36. The tremendous utility of (165) stems from the fact that one can create a finite element model of the cylinder in the moving frame and compute the deformation relative to it. Thus the steady state problem becomes, essentially, a static analysis. □

8.3 Constitutive Relations

The final ingredient needed to effect a stress analysis using (165) is a constitutive relation. Based on Le Tallec and Rahier's notational scheme which leads to (165), one sees that the proper expression for the stress $\boldsymbol{S}$ is a rotated version of a standard fixed frame constitutive relation for the second Piola-Kirchhoff stress $\boldsymbol{S}_o$.

Isotropic Elastic Materials Remarkably for isotropic elastic solids the expression for $\boldsymbol{S}$ can be given in a form "independent" of $\boldsymbol{R}(t)$. Using the Representation Theorem (see e.g. Gurtin, 1981, §37) we can write

$$\boldsymbol{S}_o = k_1\boldsymbol{C}_o + k_2\boldsymbol{C}_o^2 + k_3\mathbf{1}, \tag{168}$$

where k_1, k_2, and k_3 are functions of the invariants of $\boldsymbol{C}_o$; that is to say $\mathrm{I}_{C_o} = \mathrm{tr}[\boldsymbol{C}_o]$, $\mathrm{II}_{C_o} = 1/2[(\mathrm{tr}(\boldsymbol{C}_o))^2 - \mathrm{tr}(\boldsymbol{C}_o^2)]$, and $\mathrm{III}_{C_o} = J_o^2 = \det[\boldsymbol{C}_o]$. Noting now that $\boldsymbol{S} = \boldsymbol{R}\boldsymbol{S}_o\boldsymbol{R}^T$, we have

$$\boldsymbol{S} = k_1\boldsymbol{C} + k_2\boldsymbol{C}^2 + k_3\mathbf{1}, \tag{169}$$

where k_1, k_2, and k_3 are taken as functions of the invariants I_C, II_C, and III_C since the invariants of $\boldsymbol{C}$ and $\boldsymbol{C}_o$ are identical. Thus $\boldsymbol{S}$ is given entirely in terms of the deformation from the rotating frame. Knowledge of the true reference configuration is not needed. Note further that the final expression (169) merely involves removing the subscript o's from all terms in the "standard expression" (168).

Remark 37. A similar type result also holds for the anisotropic steel belts and radial plys in a tire. While the result does not hold for anisotropic materials in general it does hold for those in a tire due to the fact that they are laid out with symmetry with respect to the rotational axis of the tire. □

Isotropic Viscoelastic Constitution In the classical work of Oden and Lin (1986) a viscoelastic model for $\boldsymbol{S}$ was provided by taking the model of Christensen (1980) for a fixed reference frame and effectively "dropping all the subscript o's." A similar situation can also be observed in Padovan's works (though not as clearly, since not all kinematic and stress quantities that appear were fully defined in these works). Christensen's original model can be written in the present notation as:

$$\boldsymbol{S}_o(t) = \hat{\boldsymbol{S}}^e(\boldsymbol{E}_o(t)) + \nu\int_{-\infty}^{t} \exp\left(-(t-s)/\tau\right)\frac{\partial \boldsymbol{E}_o}{\partial s}ds, \tag{170}$$

where t is the current time, $\boldsymbol{E}_o = 1/2(\boldsymbol{C}_o - \mathbf{1})$, $\hat{\boldsymbol{S}}^e(\cdot)$ is the Mooney-Rivlin response function, and ν and τ are material parameters. In Oden and Lin (1986) it is proposed that

$$\boldsymbol{S}(t) = \hat{\boldsymbol{S}}^e(\boldsymbol{E}(t)) + \nu\int_{-\infty}^{t} \exp\left(-(t-s)/\tau\right)\frac{\partial \boldsymbol{E}}{\partial s}ds \tag{171}$$

in the steady state spinning problem. Is is noted that such a relation is inconsistent with the development leading to the weak form (165). Using the proper relation for $\boldsymbol{S}$, (166), and (170) shows

$$\boldsymbol{S}(t) = \hat{\boldsymbol{S}}^e(\boldsymbol{E}(t)) + \nu\int_{-\infty}^{t} \exp\left(-(t-s)/\tau\right)\frac{d}{ds}\left\{\boldsymbol{R}(t)\boldsymbol{R}^T(s)\boldsymbol{E}(s)\boldsymbol{R}(s)\boldsymbol{R}^T(t)\right\}ds, \tag{172}$$

where $\boldsymbol{E} = 1/2(\boldsymbol{C} - \boldsymbol{1})$. The error involved in using (171) versus (172), involves a non-physical contribution of viscoelastic effects to the stress. In particular the orientation of the "over-stress" tensor is not properly accounted for in (171).

Remark 38. When using (172) it is noted that even though $\boldsymbol{R}(\cdot)$ explicitly appears the entire motion need not be known since $\boldsymbol{R}(\cdot)$ always appears at $\boldsymbol{R}(t)\boldsymbol{R}^T(s)$; i.e. only the relative rotation between two points in time is needed. Due to the steady nature of the problem this is explicitly give by $\exp[\boldsymbol{\Omega}(t-s)]$. □

If we consider the more general N-relaxation mechanism model of Govindjee and Simo (1992a) we obtain similar results. In particular the fixed reference frame model is given by

$$\boldsymbol{S}_o(t) = 2\frac{\partial W^\infty}{\partial \boldsymbol{C}_o}(\boldsymbol{C}_o(t)) + \text{DEV}_o(t)\left[\sum_{k=1}^{N} \boldsymbol{Q}_o^k(t)\right], \tag{173}$$

where W^∞ is the "long-time" hyper-elastic stored energy function for the material, $\text{DEV}_o(\cdot) = J_o^{-2/3}\{(\cdot) - 1/3[(\cdot) : \boldsymbol{C}_o]\boldsymbol{C}_o^{-1}\}$, and $\boldsymbol{Q}_o^k$ are a set of viscoelastic second Piola-Kirchhoff over-stresses. More specifically

$$\boldsymbol{Q}_o^k(t) = \int_{-\infty}^{t} \exp\left(-(t-s)/\tau_k\right) \frac{d}{ds}\left\{\text{DEV}_o(s)\left[2\frac{\partial W^k}{\partial \boldsymbol{C}_o}(\boldsymbol{C}_o(s))\right]\right\} ds, \tag{174}$$

where τ_k are the relaxation times and the $W^k(\cdot)$ are stored energy functions associated with each relaxation mechanism. In this setting, if $W^e(\cdot)$ and $W^k(\cdot)$ are isotropic, then one has that

$$\boldsymbol{S}(t) = 2\nabla W^e(\boldsymbol{C}(t)) + \text{DEV}(t)\left[\sum_{k=1}^{N} \boldsymbol{Q}^k(t)\right], \tag{175}$$

where $\text{DEV}(\cdot) = J^{-2/3}\{(\cdot) - 1/3[(\cdot) : \boldsymbol{C}]\boldsymbol{C}^{-1}\}$ and

$$\boldsymbol{Q}^k(t) = \int_{-\infty}^{t} \exp\left(-(t-s)/\tau_k\right) \frac{d}{ds}\left\{\boldsymbol{R}(t)\boldsymbol{R}^T(s)\text{DEV}\left[2\nabla W^k(\boldsymbol{C}(s))\right]\boldsymbol{R}(s)\boldsymbol{R}^T(t)\right\} ds. \tag{176}$$

Thus in this more general setting one also has the result that the appropriate stress for the weak form can be computed without a knowledge of the entire motion of the cylinder.

Viscoelastic Steady State Space-Time Mapping The viscoelastic model requires a history of the particle strains over time. For the steady state problem, the time domain can be held fixed and computations based on a particle's history are mapped onto an angular position as proposed first by Lynch (1969). In other words the history data at a past time s for a material particle currently at say an angle β can be found in the steady case by looking at the state of the material located (at the current time t) at the angle $\beta - ||\boldsymbol{\Omega}||(t-s)$. Thus, the history of a particle is shown to be directly related to the annulus of particles within a fixed radius in the moving reference frame.

Applying this result to the viscoelastic contribution (176) gives:

$$
\begin{aligned}
\boldsymbol{Q}^k(\beta) = \mathrm{DEV}(\beta)\left[2\frac{\partial W^k}{\partial \boldsymbol{C}}(\boldsymbol{C}(\beta))\right] - \\
\frac{1}{||\boldsymbol{\Omega}||\tau_k}\left\{\int_{-\infty}^{\beta}\exp\left(-\frac{\beta-\alpha}{||\boldsymbol{\Omega}||\tau_k}\right)\boldsymbol{R}(\beta)\boldsymbol{R}^T(\alpha)\right. \\
\left.\mathrm{DEV}(\alpha)\left[2\frac{\partial W^k}{\partial \boldsymbol{C}}(\boldsymbol{C}(\alpha))\right]\boldsymbol{R}(\alpha)\boldsymbol{R}^T(\beta)d\alpha\right\}.
\end{aligned}
\tag{177}
$$

Viscoelastic BVP solution Relation (165) can in principle be solved by a straight-forward Newton method. However since (177) yields a non-local constitution the linearization of the viscous stresses is non-standard and does not easily fit typical FEM code architectures. One proposal is to use a quasi-Newton methodology where the viscoelastic stresses $\mathrm{DEV}(t)\left[\sum_{k=1}^{N}\boldsymbol{Q}^k(t)\right]$, are held fixed during a series of Newton iterations on (165). They are then updated based on the final iterate through an integration of (177) and a new Newton sequence is initiated. This proceeds until convergence. To speed matters up, one can use a single point Euler integration rule to integrate (177),

$$
\begin{aligned}
\boldsymbol{Q}^k(\beta) = \mathrm{DEV}(\beta)[2\frac{\partial W^k}{\partial \boldsymbol{C}}(\beta)] - \\
\frac{\Delta\beta}{||\boldsymbol{\Omega}||\tau_k}\left[\sum_{i=1}^{\bar{N}}\exp\left(-\frac{\beta-\beta_i}{||\boldsymbol{\Omega}||\tau_k}\right)\boldsymbol{R}(\beta)\boldsymbol{R}^T(\beta_i)\right. \\
\left.\mathrm{DEV}(\beta_i)\left[2\frac{\partial W^k}{\partial \boldsymbol{C}}(\boldsymbol{C}(\beta_i))\right]\boldsymbol{R}(\beta_i)\boldsymbol{R}^T(\beta)\right],
\end{aligned}
\tag{178}
$$

where $\beta_i = \beta - i\Delta\beta$, $\Delta\beta$ is the angular resolution of the mesh, and $\bar{N}$ is the number of points used in the Euler approximation. Review of this equation shows that the viscoelastic over-stress for a point is a weighted sum of all previous points within an annulus containing the point of interest. Also, the weighting of the sum has an exponential decay which may require multiple revolutions around the annulus before it decays sufficiently. Thus, $\bar{N}$ should be chosen to include all terms which yield a significant contribution to the viscoelastic stress. We note that each term in the series sum can be mapped directly to a circumferential element in the finite element model.

Remark 39. One alternative to using this type of "Gauss-Siedel" method of solving the equations to simultaneously linearize the viscous over-stresses with the displacements but to only include the contributions from directly neighboring element. This way the major influences are accounted for without creating a large bandwidth problem. This technique was shown to be effective by engineers at HKS.

8.4 Finite nonlinear viscoelasticity

The use of the thermodynamically consistent finite nonlinear viscoelasticity model is a bit less straightforward in this setting. First, one must cast the evolution equation, $-\frac{1}{2}\mathcal{L}_v \boldsymbol{b}^e \boldsymbol{b}^{e-1} = \mathcal{V}^{-1}$:

τ^{NEQ}, into the rotating frame. To do this, note the pull back relations

$$\mathcal{L}_v b^e = F_o \overline{C_o^{v-1}}^{\boldsymbol{\cdot}} F_o^T \quad \text{and} \quad b^e = F_o C_o^{v-1} F_o^T . \tag{179}$$

Now employ the following notational convention for rank-2 tensor $(\cdot)_o = R^T(\cdot)R$. Under these observations, the evolution equation can be re-written as:

$$C^{v-1}\Omega - \Omega C^{v-1} + \text{GRAD}[C^{v-1}]\Omega X = -2F^{-1}(\mathcal{V}^{-1} : \tau^{NEQ})b^e F^{-T} . \tag{180}$$

The integration (180) is a bit tricky and the proper procedure was first published by Le Tallec and Rahier (1994). The basic idea is to note that (180) is a first order PDE in space and that it can be effectively treated using an upwinding scheme around the tire. One begins with a weak enforcement of the evolution law and the assumption that the viscoelastic internal variables C^{v-1} are approximated as piecewise constants in each element. Thus in integral form we have

$$\int_{\mathcal{B}} C^{v-1}\Omega - \Omega C^{v-1} + \text{GRAD}[C^{v-1}]\Omega X = \int_{\mathcal{B}} -2F^{-1}(\mathcal{V}^{-1} : \tau^{NEQ})b^e F^{-T} \tag{181}$$

The term $\text{GRAD}[C^{v-1}]\Omega X$ is treated using an efficient upwinding scheme outlined by Le Tallec and Rahier. First note that

$$\int_{\mathcal{B}_e} \text{GRAD}[C^{v-1}]\Omega X = \int_{\mathcal{B}_e} \text{DIV}[C^{v-1} \otimes \Omega X] = \int_{\partial\mathcal{B}_e} C^{v-1}(N\Omega X) , \tag{182}$$

where $\mathcal{B}_e$ in a single element e in the mesh. Now noting that $N\Omega X$ is zero on all "crowns" which gives the relation:

$$\text{GRAD}[C^{v-1}]\Omega X = \frac{1}{\text{vol}(\mathcal{B}_e)} \left[C^{v-1}_{\text{upwind}} - C^{v-1} \right] \int_{\text{upwind face}} N\Omega X \tag{183}$$

Remark 40. One interesting observation about this model and integration method is that it leads to less viscoelastic coupling in the tangent than in the standard convolution integral form. But this happens at the expense of finding the viscoelastic variables at the global FEA level. □

9 Damage Modeling: Mullins Effect

The modeling of damage via continuum models is a popular though somewhat dangerous endeavor. The danger arises if one attempts to model actual material failure using continuum models. The simple problem is that the length scales are incorrect; energy is dissipated per unit volume in a continuum model instead of per unit area as happens typically at final system failure. If one does not push the material failure so far then the notion of continuum damage modeling is a reasonable exercise with useful engineering application. In what follows we outline a basic, yet general, model that is appropriate for modeling Mullins effect and discuss how one treats it numerically.

9.1 Continuum Damage Model

There are great variety of damage models for Mullins effect and what is described below has its origins in the works of Govindjee and Simo (1991) and Govindjee and Simo (1992b). The important feature of these models, which are similar to the model of Ogden and Roxburgh (1999) is that the damage parameters enter into the resulting stress-strain relations in a non-separable fashion from the principal stretches. Note that this is in sharp contrast to "one-minus-d" type models which assume separability of the stress response. The non-separability is crucial to achieving the correct shape of the loading unloading curves. It is also a direct consequence of the statistical mechanics of the damaging networks as shown in Govindjee and Simo (1992b).

The most basic model equations which can be usefully employed have the following structure. First one has a free energy function in terms of the deformation through $\boldsymbol{C}$ and a damage variable d. For concreteness we assume isotropy and an Ogden framework for the energy density.

$$W(\boldsymbol{C}, d) = U(J) + \sum_{A=1}^{3} w(\widetilde{\lambda}_A, d)\,, \tag{184}$$

where the $\widetilde{\lambda}_A = J^{-1/3}\lambda_A$ are the isochoric stretches as defined earlier and the functions $w(\cdot)$ are defined by

$$\frac{\partial w}{\partial \widetilde{\lambda}_A} = [1 + \chi(\widetilde{\lambda}_A, d)] \sum_{p=1}^{M} m_p \widetilde{\lambda}_A^{\alpha_p - 1}\,. \tag{185}$$

The function $\chi(\cdot)$ is a damage function which most notably depends on the stretch. M is the number of Ogden terms in the chosen model and m_p and α_p are the Ogden parameters. The damage criteria is given as

$$g(\boldsymbol{C}) - h(d) \leq 0\,, \tag{186}$$

where $g(\cdot)$ is some type of norm on $\boldsymbol{C}$ and $h(\cdot)$ is often simply taken to be equal to d. The evolution of the damage can be generated in many ways. If one assumes a maximum dissipation type damage then one has

$$\frac{\partial^2 W}{\partial \boldsymbol{C} \partial d} \dot{d} = -\gamma \frac{\partial g}{\partial \boldsymbol{C}}\,, \tag{187}$$

where γ is the so-called consistency parameter and

$$\gamma \geq 0, \quad \gamma g = 0, \quad \text{and} \quad \gamma \dot{g} = 0\,. \tag{188}$$

This choice simply say that the damage variable will always track the maximum of $g(\boldsymbol{C})$ over the history of the deformation. Clearly, if $g < 0$ the condition $\gamma g = 0$ implies $\gamma = 0$. On the other hand, if damage is taking place, then $g = 0$, $\dot{g} = 0$, and $\gamma > 0$. Equation (187) can then be solved for the consistency parameter γ by contracting both sides with the tensor $\partial g / \partial \boldsymbol{C}$ to obtain

$$\gamma = -\frac{(\partial^2 W / \partial \boldsymbol{C} \partial d) : (\partial g / \partial \boldsymbol{C})}{(\partial g / \partial \boldsymbol{C}) : (\partial g / \partial \boldsymbol{C})} \dot{d}\,. \tag{189}$$

From (188) one has that during active damage of the system that $\dot{g}(\boldsymbol{C}, d) = 0$. Hence, one also has that

$$\dot{d} = -\frac{1}{\partial g / \partial d} (\partial g / \partial \boldsymbol{C}) : \dot{\boldsymbol{C}} \tag{190}$$

in the case of active loading. This is the evolution equation that one must actually solve during the process of active damage. These results can be combined up to give the consistency parameter as:

$$\gamma = \begin{cases} \frac{1}{\partial g/\partial d} \frac{(\partial^2 W/\partial \boldsymbol{C}\partial d):(\partial g/\partial \boldsymbol{C})}{(\partial g/\partial \boldsymbol{C}):(\partial g/\partial \boldsymbol{C})} (\partial g/\partial \boldsymbol{C}) : \dot{\boldsymbol{C}} & \text{if } g = 0 \text{ and } \dot{g} = 0 \\ 0 & \text{otherwise}. \end{cases} \tag{191}$$

The material tangent expression can also be derived fairly easily by first noting that:

$$\dot{\boldsymbol{S}} = \mathbb{C} : \frac{1}{2}\dot{\boldsymbol{C}}, \tag{192}$$

where $\boldsymbol{S}$ is the second Piola-Kirchhoff stress tensor. To find $\mathbb{C}$, note that we also have that

$$\dot{\boldsymbol{S}} = 4\frac{\partial^2 W}{\partial \boldsymbol{C}^2}\frac{1}{2}\dot{\boldsymbol{C}} + 2\frac{\partial^2 W}{\partial \boldsymbol{C}\partial d}\dot{d}. \tag{193}$$

If the damage is not evolving then $\dot{d}$ is zero and the tangent expression is identical to that of the hyperelastic material (at fixed level of damage). If on the other hand the damage is evolving one has by combining (187) and (191) that

$$2\frac{\partial^2 W}{\partial \boldsymbol{C}\partial d}\dot{d} = -4\frac{1}{\partial g/\partial d}\frac{(\partial^2 W/\partial \boldsymbol{C}\partial d) : (\partial g/\partial \boldsymbol{C})}{(\partial g/\partial \boldsymbol{C}) : (\partial g/\partial \boldsymbol{C})}\left[\frac{\partial g}{\partial \boldsymbol{C}} \otimes \frac{\partial g}{\partial \boldsymbol{C}}\right] : \frac{1}{2}\dot{\boldsymbol{C}}. \tag{194}$$

Thus in the case of evolving damage

$$\mathbb{C} = 4\frac{\partial^2 W}{\partial \boldsymbol{C}^2} - 4\frac{1}{\partial g/\partial d}\frac{(\partial^2 W/\partial \boldsymbol{C}\partial d) : (\partial g/\partial \boldsymbol{C})}{(\partial g/\partial \boldsymbol{C}) : (\partial g/\partial \boldsymbol{C})}\left[\frac{\partial g}{\partial \boldsymbol{C}} \otimes \frac{\partial g}{\partial \boldsymbol{C}}\right]. \tag{195}$$

9.2 Finite Element Issues

The solution of finite element problems utilizing damaging material models as described above is straightforward. First one assumes that the given body forces $\boldsymbol{b}$, surface tractions $\boldsymbol{t}$, and imposed displacements $\bar{\varphi}$ are known functions of space and time. The solution of the governing equilibrium equations is then performed over an interval of time $\mathbb{I} = [t_o, t_f]$ in an incremental fashion at particular times $t^{(k)} = t_o + k\Delta t$ where $\Delta t > 0$ is a given time step. At each instant $t^{(k)}$ one needs to solve the weak equilibrium equations

$$G(\boldsymbol{\varphi}^{(k)}, d^{(k)}; \boldsymbol{\eta}) = \int_{\mathcal{B}} \boldsymbol{P}(\boldsymbol{\varphi}^{(k)}, d^{(k)}) : \text{GRAD}[\boldsymbol{\eta}] - \int_{\mathcal{B}} \boldsymbol{b} \cdot \boldsymbol{\eta} - \int_{\mathcal{B}_t} \bar{\boldsymbol{t}} \cdot \boldsymbol{\eta} = 0, \tag{196}$$

where $\boldsymbol{\eta}$ is a suitable variation. In (196), one needs to find the damage field $d^{(k)} = d(\boldsymbol{x}, t^{(k)})$ in addition to the motion of the body $\boldsymbol{\varphi}^{(k)} = \boldsymbol{\varphi}(t^{(k)})$. It is presumed that the complete solution at the prior time $t^{(k-1)}$ is known.

To effect the solution one locally integrates the damage evolution equations at each Gauss point; in this way only the motion of the body appears as a global finite element variable. If one has that $h(\cdot)$ is a non-decreasing and invertible then it is possible to locally integrate the damage evolution equations exactly without recourse to time integration formulae such as Backward Euler and other schemes popular in the plasticity and damage mechanics literature. This situation is a strong point in favor of using strain based damage criteria as opposed to stress based criteria (i.e. $g(\cdot)$ as a function of stress like variables). The basic outline of the local (exact) integration of the evolution equations is as follows.

1. Assume $\boldsymbol{C}^{(k)}$ and $d^{(k-1)}$ are given.
2. If $h^{-1}(g(\boldsymbol{C}^{(k)})) \leq d^{(k-1)}$, then no damage.
   (a) $d^{(k)} = d^{(k-1)}$
   (b) Stress $\boldsymbol{S} = 2\partial W(\boldsymbol{C}^{(k)}, d^{(k)})/\partial \boldsymbol{C}$
   (c) Tangent modulus $\mathbb{C} = 4\partial^2 W(\boldsymbol{C}^{(k)}, d^{(k)})/\partial \boldsymbol{C}^2$
3. Else, damage is actively taking place
   (a) $d^{(k)} = h^{-1}(g(\boldsymbol{C}^{(k)}))$
   (b) Stress $\boldsymbol{S} = 2\partial W(\boldsymbol{C}^{(k)}, d^{(k)})/\partial \boldsymbol{C}$
   (c) Tangent modulus $\mathbb{C}$ determined by (195) evaluated at $\boldsymbol{C} = \boldsymbol{C}^{(k)}$ and $d = d^{(k)}$

One big advantage of this formulation is that there are no nonlinear equations to locally solve for the damage variables and further the continuum tangent expressions are the correct expressions to be used in the finite element tangent matrix assembly. Thus the "consistent" tangent expression here is also the "continuum" tangent expression.

10 Release rates

A common problem in the design of rubber components is to determine the likelihood of tearing or fracture of a rubber part. The relevant quantity to be computed is known as the "energy release rate". Loosely this quantity represents the amount of energy available to the tip of a crack to move forward and it is given in terms of energy per unit area. If this available energy exceeds the resistance the material can provide (a material parameter to be measured experimentally) then the crack will move forward. Energy release rate (crack tip driving force) was a quantity introduced by Eshelby (1956) to characterize the driving force upon a singularity in an elastic field; see also Eshelby (1975). This work is intimately related to the pioneering fracture studies of Ingles (1913), Griffith (1920), and Irwin (1948). In the context of fracture, this driving force is often associated with the J-integral criteria; see Rice (1968) and Budiansky and Rice (1973). For the particular case of fracture of elastomers one typically attributes the main results to Rivlin and Thomas (1953). Finite element based methods for doing such computations have been present for many years and are well entrenched in commercial finite element packages. Most commonly, commercial programs utilize the method of domain integrals between crack tip contours; see the review article of Li et al. (1985). Here we will present a more recent development that is equivalent but which emanates from a different viewpoint. The material to be described has its origins in the work of Maugin and Tirmarco (1992) as interpreted by Steinmann (2000) and Steinmann et al. (2001).

10.1 Basic energy-momentum balance

The basis of Steinmann's method for computing energy release rates starts with the basic definition of crack tip driving forces as:

$$\boldsymbol{G} = \lim_{\epsilon \to 0} \int_{\Gamma(\epsilon)} \boldsymbol{\Sigma} \boldsymbol{N} \, d\Gamma(\epsilon) \,, \tag{197}$$

where $\Gamma(\epsilon)$ is a contour around a crack tip of "radius" ϵ and $\boldsymbol{N}$ is the unit normal to this contour. The tensor $\boldsymbol{\Sigma} = W(\boldsymbol{F})\mathbf{1} - \boldsymbol{F}^T \boldsymbol{P}$ in the quasi-static elastic case and is commonly known as the Eshelby energy momentum tensor. This expression is valid for elasticity problems and it can change depending upon the material model at hand. Physically, (197) computes the amount of energy moving through the contour $\Gamma(\epsilon)$. Thus in the limit $\epsilon \to 0$ one is left with the amount of energy available for a crack tip to move. Another way of interpreting this result is that what one is really after is the singular part of Eshelby's energy-momentum tractions.

Following Maugin and Tirmarco (1992), the strong form balance equation that is used to determine the energy-momentum tensor is found by pulling-back linear momentum balance to the material manifold; i.e.,

$$\boldsymbol{F}^T \left\{ \mathrm{DIV}[\boldsymbol{P}] + \bar{\rho}\boldsymbol{B} - \bar{\rho}\ddot{\boldsymbol{\phi}} \right\} = 0 \,. \tag{198}$$

In the hyperelastic case where the 1st Piola-Kirchhoff stress tensor is given by $\boldsymbol{P} = \partial W / \partial \boldsymbol{F}$ one has that:

$$\begin{aligned}
0 &= \boldsymbol{F}^T \left\{ \mathrm{DIV}[\boldsymbol{P}] + \bar{\rho}\boldsymbol{B} - \bar{\rho}\ddot{\boldsymbol{\phi}} \right\} & (199)\\
&= \mathrm{DIV}[\boldsymbol{F}^T\boldsymbol{P}] - \mathrm{GRAD}[\boldsymbol{F}^T] : \boldsymbol{P} + \boldsymbol{F}^T\bar{\rho}\boldsymbol{B} - \boldsymbol{F}^T\bar{\rho}\ddot{\boldsymbol{\phi}} & (200)\\
&= \mathrm{DIV}[-W\mathbf{1} + \boldsymbol{F}^T\boldsymbol{P}] + \mathrm{GRAD}[W]\mathbf{1} - \mathrm{GRAD}[\boldsymbol{F}^T] : \boldsymbol{P} + \boldsymbol{F}^T\bar{\rho}\boldsymbol{B} - \boldsymbol{F}^T\bar{\rho}\ddot{\boldsymbol{\phi}} & (201)\\
&= \mathrm{DIV}[-W\mathbf{1} + \boldsymbol{F}^T\boldsymbol{P}] + \boldsymbol{P} : \mathrm{GRAD}[\boldsymbol{F}]\mathbf{1} - \mathrm{GRAD}[\boldsymbol{F}^T] : \boldsymbol{P} + \boldsymbol{F}^T\bar{\rho}\boldsymbol{B} - \boldsymbol{F}^T\bar{\rho}\ddot{\boldsymbol{\phi}} & (202)\\
&= \mathrm{DIV}[\boldsymbol{\Sigma}] - \boldsymbol{F}^T\bar{\rho}\boldsymbol{B} + \boldsymbol{F}^T\bar{\rho}\ddot{\boldsymbol{\phi}} \,. & (203)
\end{aligned}$$

If we wish to consider viscoelasticity within this framework then we need a fully thermodynamically consistent theory in finite deformation. Such a model is furnished by Govindjee and Reese (1997), Reese and Govindjee (1998b), and Reese and Govindjee (1998a). In review, for the isothermal case, this model employs a kinematic decomposition of the deformation gradient as: $\boldsymbol{F} = \boldsymbol{F}^e\boldsymbol{F}^v$. The important deformation measures are $\boldsymbol{b}^e = \boldsymbol{F}^e\boldsymbol{F}^{eT}$ and $\boldsymbol{C}^v = \boldsymbol{F}^{vT}\boldsymbol{F}^v$. The energy ansatz states that $W(\boldsymbol{b}, \boldsymbol{b}^e) = \Psi^{EQ}(\boldsymbol{b}) + \Psi^{NEQ}(\boldsymbol{b}^e)$. Then from standard thermodynamic arguments one has that the Kirchhoff stresses are given as: $\boldsymbol{\tau} = \boldsymbol{F}(\partial W / \partial \boldsymbol{C})\boldsymbol{F}^T = \boldsymbol{\tau}^{EQ} + \boldsymbol{\tau}^{NEQ} = 2(\partial \Psi^{EQ} / \partial \boldsymbol{b})\boldsymbol{b} + 2(\partial \Psi^{NEQ} / \partial \boldsymbol{b}_e)\boldsymbol{b}^e$. The model is closed out by the evolution law $-\mathcal{L}_v \boldsymbol{b}^e \boldsymbol{b}^{e-1} = \mathcal{V}^{-1} : \boldsymbol{\tau}^{NEQ}$. We remind the reader that suitable choices for the inverse viscosity, $\mathcal{V}^{-1}$, and/or linearization of it give a number of existing (linear) models in the literature. But most importantly the model is rigorously 2nd Law satisfying in finite dimensions. For this model, we will again assume the canonical energy-momentum tensor expression $\boldsymbol{\Sigma} = J[\partial(W/J)/\partial \boldsymbol{F}^{-1}]\boldsymbol{F}^{-T} = W\mathbf{1} - \boldsymbol{F}^T\boldsymbol{P}$. Other authors often make variations on this expression to suit special needs but in the computational setting to be developed in the next sections these variations are moot. With this in hand we can pull the linear momentum balance equation back to the material manifold to give the appropriate balance law for the energy-momentum field:

$$\begin{aligned}
0 &= \boldsymbol{F}^T \left\{ \mathrm{DIV}[\boldsymbol{P}] + \bar{\rho}\boldsymbol{B} - \bar{\rho}\ddot{\boldsymbol{\phi}} \right\} & (204)\\
&= \mathrm{DIV}[\boldsymbol{F}^T\boldsymbol{P}] - \mathrm{GRAD}[\boldsymbol{F}^T] : \boldsymbol{P} + \boldsymbol{F}^T\bar{\rho}\boldsymbol{B} - \boldsymbol{F}^T\bar{\rho}\ddot{\boldsymbol{\phi}} & (205)\\
&= \mathrm{DIV}[-W\mathbf{1} + \boldsymbol{F}^T\boldsymbol{P}] + \mathrm{GRAD}[W]\mathbf{1} - \mathrm{GRAD}[\boldsymbol{F}^T] : \boldsymbol{P} + \boldsymbol{F}^T\bar{\rho}\boldsymbol{B} - \boldsymbol{F}^T\bar{\rho}\ddot{\boldsymbol{\phi}} & (206)\\
&= \mathrm{DIV}[-W\mathbf{1} + \boldsymbol{F}^T\boldsymbol{P}] + \boldsymbol{P} : \mathrm{GRAD}[\boldsymbol{F}]\mathbf{1} + \partial W / \partial \boldsymbol{F}^v : \mathrm{GRAD}[\boldsymbol{F}^v] & (207)\\
&\quad - \mathrm{GRAD}[\boldsymbol{F}^T] : \boldsymbol{P} + \boldsymbol{F}^T\bar{\rho}\boldsymbol{B} - \boldsymbol{F}^T\bar{\rho}\ddot{\boldsymbol{\phi}} & (208)\\
&= \mathrm{DIV}[\boldsymbol{\Sigma}] - \boldsymbol{F}^T\bar{\rho}\boldsymbol{B} + \boldsymbol{F}^T\bar{\rho}\ddot{\boldsymbol{\phi}} + \boldsymbol{A} : \mathrm{GRAD}[\boldsymbol{C}^v] \,, & (209)
\end{aligned}$$

where $\boldsymbol{A} = \frac{1}{2}\boldsymbol{F}^{-1}\boldsymbol{\tau}^{NEQ}\boldsymbol{b}^e\boldsymbol{F}^{-T}$.

10.2 Steinmann's Observation

Steinmann's method hinges upon two observations:

1. Looking at (197) we see that the expression will result in zero unless the integrand is singular.
2. During the typical application of the divergence theorem in generating weak balance laws one typically assumes that no singularities are present.

These two observations can be combined into the following powerful procedure. First, take the strong form balance equation for the energy-momentum field and cast it into weak form. If we use $\boldsymbol{v}$ as the test function then we will have a weak balance relation as:

$$\begin{aligned} \int_{\mathcal{B}} \mathrm{GRAD}[\boldsymbol{v}] : \boldsymbol{\Sigma} + \int_{\mathcal{B}} \boldsymbol{v} \cdot \boldsymbol{F}^T \bar{\rho} \boldsymbol{B} - \int_{\mathcal{B}} \boldsymbol{v} \cdot \boldsymbol{A} : \mathrm{GRAD}[\boldsymbol{C}^v] \\ - \int_{\mathcal{B}} \boldsymbol{v} \cdot \boldsymbol{F}^T \bar{\rho} \ddot{\boldsymbol{\phi}} - \underbrace{\int_{\partial\mathcal{B}} \boldsymbol{v} \cdot \boldsymbol{\Sigma} \boldsymbol{N}}_{\text{non-singular terms}} = 0 \end{aligned} \tag{210}$$

In the above we have assumed *no* singularities. Thus the right-hand-side (RHS) can be used to find the singular part of Eshelby's energy-momentum tractions through computation of the left-hand-side (LHS).

Remark 41. For elastic systems this weak form represents the first order optimality conditions for the stationarity of the potential energy with respect to the inverse motion. □

10.3 Finite elements

If one assumes a variation approximation as $v = \sum_a v^a M^a$, where M^a are given shape functions, then the discrete nodal energy release rates are given by:

$$\begin{aligned} \boldsymbol{G}^a = \int_{\mathcal{B}} \mathrm{GRAD}[M^a] \boldsymbol{\Sigma}^T + \int_{\mathcal{B}} M^a \boldsymbol{F}^T \bar{\rho} \boldsymbol{B} - \int_{\mathcal{B}} M^a \boldsymbol{A} : \mathrm{GRAD}[\boldsymbol{C}^v] \\ - \int_{\mathcal{B}} M^a \boldsymbol{F}^T \bar{\rho} \ddot{\boldsymbol{\phi}} - \underbrace{\int_{\partial\mathcal{B}} M^a \boldsymbol{\Sigma} \boldsymbol{N}}_{\text{non-singular terms}} \end{aligned} \tag{211}$$

The use of this weak form is quite straight-forward and does not involve in anyway actually having to solve the energy-momentum balance equation – a computation that is near impossible in any realistic situation since the energy momentum tractions are typically unknown. What one actually does is to first solve the linear momentum balance equations utilizing a completely standard finite element approach. Once this solution is at hand then everything on the RHS of (211) is at once known. Thus to determine the energy release rates one only needs to assemble the terms on the RHS. This is done using the standard finite element assembly procedure.

Remark 42. When following this procedure one get a release rate at every node in a mesh. These values are typically small except at points of singularities in the continuum problem. If there are appreciable non-zero driving forces on other nodes in the mesh then this is an indicator of poor mesh quality as it says that one can move the node (in the direction opposite of $\boldsymbol{G}^a$) and effect a lowering of the potential energy of the mesh (at least in the purely elastic case). □

Remark 43. An advantage of this type of formulation for computing release rates is that it makes more transparent how one extends "classical" methods to cases with inelastic materials, dynamics, etc. For example the extension to the case of a spinning tire in steady state analysis is quite straightforward. For this case the nodal release rate is given by

$$\begin{aligned} \boldsymbol{G}^a = \int_{\mathcal{B}} \mathrm{GRAD}[M^a]\boldsymbol{\Sigma}^T - \overbrace{\int_{\partial\mathcal{B}} M^a \boldsymbol{\Sigma}\boldsymbol{N}}^{\text{regular part only}} - \overbrace{\int_{\mathcal{B}} \bar{\rho} M^a \boldsymbol{F}^T \mathbf{1}\boldsymbol{\Omega}^2\boldsymbol{X}}^{\text{standard term}} - \overbrace{\int_{\mathcal{B}} M^a \boldsymbol{A} : \mathrm{GRAD}[\boldsymbol{C}^v]}^{\text{viscoelastic term}} \\ \underbrace{+ \int_{\mathcal{B}} \bar{\rho}\mathrm{GRAD}[M^a](\boldsymbol{F}^T\mathrm{GRAD}[\boldsymbol{u}]\boldsymbol{\Omega}\boldsymbol{X} \otimes \boldsymbol{\Omega}\boldsymbol{X})^T + \int_{\mathcal{B}} \bar{\rho} M^a \mathrm{GRAD}[\boldsymbol{F}^T]\mathrm{GRAD}[\boldsymbol{u}]\boldsymbol{\Omega}\boldsymbol{X}\boldsymbol{\Omega}\boldsymbol{X}}_{\text{finite deformation terms}} \end{aligned} \tag{212}$$

□

References

Ball, J. M. (1977a). Convexity conditions and existence theorems in nonlinear elasticity. *Archives for Rational Mechanics and Analysis* 63:337–402.

Ball, J. (1977b). Constitutive inequalities and existance theorems in nonlinear elastostatics. In Knops, R., ed., *Nonlinear analysis and mechanics: Heriot-Watt Symposium*, volume 1, 187–241.

Bass, J. (1987). Three-dimensional finite deformation, rolling contact of a hyperelastic cylinder: Formulation of the and computational results. *Computers and Structures* 26:991–1004.

Bergstrom, J., and Boyce, M. (1998). Constitutive modeling of the large strain time-dependent behavior of elastomers. *Journal of the Mechanics and Physics of Solids* 46:931–954.

Brezzi, F., and Fortin, M. (1991). *Mixed and Hybrid Finite Element Methods*. Springer-Verlag.

Budiansky, B., and Rice, J. (1973). Conservation laws and energy-release rates. *Journal of Applied Mechanics* 40:201–203.

Chadwick, P. (1974). Thermo-mechanics of rubberlike materials. *Philosophical Transactions of the Royal Society of London Series A* 276:371–403.

Chadwick, P. (1975). Applications of an energy-momentum tensor in elastostatics. *Journal of Elasticity* 5:249–258.

Christensen, R. (1980). A nonlinear theory of viscoelasticity for application to elastomers. *Journal of Applied Mechanics* 47:762–768.

Ciarlet, P. (1988). *Mathematical Elasticity, Volume 1: Three Dimensional Elasticity*. Amsterdam: North-Holland.

Dacorogna, B. (1989). *Direct Methods in the Calculus of Variations*. Springer-Verlag.

Eshelby, J. (1956). The continuum theory of lattice defects. In Seitz, F., and Turnbull, D., eds., *Solid State Physics, Advances in Research and Applications*, volume III. Academic Press. 79–144.

Eshelby, J. (1975). The elastic energy-momentum tensor. *Journal of Elasticity* 5:321–335.

Eterovic, A., and Bathe, K. (1990). A hyperelastic-based large strain elasto-plastic constitutive formulation with combined isotropic-kinematic hardening using the logarithmic stress and strain measures. *International Journal for Numerical Methods in Engineering* 30:1099–1114.

Euler, L. (1744). Methodus inveniendi lineas curvas. *Opera omnia I* 24:264–266. (reprinted 1952).

Flory, P. (1961). Thermodynamic relations for high elastic materials. *Transactions of the Faraday Society* 57:829–838.

Fried, I. (1974). Finite element analysis of incompressible materials by residual energy balancing. *International Journal of Solids and Structures* 10:993–1002.

Golub, G., and Van Loan, C. (1996). *Matrix Computations*. Johns Hopkins University Press, 3rd edition.

Govindjee, S., and Mihalic, P. (1996). Computational methods for inverse finite elastostatics. *Computer Methods in Applied Mechanics and Engineering* 136:47–57.

Govindjee, S., and Mihalic, P. (1998a). Computational methods for inverse deformations in quasi-incompressible finite elasticity. *International Journal for Numerical Methods in Engineering* 43:821–828.

Govindjee, S., and Mihalic, P. (1998b). Viscoelastic constitutive relations for the steady spinning of a cylinder. Technical Report UCB/SEMM-98/02, University of California Berkeley, Department of Civil Engineering.

Govindjee, S., and Reese, S. (1997). A presentation and comparison of two large deformation viscoelasticity models. *ASME Journal of Engineering Materials and Technology* 119:251–255.

Govindjee, S., and Simo, J. (1991). A micro-mechanically based continuum damage model for carbon black filled rubbers incorporating mullins' effect. *Journal of the Mechanics and Physics of Solids* 39:87–112.

Govindjee, S., and Simo, J. (1992a). Mullins' effect and the strain amplitude dependence of the storage modulus. *International Journal of Solids and Structures* 29:1737–1751.

Govindjee, S., and Simo, J. (1992b). Transition from micro-mechanics to computationally efficient phenomenology: Carbon black filled rubbers incorporating mullins' effect. *Journal of the Mechanics and Physics of Solids* 40:213–233.

Govindjee, S. (1999). Finite deformation inverse design modeling with temperature changes, axis-symmetry and anisotropy. Technical Report UCB/SEMM-1999/01, University of California Berkeley, Department of Civil Engineering.

Griffith, A. (1920). The phenomena of rupture and flow in solids. *Philosophical Transactions of the Royal Society of London* 221A:163–197.

Gurtin, M. (1981). *An Introduction to Continuum Mechanics*. Academic Press, Inc.

Herrmann, L. R., and Peterson, E. F. (1968). A numerical procedure for visco-elastic stress analysis. *Proc. 7th Meeting of ICRPG Mechanical Behavior Working Group* Orlando, FL.

Ingles, C. (1913). Stresses in a plate due to the presence of cracks and sharp corners. *Transactions of the Institute of Naval Architects* 55:219–241.

Irwin, G. (1948). *Fracture Dynamics*. American Society for Metals. 147–166.

Johnson, C. (1987). *Numerical Solutions of Partial Differential Equations by the Finite Element Method*. Cambridge University Press.

Kennedy, R., and Padovan, J. (1987). Finite element analysis of steady and transiently moving/rolling nonlinear viscoelastic structure–ii. shell and three-dimensional simulations. *Computers and Structures* 27:249–257.

Koishi, M., and Govindjee, S. (2001). Inverse design methodology of a tire. *Tire Science and Technology Journal* 29:155–170.

Le Tallec, P., and Rahier, C. (1994). Numerical models of steady rolling for non-linear viscoelastic structures in finite deformations. *International Journal for Numerical Methods in Engineering* 37:1159–1186.

Li, F., Shih, C., and Needleman, A. (1985). A comparison of methods for calculating energy release rates. *Engineering Fracture Mechanics* 21:405–421.

Lynch, F. D. S. (1969). A finite element method of viscoelastic stress analysis with application to rolling contract problems. *International Journal for Numerical Methods in Engineering* 1:379–394.

Maugin, G., and Tirmarco, C. (1992). Pseudomomentum and material forces in nonlinear elasticity: variational formulations and applications to brittle fracture. *Acta Mechanica* 94:1–28.

Oden, J., and Lin, T. (1986). On the general rolling contact problem for finite deformations of a viscoelastic cylinder. *Computer Methods in Applied Mechanics and Engineering* 57:297–367.

Ogden, R., and Roxburgh, D. (1999). A pseudo-elastic model for the mullins effect in filled rubber. *Proceedings of the Royal Society of London* A455:2861–2877.

Ogden, R. W. (1984). *Non-linear Elastic Deformations*. Ellis Horwood Limited.

Padovan, J., and Paramodilok, O. (1985). Transient and steady state viscoelastic rolling contact. *Computers and Structures* 20:545–553.

Padovan, J. (1987). Finite element analysis of steady and transiently moving/rolling nonlinear viscoelastic structure–i. theory. *Computers and Structures* 27:249–257.

Poincaré, H. (1929). *The Foundations of Science*. The Science Press. (Translation: B. Halstep).

Reese, S., and Govindjee, S. (1998a). Theoretical and numerical aspects in the thermo-viscoelastic material behaviour of rubber-like polymers. *Mechanics of Time-Dependent Materials* 1:357–396.

Reese, S., and Govindjee, S. (1998b). A theory of finite viscoelasticity and numerical aspects. *International Journal of Solids and Structures* 35:3455–3482.

Rice, J. (1968). A path independent integral and the approximate analysis of strain concentration by notches and cracks. *Journal of Applied Mechanics* 35:379–386.

Rivlin, R., and Thomas, A. (1953). Rupture of rubber 1. Characteristic energy for tearing. *Journal of Polymer Science* 10:291–318.

Shield, R. (1967). Inverse deformation results in finite elasticity. *Zeitschrift für Agnew. Math. Phys.* 18:381–389.

Sidoroff, F. (1974). Un modèle viscoélastique non linéaire avec configuration intermédiaire. *Journal de Mécanique* 13:679–713.

Simo, J. C., and Hughes, T. J. R. (1998). *Computational Inelasticity*. Springer-Verlag.

Simo, J., and Taylor, R. (1991). Quasi-incompressible finite elasticity in principal stretches. Continuum basis and numerical algorithms. *Computer Methods in Applied Mechanics and Engineering* 85:273–310.

Simo, J. C. (1987). On a fully three-dimensional finite-strain viscoelastic damage model: Formulation and computational aspects. *Comp. Meth. Appl. Mech. Engng.* 60:153–173.

Simo, J. (1992). Algorithms for static and dynamic multiplicative plasticity that preserve the classical return mapping schemes of the infinitesimal theory. *Computer Methods in Applied Mechanics and Engineering* 99:61–112.

Steinmann, P., Ackermann, D., and Barth, F. (2001). Application of material forces to hyperelastic fracture mechanics. II. Computational setting. *International Journal of Solids and Structures* 38:5509–5526.

Steinmann, P. (2000). Application of material forces to hyperelastic fracture mechanics. I. Continuum mechanical setting. *International Journal of Solids and Structures* 37:7371–7391.

Taylor, R. L., Pister, K. S., and Goudreau, G. L. (1970). Thermomechanical analysis of viscoelastic solids. *Int.J.Numer.Meth.Eng* 2:45–49.

Truesdell, C. (1953). A new chapter in the theory of the elastica. In *Proceedings of the First Midwestern Conference on Solid Mechanics*, 52–55.

Weber, G., and Anand, L. (1990). Finite deformation. constitutive equations and a time integration procedure for isotropic hyperelastic-viscoplastic solids. *Computer Methods in Applied Mechanics and Engineering* 79:173–202.

Zienkiewicz, O., and Taylor, R. (2000). *The Finite Element Method, Volume 2*. Butterworth and Heinemann, 5th edition.

Laminated Rubber Bearings for Seismic Applications

Franco Bettinali[1] and Alberto Dusi[2]

[1] CESI, Milan, Italy
[2] Consultant in Milan, Italy

Abstract. In this chapter a short presentation of the general principles of the base isolation technique and of the demonstration of its efficiency is presented. A few real applications as well as the results coming from the installed monitoring systems are shown.

1 Natural Rubber History

History records that the Egyptians were the first to use natural rubber. In 1839, Charles Goodyear and Thomas Hancock independently developed the process of vulcanization, which transforms natural rubber from a soft and tacky substance to a strong and stable material. As early as 1850, bearing springs were being made for railway wagons. In 1889, compression pads of natural rubber were laid on top of piers supporting a railway viaduct in Melbourne, Australia, which is still standing today. These three applications, namely anti-vibration mounts, impact absorbers, and flexible couplings, are still major uses of elastomeric materials.

Elastomers, otherwise known as elastic polymers, have long been recognized as potentially useful in replacing metal couplings and bearings. Early elastomeric designs were typically of simple geometry and subjected to low bearing stresses on the order of 0.2 to 0.7 MPa (30 to 100 psi). The concept of highly confined elastomeric configurations using metal reinforcements allowed bearing stresses to approach 7 MPa (1,000 psi) as a result of applied compressive loads (such as those applied to common bridge bearing pads for highway construction). By the early 1960s, serious consideration and application of highly loaded composite bearings to the oil field and aerospace industries was being explored. Today, elastomers have gained widespread acceptance in a variety of industries and applications. Bearing designs currently in use are capable of sustaining average compressive stresses of 140 MPa (20,000 psi).

Elastomers encompass all families of natural rubber and synthetic compounds exhibiting similar characteristics. These materials are characterized as being elastic, tough, versatile, adaptable, relatively impermeable to water and air, weather resistant (when properly compounded), and maintenance free. They are unique in that they are one of the few materials that can undergo high strains and still re-acquire their initially undeformed shape when load is removed. Elastomers have the ability to simultaneously store and dissipate energy given their viscoelastic properties (time dependent large strain behaviour). They are light in weight, relatively inexpensive, and can be vulcanized and bonded easily to other materials. In terms of energy storage, they are much more efficient than comparable steel springs. Unlike metals, elastomers exhibit some degree of damping, which allows the material to efficiently sustain and control vibration loading and displacements.

The development of elastomer-to-metal bonding systems now permits modes of deformation other than, or in addition to, compression. Specific applications of flexible composite bearings include: vibration isolation mountings for buildings and equipment on naval ships and submarines, forging hammer foundation springs; rocket nozzles, helicopter rotors, dock hinges, and wind turbine bearings; down-hole packers, blow-out preventers, seals, drilling riser flex joints, tensioner struts, and other offshore oil field system components. Construction, oil field, and aerospace applications use the same basic design methods for elastomeric elements even though the resulting hardware is significantly different in size and ruggedness.

Over time, many elastomeric bearings have been successfully developed for a variety of applications, which has led to increased confidence in basic design procedures established through experience and analysis. Examples of different configurations of laminated bearings are presented in the following as well as Finite Element (FE) models, both in the undeformed and deformed configurations.

2 Principles of Seismic Isolation

In a conventional structure the ground acceleration is amplified from the foundation to the higher floors. The building could be damaged in its main structural parts but, even if it correctly designed and no problems occur, the contents could suffer non-negligible damage.

The seismic isolation concept consists in the separation of the building from its foundations with connections guaranteed by special devices devoted to the filtering of energy. In such a way the movement takes place at the isolation level and the resulting floor accelerations are low. Thanks to the limited deformations and distortions induced in the structure the occupants and the contents are maintained in a safe condition. In practice the isolated buildings are expected to remain elastic during the earthquake without damage and the deformation and damping is provided by the seismic devices. Figures 1 and 2 depict the difference between a conventional and a base-isolated building during an earthquake.

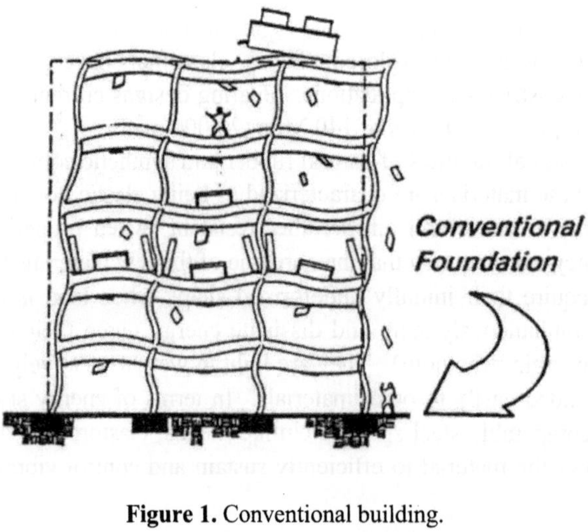

Figure 1. Conventional building.

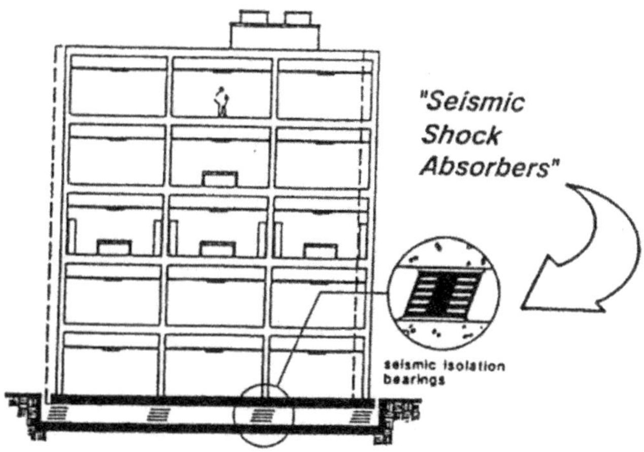

Figure 2. Seismic isolated building.

3 Examples of the Application of Seismic Isolation and of the Filtering Efficiency

3.1 USC Hospital, USA

The USC hospital, located in Los Angeles, USA, suffered the effects of the Northridge earthquake (January 17th, 1994). The epicentre was located 36 km from the building, which, with 8 floors, was designed and built in accordance with a base isolation system having an expected maximum displacement of 26 centimetres. During the event the maximum displacement recorded, at the base isolation level, was well below this figure (4 centimetres).

3.2 MGC Twin Buildings in Sanda, Japan

The twin buildings, located in Sanda near Kobe JAPAN, suffered the effects of the Great Hanshin-Awaij (Kobe) earthquake (January 17th, 1995). The epicentre was located 37 km distant from the buildings. One of the twin buildings, 3 floors in height, was designed and built with a base isolation system (4+4 high damping rubber bearings) for an expected maximum displacement of 15 centimetres. During the event the maximum displacement recorded, at the base isolation level, was below this figure (12 centimetres).

4 Effects of Base Isolation

The effects of a base isolation system are clearly evident in the following figures. By entering in the response spectrum with a longer period and a higher equivalent viscous damping it is possible to achieve a reduction in the acceleration and, as a consequence, in the inertial forces and mechanical stresses. On the other hand the displacements increase (it has to be noted that the whole superstructure moves quite rigidly on the isolation level). The response spectrum is the maximum acceleration (or displacement) recorded by a set of simple oscillators having different natural periods during a seismic event.

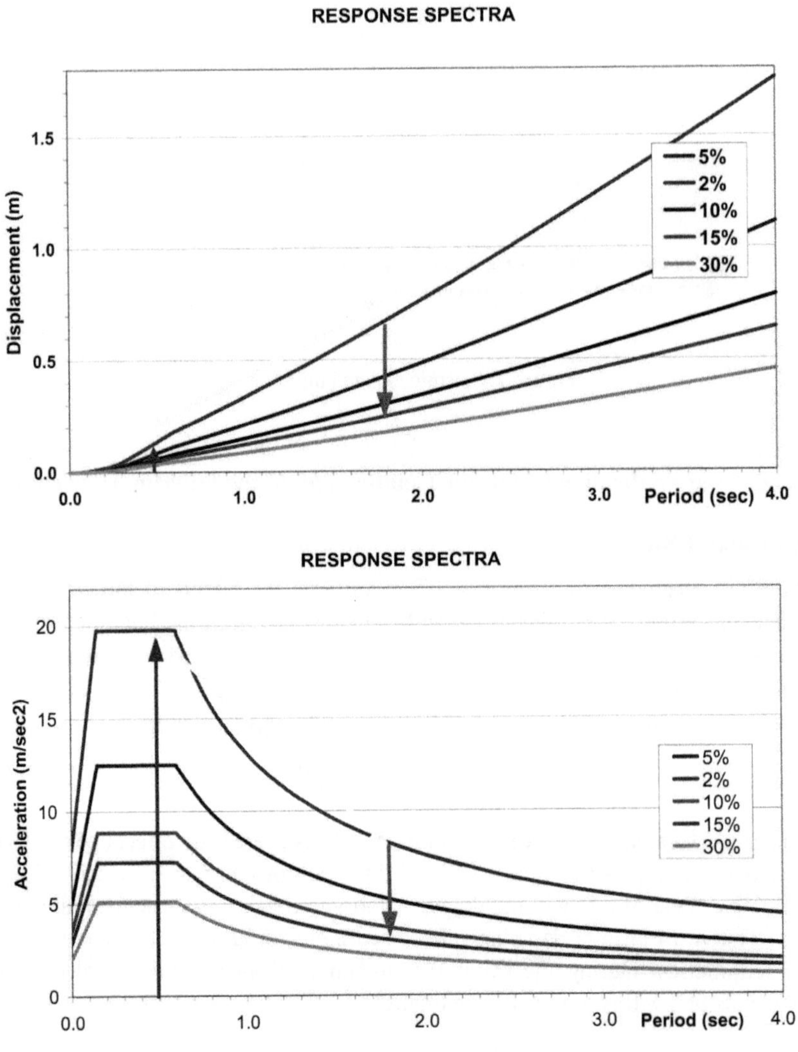

Figure 3. Conventional response spectra, displacement (top) and acceleration (bottom) for different damping coefficients (% of critical).

5 Dimensioning of a Base Isolation System

Under the assumption that the rubber device behaves as a perfect spring (acceptable in the design stage) and that the superstructure is rigid if compared to the isolation system (very commonly the case in normal buildings), it is possible to obtain the stiffness (K) of the rubber springs through the formula $T = 2\pi (M/K)^{1/2}$, where T is the isolation period and M the total mass of the superstructure. A typical isolator is shown in Figure 4.

Figure 4. Installed prototype (top) and one under test at Enel Hydro laboratories.

6 Ageing Effects

Ageing effects and their control are of the utmost importance in practical applications, and the question of ageing is very often one of the first items addressed in a technical meeting with a client or a designer having a standard background. An example of the quantitative influence of ageing is shown in Figure 5 (the results have been obtained within the framework of an EU project, contract no. BRPR-CT96-0141).

As it is possible to observe after periods of 4 and 8 months of accelerated ageing (obtained by using higher room temperatures that allow an equivalence to many years of installation in normal environmental conditions) the stiffness in the design range (shear strain equal to 1) is increased but not dramatically. On the other hand the limit resistance of the device (safety margin to rupture) is also increased.

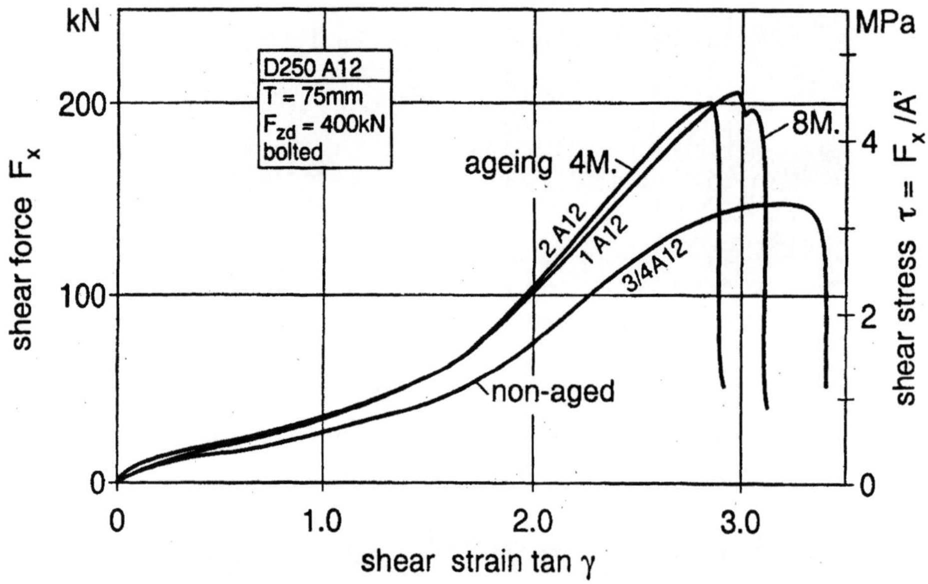

Figure 5. Shear force vs. shear deformation in aged and virgin devices.

7 Finite Element Analysis – State of the Art

Historically, the use of Finite Element Analysis (FEA) on elastomers has been limited to very specific applications. This was primarily driven by the fact that it was cheaper to build and test actual parts then to accurately evaluate them using FEA. However, as these components became increasingly more complex and subsequently more expensive, testing became cost prohibitive. This fact, coupled with rapid technological improvements in computer hardware and software, has underscored the need for and viability of FEA as a design tool.

Until recently, computer hardware capabilities limited the analyst's ability to fully examine complex elastomeric components, requiring that certain compromises had to be made. Consequently, FE models were too coarse to accurately assess local strains in the elastomer. Approximations were made to appraise the impact of non-symmetric loading on axisymmetric models and full 3D analysis was not a practical option for many companies. Most of this analysis was based on linear material models and small strain assumptions. Highly non-linear materials and complex loading environments typically resulted in poor approximation of actual performance data. Knowledge gained from past experience in modelling elastomeric bearings suggests the need for further refinement in the analytical methodology. Unlike metals, the analysis of elastomers is critically dependent on the mesh and element selection employed in the FE model. Additionally, the analysis is very sensitive to the material model formulation and to the magnitudes of the applied loads. These issues are explored in some detail in the following paragraphs.

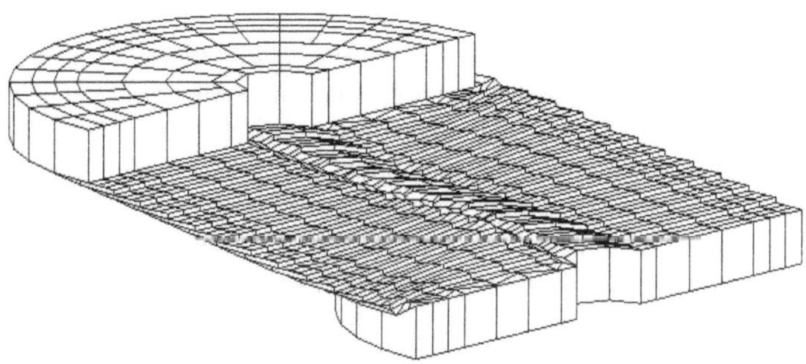

Figure 6. Numerical model deformed for a device under horizontal and vertical design loads.

7.1 Optimisation of the FE Model Geometry

In order to successfully model elastomers, the analyst must properly anticipate the final state of deflection and develop the mesh accordingly. Although this is also true of linear materials, application of large deformations to non-linear materials underscores the criticality of this point. Based on experience, linear elements do not adequately represent elastomeric materials. It is recommended that quadratic elements be used at a minimum to replicate the strain characteristics of the material. Additionally, representing elastomeric pads with only one layer of elements (irrespective of type) is sometimes found to be inadequate in replicating the bulge characteristics of the material. Proper elemental aspect ratios must be more strictly adhered to when modelling elastomeric components. Improper models can result in an unacceptable level of error between predicted and actual stress and strain results.

7.2 2D and 3D Analysis

Early analysis principally relied only on 2D models. Eventually, methods were developed to apply non-symmetric loading to axisymmetric models. Today's computers are capable of analyzing large and complex 3D FE models. This is particularly important when considering a combined loading environment with large deformation analysis and highly non-linear materials.

7.3 Mesh Convergence

Analyses have shown that results from coarsely modelled elastomers compare poorly with actual data. Given this fact, it is important to perform a mesh convergence study to ascertain the mesh density needed to achieve viable analytical results. This can easily be determined for metallic components, but it is more difficult to depict the correct distribution of strain in the elastomeric sections of a flexible bearing since there are practical limitations that restrict the level of mesh density. However, these restrictions should evaporate with more capable computer software and hardware.

7.4 Derivation of the Material Model

Until a few years ago, linear analysis was employed universally when considering flexible composite structures. Given that this approach relies on small strain theory, it can be inaccurate when compared to the actual state of stress and strain in the model. Large deformation, non-linear analysis demands a material definition that will adequately replicate non-linear response. This can be accomplished by developing an equation that will fit generated uniaxial tension, compression, and shear test data. Several methods are available to calibrate the material model. From these data, corresponding stress and strain data can be developed for input into ABAQUS, which will develop appropriate coefficients to duplicate the test data. ABAQUS assumes that elastomeric materials are elastic, isotropic, and approximately incompressible. The strain energy potential U can be defined in terms of the strain invariants I_1 and I_2 and J, as follows (ABAQUS manual, 1995): $U = U(I_1, I_2, J)$.

When the volume remains constant, the region of possible deformations is defined by uniaxial tension and uniaxial compression. Experience has shown that data from more than one test is required to reproduce an accurate material model. Relying solely on one test, while accurate for a specific condition, may give poor results for an alternate load application.

7.5 Elastomeric Material Modelling Difficulties

Elastomers by their very nature are considered to be hyperelastic materials. Hyperelasticity postulates that all of the work done when the material is deformed is stored as strain energy. Since hyperelastic materials are nearly incompressible, the use of a 'hybrid' formulation to account for this phenomenon is required (ABAQUS manual, 1995). There are other aspects of elastomeric material behaviour that are difficult to quantify, such as viscoelastic characteristics. Understanding the uniqueness of these properties is paramount in conceptualising elastomeric response. When an elastomer is subjected to a load, it deforms. As this process occurs, the molecular chains rub against each other and energy is dissipated as heat. This dissipation is known as damping and can be characterized as coulomb damping, hysteresis, or viscous damping. All three are forms of viscoelasticity and difficult to replicate with a material model. Stress relaxation within the material and the Mullins effect (i.e. variable stress-strain response based on load history) are issues that add to the difficulty of accurately modelling an elastomeric structure.

7.6 Loads and Boundary Conditions

Non-linear large deformation problems demand that more attention be given to the application of loads and boundary constraints. In many cases, contact elements must be used if the loaded model comes into contact with a rigid surface. The analyst must be familiar enough with expected levels of deformation to employ these required elements where appropriate. The point of load application should be established with the principles of large deformations in mind, understanding that its position and orientation may be changed substantially as load is applied.

Appendix. Simplified Non-linear Model Exponential Constitutive Law

Since 1993 Enel Hydro have been evaluating the applicability of seismic base isolation to industrial plants and devoted particular attention to HDLRB devices because of their very promising features and their worldwide diffusion. The proposed model assumes a decreasing exponential constitutive law $G(\gamma)$ for the rubber compound; this behaviour has been implemented in the ABAQUS code through a parallel scheme of elastoplastic elements. The model is an improvement of a previous one developed by Serino et al. (1993) and described in Bettinali et al. (1997). A peculiar feature of the model developed is its ability to reproduce the complex behaviour of HDLRB devices using only very simple information concerning the rubber compound, without the need for experimental data relevant to tests on actual devices. Static tests were carried out on single bearings to identify the force-displacement relationship, while dynamic tests were carried out using a 6-degree-of-freedom shaking table to reproduce seismic events on mock-ups connected to the ground by means of 4 or 6 HDLRB devices. Results obtained from the numerical model show very good agreement with experimental displacements and accelerations, thus demonstrating the model's ability to provide reliable results using only cheap and simple tests made on rubber specimens.

Recently, biaxial quasi-static tests have also been carried out at Enel Hydro (BU Ismes), and the comparison between numerical and experimental results indicates that the numerical model can also reproduce the biaxial behaviour of HDLRB. A slightly more sophisticated model, taking into account the high strain-hardening for uniaxial behaviour, has also been developed and tested. The model described for seismic isolators is able to represent the non-linear hysteretic behaviour as well as transverse biaxial interaction and is therefore sufficiently sophisticated to correctly predict the time history dynamic response of base isolated structures. Being based on the parallel elasto-plastic concept, the proposed model is at the same time sufficiently versatile to be easily implemented in existing non-linear finite element computer programs.

A1. Experimental Observations

Experimental tests on HDLRB devices are carried out by imposing horizontal relative displacements between the base and the top face, and by recording the value of the force required. During these tests a vertical load equal to the design load is applied. Experimental tests have shown that

(i) for values of the shear strain up to approximately 100%, the tangential modulus G_t of the bearing decreases exponentially with increasing shear strain from the initial value G_0 to the asymptotic value G_∞;

(ii) each time the relative motion between the two ends of the bearing changes sign, the tangential modulus regains its maximum value;
(iii) a kinematic hardening behaviour was observed, i.e. hysteretic loops remain inside an envelope that represents the isolator behaviour;
(iv) under the hypothesis of small thickness of each rubber layer, the rubber bearing behaviour depends only on the normal area and on the total thickness of rubber layers;
(v) the energy dissipated in a cycle depends on the maximum deformation, but is almost independent of the rate of strain applied, at least in the frequency range characteristic of an isolated structure (i.e. the amount of damping cannot be represented by means of an equivalent viscous parameter).

A2. Numerical Model

A linear model cannot reproduce in a realistic way the behaviour of HDLRB. The proposed approach, described in Serino et al. (1993), is a non-linear hysteretic model with the following assumptions:
(i) tangent modulus with decreasing exponential law, defined through three parameters related to the compound, i.e. G_∞ is the G value at design shear strain, G_0 is the G value at $\gamma = 0$, and b is an exponent multiplier;
(ii) overall bearing behaviour defined through geometric data: the normal area of the bearing and the total rubber thickness;
(iii) no viscous effects, i.e. velocity does not affect the force-displacement relationship.
The assumed G_t relationship can be expressed by the following two mathematical expressions, valid for a loading and unloading curve, respectively:

$$G_t(\gamma) = \frac{d\tau}{d\gamma} = G_\infty + ae^{-b(\gamma - \gamma_{\min})},$$

$$G_t(\gamma) = \frac{d\tau}{d\gamma} = G_\infty + ae^{-b(\gamma_{\max} - \gamma)},$$

where $a = G_\infty - G_0$, γ_{min} = starting strain value for increasing load phase, and γ_{max} = starting strain value for decreasing load phase.

It is worth noting that the model requires the correct selection of the three rubber parameters only (G_∞, a, b) which can be simply obtained from tests on rubber specimens. Hysteretic curves with continuously decreasing stiffness can easily be modelled using the parallel modelling concept. The above $G(\gamma)$ relationship has been reproduced by means of a set of elasto-plastic elements working in parallel; the parameters of each element in parallel must be calibrated to correctly reproduce the experimental bearing behaviour. The basic idea is to consider simple elastic-perfectly plastic and purely elastic elements connected in parallel, i.e. all having the same deformation but each carrying a different force. The total force acting on the assemblage is then obtained by summing the forces acting on each element. Parallel modelling is not new and has been used in the past for representing the mechanical behaviour of inelastic materials. The idea of effective stress introduced by Terzaghi to describe the behaviour of wet soils is essentially a parallel model, as water and soil fabric both carry the total stress. The concept was first formalised by Mroz (Mroz 1963) in treating non-associated flow plasticity and then used by Owen et al. (1974) and Pande et al. (1977) to achieve a better representation of actual material behaviour through a number of overlays of simple models. Nelson and Dorfmann (1995) have recently used parallel mathematical models in incremental elasto-plasticity to represent strain hardening in metals and to develop models of frictional materials such as soils, rock and concrete.

The concept applied to seismic isolators is very simple: the set of 'n' elasto-plastic elements work all together in their elastic range at the beginning of the load history; when the relative displacement reaches a threshold value, the first elasto-plastic element becomes plastic, so that the overall stiffness becomes smaller, being reduced by a quantity equal to the stiffness of the first element. When the relative displacement increases, the remaining $n-1$ elasto-plastic elements become plastic one at a time, and the resultant behaviour appears as a non-linear spring with decreasing stiffness. When the relative velocity changes sign, all the elasto-plastic elements come back to their elastic state and the global stiffness is again at its maximum value. This scheme has been implemented through a USER MATERIAL subroutine for the ABAQUS finite element code; the HDLRB behaviour can be reproduced simply by assigning this user defined material to a truss or to a solid element having the geometric characteristics of the actual bearing. The proposed model is capable of taking into account coupling between forces and displacements lying in the horizontal plane and can therefore be used to reproduce the dynamic behaviour of an isolated system subjected to triaxial seismic events, assuming that HDLRB devices behave like a rigid element in the vertical direction.

A3. Static Monoaxial Tests

Static tests have been set up to investigate the behaviour of the proposed numerical model in the case of multiple loading cycles with different amplitudes. A single bearing, having a diameter equal to 125 mm and a total rubber height equal to 33 mm, composed of a 'hard' compound with $G = 0.8$ MPa, has been tested by imposing a history of relative displacement characterised by a maximum value of 30 mm and by a series of smaller cycles with amplitudes of 1, 2, 5 and 10 mm. Such a displacement history has been designed to make evident the kinematic hardening phenomena. Comparison between experimental data and numerical results is shown in Figure A1. The agreement appears quite good, highlighting a remarkable fitting of the real behaviour of the bearing even if the smallest loops of the numerical model have areas (which corresponds to the dissipated energy) smaller than the experimental ones.

A4. Seismic Tests

The numerical model of elastomeric bearings has been used to reproduce the dynamic behaviour of real base isolated structures subjected to seismic excitation. Different tests have been carried out at Enel Hydro (BU Ismes) using a shaking table. Among other tests, shaking table experiments have been conducted on an isolated mock-up consisting of a concrete block, weighing 400 kN, supported on four bearings manufactured with rigid compound. The Friuli 1976 earthquake (Tolmezzo 3D records) has been applied as triaxial excitation with a time scale of ½ to the shaking table. Comparison between numerical results and experimental data are shown in Figure A2 for the north-south record: the agreement in terms of acceleration and displacement is very satisfactory, taking also into account the presence of coupling effects. The second seismic test considered in this chapter is relevant to a four-storey steel frame, loaded with a total mass of 245 kN supported on 6 bearings. A scaled (9 dB) Campano-Lucano 1981 earthquake (Calitri records) has been applied to the shaking table.

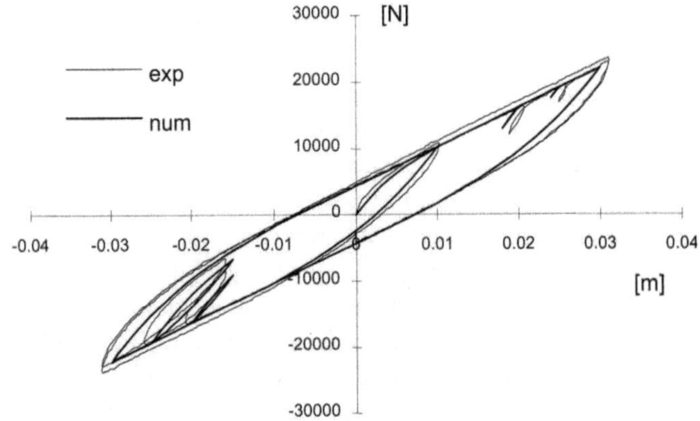

Figure A1. Experimental and numerical hysteresis loops for a rigid HDLRB ($G = 0.8$ MPa).

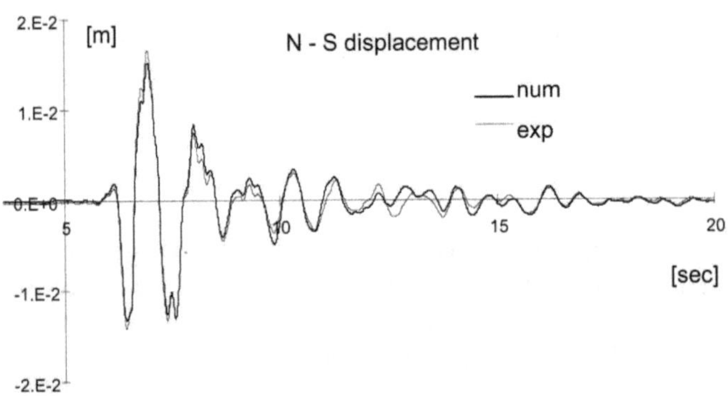

Figure A2. Experimental and numerical relative displacements, 400 kN isolated mock-(1976 Tolmezzo NS record).

Results in terms of horizontal relative displacement of bearings are compared with the experimental data in Figure A3. This last case is closer to a real application, because all the three components of the isolation problem are present: the excitation, the isolation system and the elastic supported structure with its own dynamic behaviour. The comparison between experimental data and numerical results can be considered satisfactory..

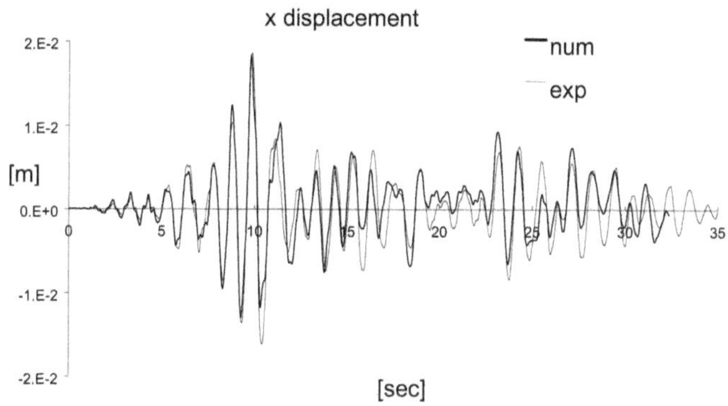

Figure A3. Experimental and numerical relative X displacement steel frame isolated mock-up 1981 Calitri record.

A5. Biaxial Tests

To get a deeper insight in the biaxial behaviour of HDLRB, a set of biaxial tests has been carried out. Low frequency (0.1 Hz) time histories of sinusoidal displacements have been imposed simultaneously in the two horizontal x and y directions of a HDLRB. Three different amplitudes and three different phase lags between the horizontal components have been considered for different tests; vertical load was simultaneously applied and controlled. Typical paths for the three assumed phase lags are reported, for a single cycle, in Figure A4; while the path relevant to 15 degrees is similar to a monoaxial excitation, the 90 degrees path represents the most heavy biaxial working condition for the isolation device. In fact, following a circular path, the isolator is subjected to a constant shear strain amplitude, only the direction of the shear strain being variable. The energy dissipation occurring during the tests with the 90 degrees phase lag is also qualitatively different from those relevant to 15–30 degrees phase lag because of the absence of variation in the strain amplitude. External work supplied during the different tests is reported in Figure A5. While for tests with 0–15–30 degrees phase lags, the time history of external work is periodic (i.e. there is a certain part of the external work that is recovered), for the 90 degrees test the time history of external work is an increasing linear function of time.

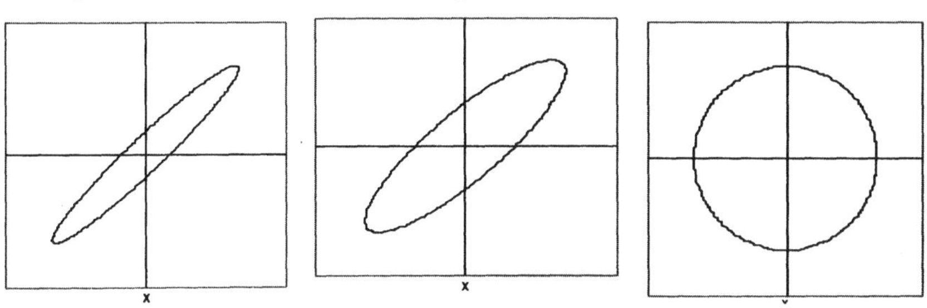

phase lag = 15 degrees phase lag = 30 degrees phase lag = 90 degrees

Figure A4. Typical paths for assumed phase lags.

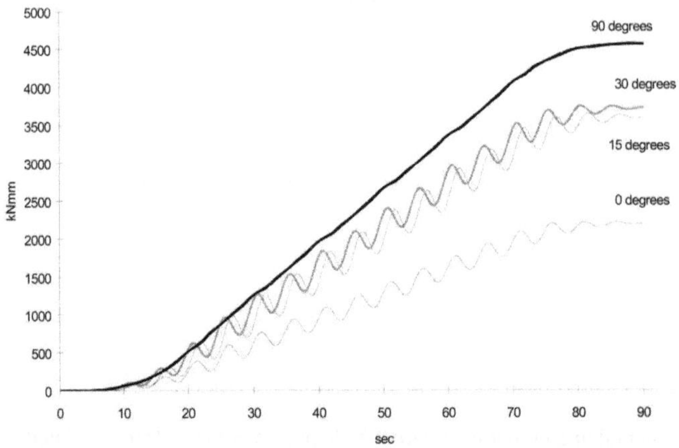

Figure A5. External work as function of time during biaxial tests.

The amount of dissipated energy directly depends on the relative phase between force vector F and incremental displacement ds:

$$E_{diss} = \oint F \cdot ds \cdot$$

Analysis of experimental data relevant to tests with the circular path showed an almost constant relative phase between force and displacement vectors equal to 11–15 degrees. The numerical simulation of these tests, carried out using the developed user subroutine, gave as a result a constant phase of about 10 degrees, demonstrating that the implemented model is able to reproduce with good accuracy the observed biaxial behaviour. The comparison between experimental data and numerical results for the test with 25% shear strain and phase lag equal to 15 degrees is reported in Figure A6, together with the absolute value of the imposed displacement. As can be noted, the agreement is qualitatively and quantitatively good, with minimum relative phases around 5 degrees corresponding to the minimum values of displacement cycles.

As an example, for the same 15 degrees test, experimental and numerical graphs of applied force modulus versus imposed displacement modulus are reported in Figure A7. Again, the agreement can be considered satisfactory.

A6. Model Development for High Shear Strains

The parallel model examined is able to reproduce the observed experimental behaviour of HDLRB as long as the maximum shear strain does not exceed a limit value around 100%. At larger shear strains in the rubber (above 150%) an increase of the tangent modulus is observed due to the strain crystallisation of the rubber matrix. Since the model is implemented under the assumption of exponential stiffness decrement along with shear strain, the high strain hardening phenomenon cannot be taken into account.

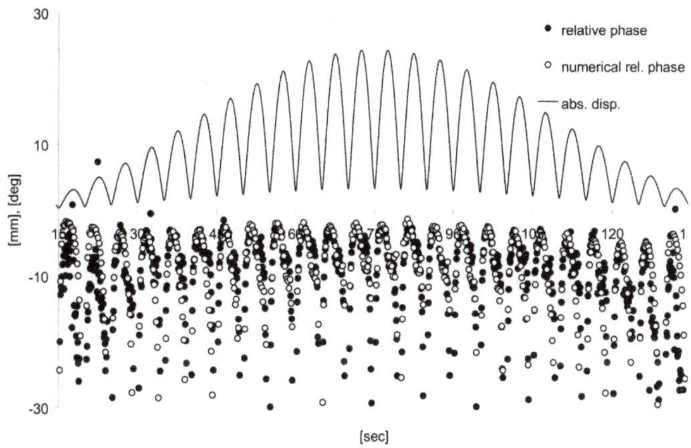

Figure A6. Experimental and numerical phase lags for biaxial test at 25% shear strain, 15 degrees phase lag.

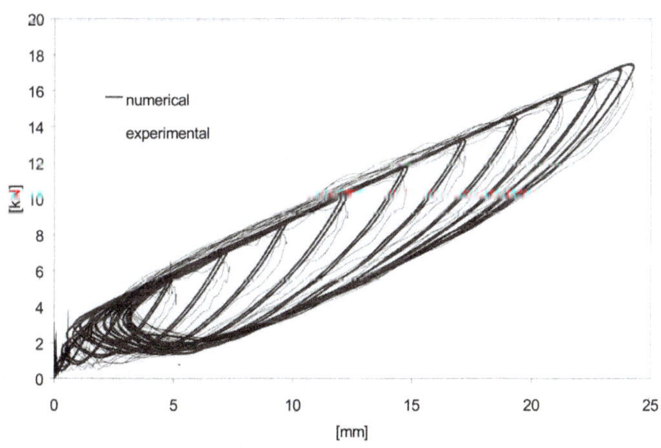

Figure A7. Experimental and numerical applied force modulus vs. displacement, biaxial test, 25% shear strain, 15° phase lag.

A more refined constitutive model was therefore set up to allow for the reproduction of the high shear strain behaviour. This was obtained by putting in parallel with the original model a second one, whose $G_h(\gamma)$ function has the following expression (c and d define the slope and curvature of the exponential law):

$$G_h(\gamma)=c(e^{d|\gamma|}-1)\,.$$

The above expression was implemented using again a set of parallel elements. Figures A8, A9 and A10 show comparisons between experimental hysteresis loops (obtained from quasi-static monoaxial tests) and numerical results. Figure A8 is relevant to static tests with the same compound used as in the example reported in Figure A1, but in this case the maximum shear strain is approximately 150%: the agreement is remarkable. It has to be stressed that the new model can reproduce the experimental behaviour for a wide range of hysteresis loops, ranging from the smallest one (20 mm) to the largest one (200 mm). It is also worth noting that both the two main effects of hardening are well reproduced: the increasing stiffness for strains above 100%, and the increasing 'thickness' of loops for zero strain.

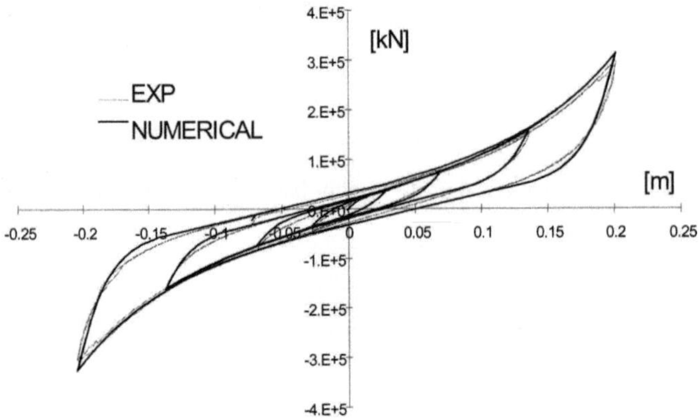

Figure A8. Experimental and numerical hysteresis loops for a rigid HDLRB ($G = 0.8$ MPa).

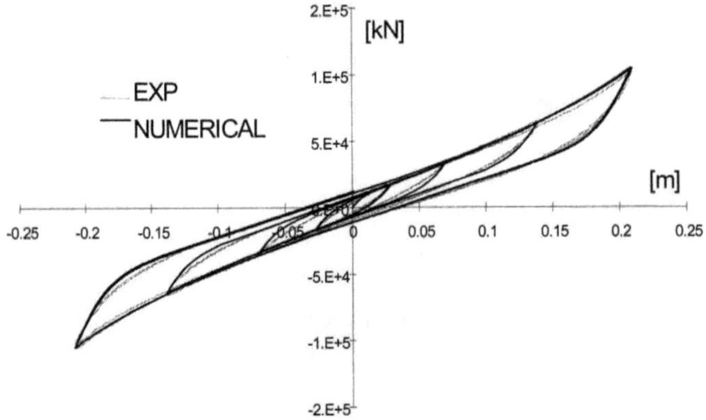

Figure A9. Experimental and numerical hysteresis loops for a soft HDLRB ($G = 0.4$ MPa).

A test with a bearing made from a 'soft' compound is reported in Figure A9; again the agreement is satisfactory (even if the hardening effect is less evident) for a wide range of loop amplitude, ranging from 20 to 200 mm. The last test performed is reported in Figure A10. A

HDLRB made with the soft compound has been tested with a maximum shear strain equal to 250%. Even if the agreement is less satisfactory, compared to the previous test analysed, it can be said that the model is adequate to catch the recorded behaviour, and thus to well simulate both the stiffness and the energy dissipation capacity of the isolator.

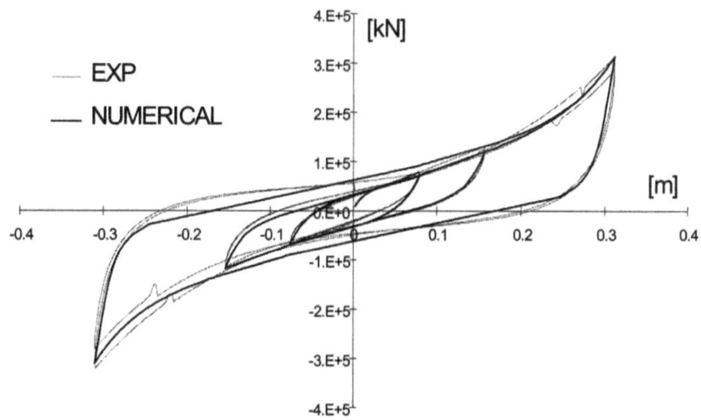

Figure A10. Experimental and numerical hysteresis loops for a soft HDLRB ($G = 0.4$ MPa)

A7. Finite Element 3D Models

In the case of project optimization, a three-dimensional Finite Element analysis of the device is necessary. Experience has been gained in the past by investigating the behaviour of the bearings under vertical load and shear strain by means of the commercial code at our disposal (ABAQUS). The mechanical behaviour of rubber-like materials has been described by an elastic, isotropic and approximately incompressible model; therefore, the governing constitutive equations are derived from a strain energy function U. In particular among the different forms of the strain energy, the polynomial form is chosen (Rebelo, 1991):

$$U = \sum_{i+j=1}^{N} C_{ij}\left(\bar{I}_1 - 3\right)^i \left(\bar{I}_2 - 3\right)^j + \sum_{i=1}^{N} \frac{1}{D_i}\left(J_{el} - 1\right)^{2i},$$

where I_1 and I_2 are the first and the second invariants of the deviatoric strain, J is the elastic volume ratio, C_{ij} and D_i are material constants, while N is the order of the energy function. The values of the constants D_i determine the compressibility of the material and are set equal to zero for fully incompressible materials. The choice of N (generally $N = 1,2,3$) provides polynomial forms of strain energy more or less complex. The coefficients in U have been obtained from simple states of deformations and stress experiments, uniaxial, biaxial and shear (Ogden, 1984; Rebelo, 1991 and Forni et al., 1995) and carried out at Enel Hydro laboratories (Bettinali et al., 1996). In the numerical analyses the material has usually been assumed to be incompressible

($D_i = 0$) and a polynomial N=2 form was adopted. Steel shims have been assumed elastic-plastic.

Finite Element Discretization, The numerical problems due to an incompressible rubber-like materials has been overcome by a mixed finite element formulation. Hybrid elements with pressure stress independently interpolated from the displacement field, make the numerical formulation of the variational problem well-behaved. An updated Lagrangian formulation has been adopted. Device geometry and loading conditions make the problem axisymmetric; therefore, only one half of the bearing need be modeled. The recess attachment system is difficult to model because of the sliding of a deformable body on a rigid one. Figure A11 depicts a typical FE mesh, while Figure A12 shows a bolted bearing under compression with shear at 300% and Figure A13 shows a comparison of the experimental results with the FE simulation. A corresponding comparison for a recessed bearing is shown in Figure A14.

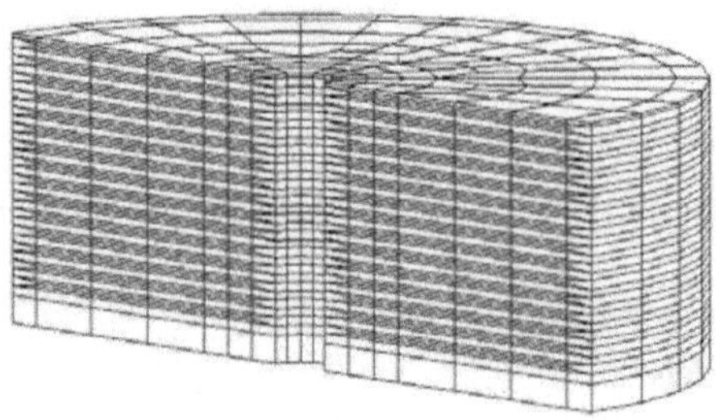

Figure A11. Example of a 3D Finite Element mesh.

Figure A12. Compression and shear test at 300% (bolted).

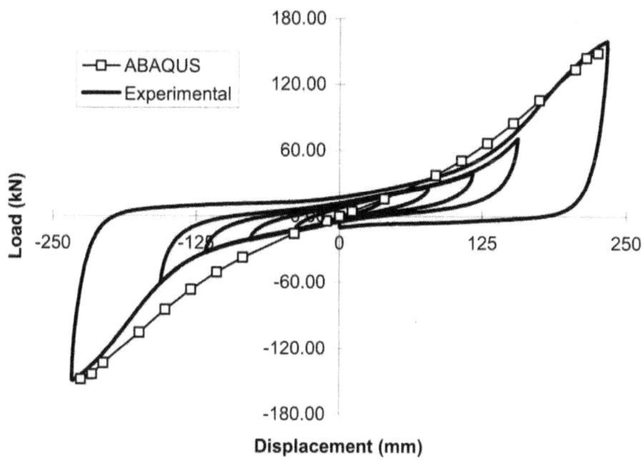

Figure A13. Experimental and numerical force-displacement comparison (compression and 300% shear strain): bolted.

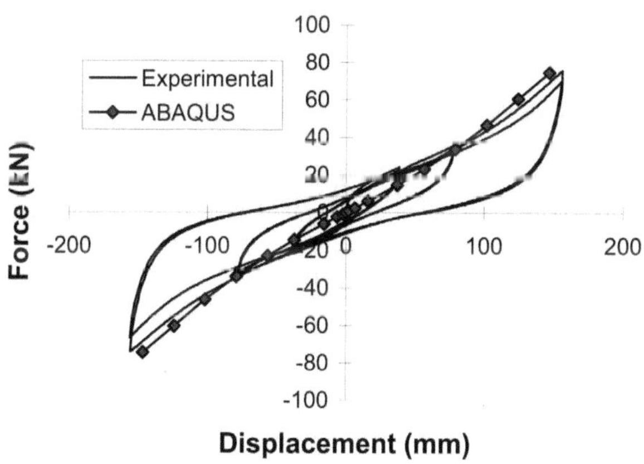

Figure A14. Experimental and numerical force-displacement comparison (compression and 200% shear strain): recess.

References

ABAQUS Manual, Version 5.5. (1995). Hibbitt, Karlsson & Sorensen Inc., Providence, Rhode Island, USA.

Bettinali, F., Dusi, A., Bonacina, G., and Rebecchi, V. (1997). Seismic behaviour of HDLRB through a simplified non linear model with exponential constitutive law. In *Proceedings of the International Conference on New Technologies in Structural Engineering, Lisbon, Portugal.*

Bettinali, F., Cazzuffi, D., Dusi, A., and Fede, L. (1996). *Valutazione delle Caratteristiche di Mescole in Gomma Naturale Utilizzate per la Costruzione di Isolatori Sismici,* ENEL S.p.A.-CRIS internal report n° 5187 (in italian).

Bonacina, G., Serino, G., and Spadoni, B. (1992). Implication of shaking table tests in the analysis and design of base isolated structures. In *Proceedings of the Tenth World Conference on Earthquake Engineering,* Vol. 4, 2405–2410.

Forni, M., Martelli, A., Dusi, A., and Castellano, G. (1995). *Hyperelastic Models of Steel-Laminated Rubber Bearings for the Seismic Isolation of Civil Buildings and Industrial Plants.* In *Proceedings of the ABAQUS Users' Conference,* Paris, France.

Kelly, J. M. (1997). *Earthquake Resistant Design with Rubber.* Springer.

Mroz, Z. (1963). Non-Associated flow rules in plasticity. *Journal de Mecanique* 2 :21–42.

Nelson, R.B., and Dorfmann, A. (1995). Parallel elastoplastic models of inelastic material behaviour. *Journal. of Engineering Mechanics* 121:1089–1097.

Owen, D.R.J., Prakash, A., and Zienkiewicz, O.C. (1974). Finite Element analysis of nonlinear composite materials by use of overlay systems. *Computers & Structures* 4 :1251–1267.

Ogden, R.W. (1984). *Non-Linear Elastic Deformations.* Ellis Horwood Limited, Chichester.

Pande, G.N., Owen, D.R.J., and Zienkiewicz, O.C. (1977). Overlay models in time-dependent nonlinear material analysis. *Computer & Structures* 7 :435–443.

Rebelo, N., (1991). *Analysis of Rubber Compound with ABAQUS,* In *Proceedings of the Second National Congress of ABAQUS Users' Group Italia.*

Serino, G., Bonacina, G., and Bettinali, F. (1993). Proposal and experimental validation of analytical models for seismic and vibration isolation devices in nuclear and non-nuclear facilities. In *Proceedings of SMiRT-12.* Elsevier.

Skinner, R.I., Robinson, W.H., and McVerry, G.H. (1995). *An Introduction to Seismic Isolation.* Wiley.

MIX
Papier aus verantwortungsvollen Quellen
Paper from responsible sources
FSC® C105338

If you have any concerns about our products,
you can contact us on
ProductSafety@springernature.com

In case Publisher is established outside the EU,
the EU authorized representative is:
Springer Nature Customer Service Center GmbH
Europaplatz 3, 69115 Heidelberg, Germany

Printed by Libri Plureos GmbH
in Hamburg, Germany